# Decision Support Systems for Weed Management

Guillermo R. Chantre • José L. González-Andújar
Editors

# Decision Support Systems for Weed Management

Springer

*Editors*
Guillermo R. Chantre
Departamento de Agronomía
Universidad Nacional del Sur -
CERZOS/CONICET
Bahía Blanca, Buenos Aires, Argentina

José L. González-Andújar
Department of Crop Protection
Institute for Sustainable Agriculture
Córdoba, Córdoba, Spain

ISBN 978-3-030-44401-3 ISBN 978-3-030-44402-0 (eBook)
https://doi.org/10.1007/978-3-030-44402-0



This Springer imprint is published by the registered company Springer Nature Switzerland AG
The registered company address is: Gewerbestrasse 11, 6330 Cham, Switzerland

*This book is dedicated to León*

# Foreword

It is about time! Or, perhaps more accurately, it is about "timing." Safe and successful short- and long-term weed management is highly dependent upon when weed seeds lose or gain dormancy, when they germinate, when seedlings emerge, how fast plants grow, when flowers and seeds form, differential sensitivities to disruption of growth and development during all phases of plant life cycles, and the fickle nature of herbicide fate. No farmer can understand all of these dependencies for even a single weed species. Nor, for that matter, can any individual weed scientist. Failure to comprehend and predict these dependencies helps explain why weeds remain common and usually unwanted residents of agricultural fields even after decades of intense efforts at controlling them. Indeed, by the year 2020, many species of weeds have evolved resistance to various forms of weed control, and they now are not just common, but rampantly abundant in some fields. The sheer volume of literature in Weed Science published during the past two decades pertaining to resistance underscores the fact that this problem is increasing, not diminishing.

Even though no individual person understands all of the variables that affect any weed, groups of weed scientists can come close to doing so. These groups of scientists can collaborate, conceptualize, experimentally test, and develop models that attempt to mimic weed behavior and control. Although models of weed growth and management were initiated many years ago, only some scientific groups continued pursuing this line of research to the present. Many other groups, however, curtailed modeling activities with the advent of genetically modified herbicide tolerant crops. Creation of herbicide tolerant crops represented truly remarkable scientific achievements, and these achievements revolutionized weed management beginning in the mid-1990s in countries that allowed GM crops to be grown. Unfortunately for farmers and weed scientists in those same countries, evolution also is quite remarkable. Selection for weed resistance to herbicides used in GMO-based cropping systems occurred faster and was more widespread than anyone had anticipated. This was, indeed, a sobering development for Weed Science.

Weed resistance to herbicides is not confined to GMO-based cropping systems. Weeds evolve resistance to herbicides whenever and wherever overreliance on herbicides occurs, even in countries that banned GM crops. Consequently, the need for understanding weed biology and management is worldwide in scope, and it is never-ending, as weeds will continue evolving as new cropping systems and weed control techniques are developed and implemented.

Fortunately, small pockets of weed scientists scattered across the globe recognized the continued need for weed models even during the GMO revolution. The continued efforts, intellect, and dedication of those groups are reflected in this book, Decision Support Systems for Weed Management. The book is divided into four parts, each with multiple chapters: (1) Modelling: A Brief Introduction to Decision Support Systems, (2) Bio-Ecological and Site-Specific based models, (3) Environmental Risk Modelling, and (4) Weed Management Decision Support Systems: Study Cases. These parts explain to readers the general and technical aspects of modeling and its utility in Weed Science; historical and recent advances in the modeling of weed behavior and dynamics, crop–weed interactions, and site-specific phenomena; assessments of unintended consequences of weed management, especially herbicide fate and effects; the utility of several highly functional DSS models developed in Australia, Europe, and Latin America. These are truly exciting developments.

In my view, the individual chapters, its sections, and the book as a whole represent the twenty-first-century basis for integrated weed management. In other words, adoption of the concepts, if not the specific models, described in this book will help lead to the sustainable cropping systems that agriculture must have in the future. It is about time!

Frank Forcella,

University of Minnesota
St Paul, MN, USA

# Preface

Weed management decision support systems (DSS) are increasingly important computer-based tools for modern agriculture. Nowadays, extensive agriculture has become highly dependent on external inputs, and both economic costs and the negative environmental impact of agricultural activities demand knowledge-based technology for the optimization and protection of nonrenewable resources. In this context, weed management strategies should aim to maximize economic profit by preserving and enhancing agricultural systems resources. Although previous contributions focusing on weed biology and weed management provide valuable insight on many aspects of weed species ecology and practical guides for weed control, no attempts have been made to highlight the forthcoming importance of DSS in weed management. This book is a first attempt to integrate "concepts and practice" providing a novel guide to the state of the art of DSS and the future prospects, which hopefully would be of interest to higher-level students, academics, and professionals in related areas.

Buenos Aires, Argentina
Córdoba, Spain

Guillermo R. Chantre
José L. González-Andújar

# Contents

# Part I
# Modelling: A Brief Introduction to Decision Support Systems

# Chapter 1
# Mathematical Models

**Niels Holst**

**Abstract** Decision support systems (DSSs) rely on computational machinery in which mathematical models often constitute an important part. In this chapter, it is discussed which kinds of models are best suited for different kinds of DSSs. The practical steps involved in model construction are outlined, keeping in mind that model construction is a process that must be integrated into the larger software development project launched to construct the whole DSS. You are invited into the modeller's workshop, as you follow the considerations involved in formulating a simple model of weed emergence. Two case studies close the chapter, demonstrating models of the population dynamics of annual weeds in a crop rotation and of an invasive weed. R scripts for all models can be found in the book's online appendix. It is concluded that weed modellers must be prepared to work in multidisciplinary teams and that they should be better at considering the needs of the DSS users. For purposes of quality control, the mathematical models should be published open-source, while the DSS itself might be proprietary.

**Keywords** Decision support systems · Model construction · Software development · Weed population dynamics · Invasive weed · Weed modeller · R scripts

**Electronic supplementary material:** The online version of this chapter (https://doi.org/10.1007/978-3-030-44402-0_1) contains supplementary material, which is available to authorized users.

N. Holst (✉)
Department of Agroecology, Aarhus University, Slagelse, Denmark
e-mail: niels.holst@agro.au.dk


G. R. Chantre, J. L. González-Andújar (eds.), *Decision Support Systems for Weed Management*, https://doi.org/10.1007/978-3-030-44402-0_1

## 1.1 Introduction

We build models to grasp the world and to manage our lives and surroundings. Whether in science or in everyday life, we express ourselves, we rationalise and we communicate by concepts that reflect our perspective on reality. We all have models of the world in our minds, whether we are humans—or bats (Nagel 1974). When we express models in the language of mathematics, we take our more or less fluffy concepts and dip them in the acid of mathematics. Whatever is left stands clearly written in equations. Then truly, *what can be said at all can be said clearly* (Wittgenstein 1922).

The disciplines of mathematical modelling and software engineering are essential to any decision support system (DSS). Mathematical models, which are constructed from mathematics and algorithms, constitute the wisdom of the DSS, while the DSS user interface makes that wisdom accessible in a language and operational mode that is convenient to the user. New DSSs are created in research and development environments by teams comprising experts on the problem domain (e.g. weed control), together with modellers and software engineers.

In a professional setting, the whole software development process is played out according to a well-defined software development protocol, such as agile development (Martin 2006). Ideally, in the early design phase of a DSS, the users and their needs are defined. Once the user problem domain has been delineated, the next step is to identify the modelling approach that will enable the development of models, which can provide information helpful to the user.

An unfortunate but common déroute in DSS development is to let the whole construction process take place in a closed forum of researchers and modellers, who believe that there is a real need for the DSS that they have in mind. When the finished DSS ultimately attracts little interest, they will blame the end users (e.g. the farmers for being too lazy to count weed seedlings and enter those numbers into the DSS). I wish to reiterate what has been said many times yet seems a surprisingly difficult advice to follow: *Before developing a model, make clear what its purpose is*. Add to that: *Before making a DSS, make certain there is an actual need for the guidance it will offer, and that end-users will pay the price in time and money needed to use the DSS*. Private companies would call it a business plan.

To make certain that an initial brainstorm will reveal the full range of possible DSS designs, the matrix in Table 1.1 can be used as a guide. The matrix is defined

**Table 1.1** A design matrix for decision support systems (DSSs) with different scopes

| | Query<br>What's the status? | Q&A<br>What should I do? | Scenarios<br>What if? |
|---|---|---|---|
| Tactics | 2.1 | 2.2 | 2.3 |
| Strategy | 3.1 | 3.2 | 3.3 |
| Policy | 4.1 | 4.2 | 4.3 |

Numbers refer to the subsections which explore these nine types of DSS further
Question and answer (Q&A)

by Conway's (1984) typology which classifies decisions at either tactical, strategic or policy level vs. the kind of the support needed, whether it's a status query, a question and answer session or an exploration of *what-if* scenarios. In the following, I will discuss which modelling approaches are most appropriate for the nine classes of DSS resulting from the combination of the two typologies in Table 1.1. Two case studies and a few recommendations conclude the chapter.

## 1.2 Models to Support Tactics

Tactical decisions are often the easiest to support with a DSS and also the easiest for which to confirm that a model provides accurate advice. Tactical decisions define a short time frame and a narrow spatial scale (e.g., weed management decisions within a given season and for a specific field). Long-term and larger-scale consequences of one's actions are deliberately ignored. The typical decision maker is a farmer or technical advisor.

### *1.2.1 Tactical Queries*

Basic queries concern weed status: Which species have emerged? At which densities? In which fields? Where in the fields? Previously, these questions were difficult to address except by personal observation, but with the advent of artificial vision and multispectral imaging, weed maps can now be drawn with increasing precision from videos captured by Global Positioning System (GPS)-enabled field equipment or from more or less autonomous rolling or flying drones. The development of mathematical models to extract patterns, such as weed species distributions, from digital images is a ripe research field driven by demands outside agronomy (e.g. military intelligence). This means that a DSS should be designed with future changes in mind; it should be easy to plug in new methods for pattern recognition as they become available.

### *1.2.2 Tactical Q&A*

When the current state of weed pressure has been assessed, whether through high-tech monitoring, visual scouting or personal experience from earlier growing seasons, the question is what to do about it? Thus, a farmer may ask whether weed control is necessary, and, if so, which herbicide/s and dosage/s will provide efficient control or minimise ecotoxicological side effects?

The model to answer such questions would be based on a database of herbicide efficacy for different weed species, maybe even parameters for dose-response relations and corrections for weed growth stage and crop. The simultaneous optimisation on several criteria might be addressed best by optimising each separately, and then let it over to the farmer to take the final decision, weighing the options.

The models defining the optimisation problem might be based on simple regression models that describe dose-response-price-environment relations. However, with all the possible combinations of weed species, crops, herbicides and non-chemical treatment options, these models quickly turn very data hungry. Sensible ways of cutting down on this combinatorial explosion should be addressed early in the process of model development.

In precision agriculture (PA), questions must be addressed at a fine spatial resolution within each field. This will make the optimisation problem more difficult, maybe difficult even to define. Numerical optimisation in itself is a classical discipline within mathematics, physics and computer science. Please, see Chap. 3 of this section for a detailed description on numerical optimisation.

### *1.2.3 Tactical Scenarios*

We most often think of scenarios as something distant and far reaching, but even within the scope of a single field in a single season, different scenarios can be envisaged at the time of weed control. Thus, a farmer may ask, which among the available control options will give the highest yield, by grain or by net income? If we get a dry spring and I do not control the weeds, what will the yield loss be? How much should the price of grain change to make one control tactic economically better than another?

With scenarios, DSS models become more demanding. Maybe the total range of possible outcomes cannot be described by regression models alone. More complex simulation models might become necessary. This will incur additional costs in terms of model development and assessment of model reliability. A DSS in scenarios mode easily gets more speculative, and the user interface more difficult to design to strike the right level of detail and functionality.

## 1.3 Models to Support Strategy

While tactical decisions are taken in the season as a reaction to imminent weed problems, strategic decisions can be made off-season usually as a simulation exercise. The scale of strategic decisions extends into weed management over several years in the same field and across all the fields belonging to a farm or a landscape. Invasive weeds and the management of weeds in natural habitats are problems that necessitate strategic (and policy) level decisions. When we are developing a DSS for strategic planning, we should be careful to recognise that weed management forms only a small part of farm management and the whole-farm organisation. We should always think carefully about the interface between the DSS and other farm management software to achieve a smooth integration and convenience of use. The typical user is an agricultural consultant.

### 1.3.1 Strategic Queries

The necessity of a strategy, rather than just simple tactics, for weed management becomes obvious when weed problems escalate above the norm. Common causes are a reduced diversity in crop rotation (in the extreme case, monoculture), an over-reliance on a small subset of herbicides with similar modes of action and, ultimately, the advent of herbicide resistance. For example, a DSS could help by identifying and predicting imminent weed outbreaks. If monitoring data on weed occurrence were logged, together with a log of field activities, then an ideal DSS could issue early warnings which could then inspire changes in weed management strategies. Models for such a DSS would incorporate weed population dynamics analysed either statistically or numerically through simulation. However, it is doubtful whether farmers/advisers really need an early warning system for weeds. Field infestations are obvious to the naked eye, and weed problems will usually announce themselves in a few hot spots before large areas suffer from the infestation. At landscape level, a DSS taking input from remote sensing could point out patches of invasive weeds.

### 1.3.2 Strategic Q&A

An aspect of weed status that is important for strategic planning yet remains difficult to ascertain is an answer to the query: What is the current prevalence of herbicide resistance? It still seems far into the future that a DSS, fed with drone-collected biomolecular characteristics of weeds, could provide this information. The models underpinning such a DSS would be in the reign of bioinformatics.

A more approachable strategic question might be: will this crop rotation control this weed? Or, if I choose this crop rotation, which weed species would be prevented and which would be promoted? Or, with this rotation of herbicides, will I prevent herbicide resistance building up? A model to answer these questions could be a rather simple simulation model working in time steps of cropping seasons. The model would consist of difference equations describing the mechanisms at a rather coarse level. Even so, it might prove difficult to find solid empirical data to estimate all model parameters. The best course is then to include parameter uncertainty in the model (e.g. by supplying min-max values for all parameters) and use proper methods to derive the resulting uncertainty in model outputs (Saltelli et al. 2008).

### *1.3.3 Strategic Scenarios*

A DSS could provide tools to design a complete weed management strategy, including crop rotations and herbicides, or the full complement of methods used in organic farming. Outputs could include economic performance, yields, weed densities, herbicide-resistance prevalence and environmental side effects. Such a DSS might acquire the flavour of a computer game, in which the user tries to win by fulfilling as many goals as possible, accepting trade-offs according to personal preferences. The model underlying this DSS will be more complex than the previous. A simulation model is clearly called for, and even more detail is needed, reflecting the detail of the scenarios and the outputs.

## 1.4 Models to Support Policy

Policy models are for decision makers at the highest organisational level. They might be decision makers at international, national or regional levels, or decision makers working for the interest of non-governmental organisations (NGOs), such as farmer organisations or nature conservation societies. For a modeller, it can be a frightening experience to develop models that will feed into decision processes affecting society at large, even though economist modellers seem less challenged by this prospect. Policy models play such a powerful role in modern society that they have been put in their own category dubbed *post-normal models* (Funtowicz and Ravetz 1993). For policy models, it is of particular importance to include uncertainty in model inputs (and consequently, in model outputs) to prevent abuse of the models by overzealous policymakers. In a democratic society, the models should be open-source since they are used to formulate arguments in the public debate.

### *1.4.1 Policy Queries*

At policy level, queries are made to identify problems and motivate the formulation of policies. Thus, one may ask, what is the current distribution of an invasive weed? What is the current use of herbicides per year, and in which crops are they applied? Queries such as these can often be translated into queries into databases. Thus, the underlying model is within the range of software engineering, maybe overlaid with descriptive statistics.

### 1.4.2 Policy Q&A

Weed management policies are formulated with an eye to political goals, preferentially those that can be formulated in terms of performance indicators: production quantity and quality, farmer economy, environmental side effects, etc. Political instruments are foremost economical (taxes, subsidies) but also include indirect measures such as education and research. This means that even the simplest question (e.g. if herbicide taxes were increased in proportion to their ecotoxicity, what are the consequences on farmer economy and the environment?) will involve several fields of knowledge (agronomy, ecology, economics, sociology). The corresponding models will tend to be rich in assumptions and parameters estimated by expert opinion. Model outputs will be equally rich and challenging to condense into information useful to the decision maker.

It is very difficult to construct a policy model with a clear rationale for which components and mechanisms should be included and at which level of detail. Model uncertainties must be included in the DSS outputs, but there is a high risk of uncertainty being caused by structural faults (i.e. the exclusion or misrepresentation of key elements and key processes), which cannot be diagnosed by formal methods but only by scientific argument. Structural faults might lead to biased outputs, as will be pointed out soon enough by political combatants. The modeller must be prepared to defend in public the scientific base of a policy model.

### 1.4.3 Policy Scenarios

When even the simplest policy question leads to models of high complexity, the modelling of policy scenarios will lead to even higher complexity. We quickly reach the limit of what can be modelled with some confidence—and within a weed research budget. The conscientious modeller confronted with a demand for a model of such immense complexity should consider to decline the order.

A problem domain bordering that of weed management is pesticide legislation and regulation. Legislators and land-use administrators are in need of information on the fate of herbicides in the environment (e.g. persistence, leakage to ground and surface water) and on the magnitude of their unwanted side effects (ecotoxicological and human toxicological). A DSS to support these policymakers would incorporate models of the physicochemical pathways of herbicides in air, soil and biota and the derived effects on exposed populations (by necessity including only a few key species from selected taxa). The information provided by such a DSS could be used to formulate laws and regulations on herbicide use, including the possible banning of a specific herbicide. Due to the vast economical interest in herbicides, represented by farmers and pesticide companies, and the skepticism of NGOs representing a variety of interests, the modeller should be prepared that the DSS will be playing part in a complex political theatre.

## 1.5 Model Development

For models that consist of a few regression equations or other statistical measures, the modelling procedure falls inside common research practice. The only challenge will be to communicate with software developers on how to embed the statistics in a DSS. For models that are simply queries into a database, the software engineer is in command and will need the weed modeller only as a consultant to assist in the proper interpretation of the data.

The really demanding models are simulation models. Since they will need to be embedded in dedicated DSS software, their implementation will become an integral part of a commercial-scale software development project. The best software design will ensure a loose coupling (Seemann 2012) between the DSS user interface and the simulation model, both kept in separate modules. This will allow independent development of the DSS and the model. Furthermore, it will allow the DSS code to be proprietary (i.e. owned by a company or institute) and the model code to be open-source and thereby open for scientific publication and public scrutiny.

Model development goes through a series of steps, generally acknowledged in the modelling community and outlined in the following: formulation, parameter estimation, verification, testing, validation, uncertainty analysis and sensitivity analysis.

### *1.5.1 Formulation*

Simulation models are formulated in the language of mathematics and logic. They should be based on the theoretical concepts of the topic and should re-use earlier models or sub-models when possible. If the model contains many interacting components, consider software engineering methods to manage the complexity (reviewed by Holst and Belete 2015).

An important part of model formulation is parameterisation. This term is most often used in the wrong sense to mean 'parameter estimation' (see next subsection). What it means properly is to 'formulate in terms of parameters'. For example, you may need a hump-shaped curve to represent a process such as seedling emergence rate through time (Fig. 1.1 top). You can formulate that as a parabolic curve using the standard parameterisation

$$y = ax^2 + bx + c \tag{1.1}$$

This parameterisation, however, has the problem that none of the parameters represent a biological feature of seedling emergence. A better parameterisation describes the curve by its start ($x_{begin}$) and end ($x_{end}$) on the $x$-axis and by its maximum on the $y$-axis ($y_{max}$). Equivalent to Eq. (1.1), we get

$$y = 4y_{max} \frac{\left(x - x_{begin}\right)\left(x - x_{end}\right)}{\left(x_{begin} - x_{end}\right)\left(x_{end} - x_{begin}\right)} \tag{1.2}$$

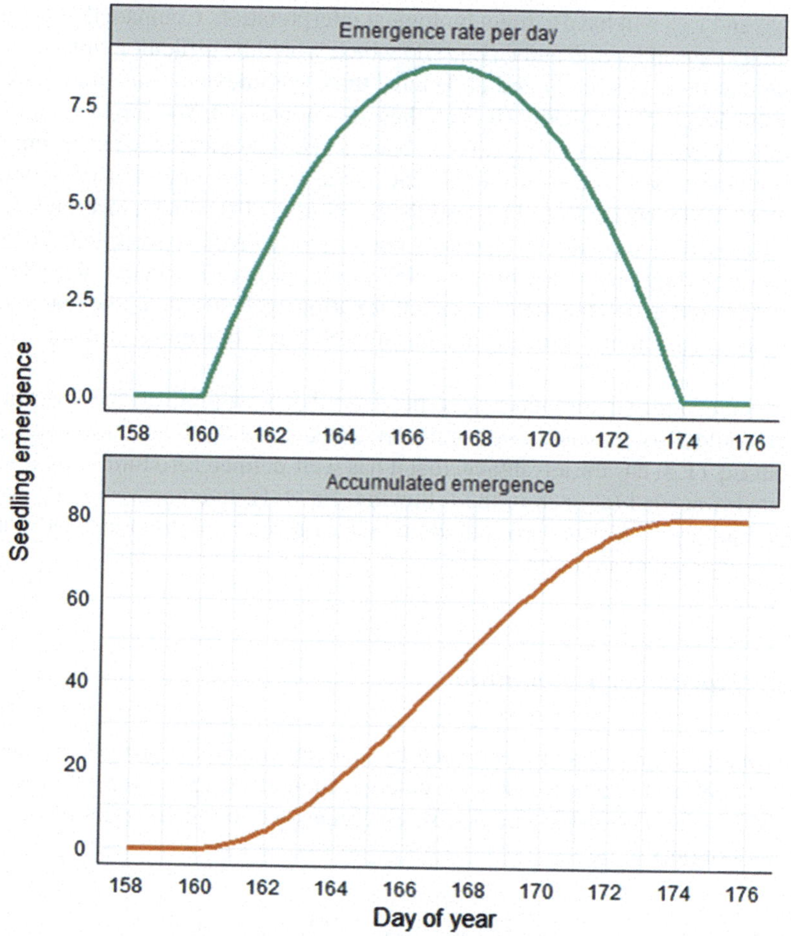


**Fig. 1.1** A hump-shaped curve to describe seedling emergence rate (Eq. 1.3, top) and accumulated seedling emergence (Eq. 1.4, bottom) with $x_{begin} = 160$, $x_{end} = 174$ and $y_{total} = 80$. Implemented in the dsswm-1-1.R script

Yet, as you begin to use this equation, you may realise that $y_{max}$ is not a convenient parameter. The area under the curve, expressing total emergence ($y_{total}$), would be a much better parameter. Hence, you proceed to integrate Eq. (1.2) and replace $y_{max}$ with $y_{total}$ and finally get

$$y = 6y_{total} \frac{\left(x - x_{begin}\right)\left(x - x_{end}\right)}{\left(x_{begin} - x_{end}\right)\left(x_{end} - x_{begin}\right)^2} \tag{1.3}$$

The benefits of this parameterisation are plenty. The user of the model (foremost yourself) can now estimate, communicate and change the parameters with a clear rationale. Moreover, in a sensitivity analysis, the uncertainty induced by parameters

$x_{begin}$, $x_{end}$ and $y_{total}$ will have a direct biological interpretation. Compare that to model uncertainty caused by *a*, *b* and *c* (Eq. 1.1) which would be difficult to interpret.

Note that both Eqs. (1.1) and (1.3) have three parameters. Thus, they have the exact same level of complexity (in fact, they are equivalent). You might want to add an additional parameter to Eq. (1.3) to obtain a skewed emergence curve, but with every parameter you add to a model, you incur an increasing debt of parameter estimation. If the curve is used to describe the course of seedling emergence in the field, there will be so many mechanisms not accounted for (weather and soil being the most important ones), that further detail is not merited. The detail of model formulation should match the detail in the information available about the real system. Modelling of the more intricate details of seed bank dynamics is dealt with in Chap. 4.

The curve (Fig. 1.1 top) has a superficial similarity with the normal distribution (which would also demand three parameters: mean, standard deviation and *y* scaling), but Eq. (1.3) has the advantage, that it has well-defined zero limits and is easily integrated if needed (as seen in the following, Eq. 1.4). In comparison, the normal distribution never reaches zero (an additional parameter would be needed), and it has no analytical integral.

### 1.5.2 Parameter Estimation

Often, model parameters are estimated by standard statistical procedures, such as linear or nonlinear regression. For the emergence model (Fig. 1.1 top), for example, you could regress observed cumulative emergence (*Y*) on the integral of Eq. (1.3) (Fig. 1.1 bottom):

$$Y = \frac{y_{total}\left(3x_{end} - x_{begin} - 2x\right)\left(x_{begin} - x\right)^2}{\left(x_{end} - x_{begin}\right)\left(x_{begin} - x_{end}\right)^2}, \tag{1.4}$$

which expands to a third-degree polynomial. The estimated polynomial coefficients can be used to calculate the three parameters: $x_{begin}$, $x_{end}$ and $y_{total}$.

In an early stage of model development, you should consider whether the DSS ought to include uncertainty. If so, each uncertain parameter must be described by a distribution (e.g. uniform between min-max values or normal defined by mean and standard deviation). Be suspicious, in particular of parameter values that are expert opinions (guesses). Ask the expert up front for parameter ranges or distributions rather than simple point values.

In the case of parameters estimated by regression of equations such as Eq. (1.4), you cannot always use the standard error of the coefficients to generate random parameter values independently. If the standard errors of the regression parameters cannot be considered independent, you must use the regression model itself to draw random values from the predicted distribution of $y$ given $x$. For the particular case of the emergence model, however, it does seems reasonable that the three parameters vary independently, though you might choose to replace $x_{\text{end}}$ with $x_{\text{begin}} + x_{\text{duration}}$, where $x_{\text{duration}}$ designates the duration of the emergence period.

Some parameters are best estimated from the model itself, a process commonly called *calibration*. This is a somewhat dubious activity: You make the model fit your expectations, usually empirical data, by fine-tuning one or several model parameters. The more parameters you calibrate, the less confident you should feel about the general applicability and robustness of the model.

### *1.5.3 Verification*

In model verification, you check that model behaviour makes sense. For this purpose, you define a series of parameter sets (often considered model *scenarios*) of increasing complexity and proceed, more or less formally, to check that model outputs look right. In other words, you check that model outputs could be true, that the behaviour of the model makes sense. Negative, zero or infinite weed densities are typical examples of model fragility discovered during verification. You proceed by mending the model as needed to pass verification. Do find the root cause of any problem and implement a scientifically sane solution. Do not thoughtlessly use this solution, often hiding in model code, *if* $(x < 0)$ *then* $x = 0$, or other hacks like it.

### *1.5.4 Testing*

Testing is an important discipline in computer science, even to the degree that the whole software development process can be centred around it (Beck 2002). Software testing is not a part of model development as such, but the testing of the DSS easily becomes intertwined with model verification. As a modeller, you should be prepared to supply the software engineers with *unit tests*: Each unit test defines the output values expected from certain input values. This makes it possible to automatise the test procedure. A less favoured method, nowadays, of software quality assurance is *debugging*, which is an unsystematic stress test traditionally carried out by the programmer.

### *1.5.5 Validation*

Validation does not mean proof of model correctness; rather, it is the comparison of model outputs with independent field data. Thus, validation aims to convince peers that there is a robust and rather accurate match between model predictions and the real world. It is wise always to make a plan for model validation in the early phase of model development. The model design should be accommodated to make a final validation possible, often by restricting the model's scope and the modeller's ambitions.

### 1.5.6 Uncertainty Analysis

If some model parameters are best described not by a single estimate but by a distribution (e.g. normal), reflecting uncertainty due to statistical error or natural (irreducible) variation, then model output will be distributed as well. Users will be familiar with uncertainty from weather forecasts which may predict, for example, a 20% chance of rain tomorrow. Likewise, a weed DSS may predict, for example, that a yield loss >10% is highly improbable due to a risk level of only 1%. Models that include uncertainty make most of their assumptions transparent as they shine through in the recommendations issued by the DSS. Since DSSs are meant to be reliable tools, one should not be shy of situations which produce a very wide range of responses. Sometimes, the future may be unpredictable in essence. That information can also be useful.

### 1.5.7 Sensitivity Analysis

Sensitivity analysis is a step following up on uncertainty analysis. In sensitivity analysis, the uncertainty in model outputs is apportioned to the inputs thus identifying those inputs, that are most decisive for model uncertainty (Saltelli et al. 2008). Sensitivity analysis is usually an academic activity related to the scientific publication of the model, but it could potentially be useful as a DSS feature. The user could be told how much certainty would be gained in DSS outputs by giving more precise estimates of, for instance, weed density or herbicide resistance.

## 1.6 Case Studies in Model Development

### 1.6.1 A Difference Equation Model for Annual Weeds

Weed populations tend to be highly dynamic; fast establishment and rapid proliferation are part of being an *r* strategist. Once established, the soil bank of seeds or shoots becomes a constant source of potential outbreaks. The long-term management of weeds is a strategic problem raising questions such as: Which level of seedling mortality will be necessary to reduce the infestation? Or, will this change of the crop rotation help to regulate the weeds? Modellers themselves have for a long time been prolific developing models to answer such questions (Holst et al. 2007). In the following, I will go through the steps of developing a classical iterative model which moves forward in steps of 1 year:

$$S = (1 - \mu_{soil})(S_{prev} - E_{prev} + P_{prev}) \tag{1.5}$$

The equation computes this year's seed bank ($S$; $m^{-2}$) from the previous year's seed bank ($S_{prev}$; $m^{-2}$), emergence ($E_{prev}$; $m^{-2}$) and seed production ($P_{prev}$; $m^{-2}$), while undergoing a certain basic seed bank mortality ($\mu_{soil}$; $y^{-1}$).

From the seed bank, a certain proportion ($\epsilon$; $y^{-1}$) will emerge as seedlings ($E$; $m^{-2}$):

$$E = \epsilon S \tag{1.6}$$

of which again a certain proportion ($\mu_{control}$; $y^{-1}$) will be killed by weed control measures:

$$N = (1 - \mu_{control})E \tag{1.7}$$

to leave some plants surviving ($N$; $m^{-2}$) to produce new seeds ($P$; $m^{-2}$):

$$P = \frac{f_\infty}{1 + \frac{f_\infty - f_1}{f_1 N}} \tag{1.8}$$

The two parameters describing fecundity are $f_1$ (seeds per plant), which is the expected number of seeds produced by one plant growing in competition with the crop only, and $f_\infty$ (seeds per $m^2$), which is the maximum number of seeds produced in competition with the crop by an infinite density of weed plants.

It is not obvious from this formulation (Eqs. 1.5–1.8) that the model is composed of *difference equations*. A mathematically more concise formulation makes this clear. Equation (1.6), for example, could be written more correctly as

$$\Delta E = \epsilon S \Delta t \tag{1.9}$$

where $\Delta E$ ($m^{-2}$) expresses the change in $E$, i.e. the difference to be added to $E$ over the time step $\Delta t = 1$ year. Note that multiplication with $\Delta t$ is necessary to make the units right; Equation (1.9) corrects Eq. (1.6) also in that sense. However, here, we will maintain the slightly incorrect formulation (Eqs. 1.5–1.8) as this is commonly found in literature. The left-out multiplications with 1 will have no effect other than to annoy finicky mathematicians, who would in any case likely prefer a differential equations formulation. To continue with Eq. (1.6) as an example, this would look like

$$\frac{dE}{dt} = \epsilon S \tag{1.10}$$

If you are mathematically skilled, the option stands open for you to build a *differential equations* model, rather than a difference equations model, but for most weed modellers, this is not the case.

Equations (1.5)–(1.7) are all linear which makes them easy to comprehend. Equation (1.8) was given a nonlinear form to take into account density dependence; fecundity per plant decreases with increasing plant density until an asymptote is reached.

Always verify that the shape of your equations makes sense in the real world. For a nonlinear equation, make a plot to ascertain its shape and check its limits both graphically and algebraically. In the case of Eq. (1.8), we get meaningful boundary conditions:

$$\begin{aligned} P &\to f_\infty \quad \text{for} \quad N \to \infty \\ P &= f_1 \quad \text{for} \quad N = 1 \\ P &\to 0 \quad \text{for} \quad N \to 0 \end{aligned} \tag{1.11}$$

Not all verification turns out as successful. For instance, Zwerger and Hurle (1989) proposed an alternative to Eq. (1.8):

$$P = Nae^{-bN} \tag{1.12}$$

for which $P \to 0$ for $N \to \infty$. At high weed density, seed production $P$ ($m^{-2}$) goes towards zero. A self-defeating weed!

A model consisting of Eqs. (1.5)–(1.8) is called an *iterative* model because you run a simulation by repeatedly computing Eqs. (1.5)–(1.8), thereby updating the four state variables of the model ($S$, $E$, $N$, $P$) iteratively in time steps of 1 year. It is a *stage-structured* model of *population dynamics* since the population is divided into separate life stages ($S$, $E$, $N$) which are simulated dynamically.

A model needs to be started from some initial state. In this case, we need initial values for $S_{prev}$, $E_{prev}$ and $P_{prev}$. More importantly, we need to estimate the values of the model parameters. For this model, there are only few parameters: $\mu_{soil}$, $\epsilon$, $\mu_{control}$, $f_\infty$ and $f_1$. The task of parameter estimation seems simple until you realise that some of the parameters are likely to depend on the crop. Moreover, they are all liable to differ between years and locations. To accommodate this inherent variability of the parameters, we will define their values as ranges rather than point estimates, some of them specific to the crop (Table 1.2).

The values in Table 1.2 are the expected average values, originally given without indication of their standard errors (Zwerger and Hurle 1989). However, both $\hat{f}$ and $\hat{\epsilon}$ will certainly vary markedly between fields due differences in soil and weather. To capture this uncertainty in the model, we pick values at random inside intervals defined as

**Table 1.2** Weed life history parameters from Zwerger and Hurle (1989), except* from CABI (2019)

| | Spring barley | Maize | Winter wheat | Any crop |
|---|---|---|---|---|
| *Alopecurus myosuroides* | | | | |
| Fecundity $(\hat{f})$ | *1000 | *1000 | *1000 | |
| Emergence $(\hat{\epsilon})$ | 0.040 | 0.034 | 0.050 | |
| Soil mortality ($\mu_{soil}$) | | | | 0.81 |
| *Avena fatua* | | | | |
| Fecundity $(\hat{f})$ | *200 | *200 | *200 | |
| Emergence $(\hat{\epsilon})$ | 0.240 | 0.240 | 0.230 | |
| Soil mortality ($\mu_{soil}$) | | | | 0.87 |
| *Fallopia convolvulus* | | | | |
| Fecundity $(\hat{f})$ | 192 | 1855 | 93 | |
| Emergence $(\hat{\epsilon})$ | 0.043 | 0.020 | 0.078 | |
| Soil mortality ($\mu_{soil}$) | | | | 0.16 |
| *Galium aparine* | | | | |
| Fecundity $(\hat{f})$ | 3 | 100 | 40 | |
| Emergence $(\hat{\epsilon})$ | 0.036 | 0.010 | 0.037 | |
| Soil mortality ($\mu_{soil}$) | | | | 0.20 |
| *Lamium purpureum* | | | | |
| Fecundity $(\hat{f})$ | 32 | 300 | 280 | |
| Emergence $(\hat{\epsilon})$ | 0.013 | 0.017 | 0.023 | |
| Soil mortality ($\mu_{soil}$) | | | | 0.16 |
| *Thlaspi arvense* | | | | |
| Fecundity $(\hat{f})$ | 60 | 630 | 330 | |
| Emergence $(\hat{\epsilon})$ | 0.073 | 0.021 | 0.043 | |
| Soil mortality ($\mu_{soil}$) | | | | 0.08 |
| *Veronica persica* | | | | |
| Fecundity $(\hat{f})$ | 150 | 200 | 150 | |
| Emergence $(\hat{\epsilon})$ | 0.079 | 0.066 | 0.030 | |
| Soil mortality ($\mu_{soil}$) | | | | 0.50 |

$\hat{f}$ : seeds per plant; $\hat{\epsilon}$ : $y^{-1}$; $\mu_{soil}$: $y^{-1}$

$$f_1 \in \left[\frac{\hat{f}}{5}; 2\hat{f}\right[$$

$$f_\infty \in \left[10\hat{f}; 100\hat{f}\right[$$

$$\epsilon \in \left[\hat{\epsilon}/5; 2\hat{\epsilon}\right[$$

Limits for random numbers are conventionally closed-open; [*a*; *b*[ designates an interval including *a* and excluding *b*.

For soil mortality ($\mu_{soil}$), we will use the point estimates (Table 1.1) without any variance. We will assume that the efficacy of weed control vary quite much picking random values, $\mu_{soil} \in [0.6; 0.9[$.

Since we let four of the parameters vary randomly, our model is a *stochastic* model; it will not always give the same result. Hence, we have to run it many times to assess the uncertainty in its predictions (Fig. 1.2). During model verification, it was found that two of the weed species were dying out (ALOMY, AVEFA). Hence, the parameter values from Zwerger and Hurle (1989) were replaced with values roughly taken from CABI (2019) (Table 1.2).

The first impression of Fig. 1.2 is that the uncertainty is much larger for some species (ALOMY, AVEFA, VERPE) than for others. Note that two units on the *y*-axis correspond to variation by a factor of 100. It could be of interest to know which of the model parameters are causing this huge variation. This could be resolved by a sensitivity analysis (Saltelli et al. 2008). Some species exhibit fluctuations clearly provoked by crop rotation (FALCO, THLAR), more clearly seen for seedling than for seed bank density. This makes sense because different weed species are known to emerge either in spring or autumn sown crops, or in both.

During the 24 years covered by this simulation, most species are attaining an equilibrium density, THLAR most quickly, AVEFA most slowly. It is difficult to imagine, however, how knowledge of the equilibrium density could be interesting from a DSS perspective. We would rather like to help the farmer to achieve the situation illustrated by GALAP for which density is decreasing in this scenario; it is a weed under control. It would be wise though to consult with weed experts and discuss whether this GALAP scenario seems realistic (a belated verification of the model).

Simulation experiments with the crop rotation and a sensitivity analysis could help suggest effective control strategies for these weed species. The model could be incorporated into a DSS, allowing the farmer/advisor to address problematic weed species through strategic means, rather than the purely tactical which entails giving up on controlling the seed bank (left-hand side of Fig. 1.2) and just limiting its expression (right-hand side of Fig. 1.2), which after all is the ultimate cause of yield loss.

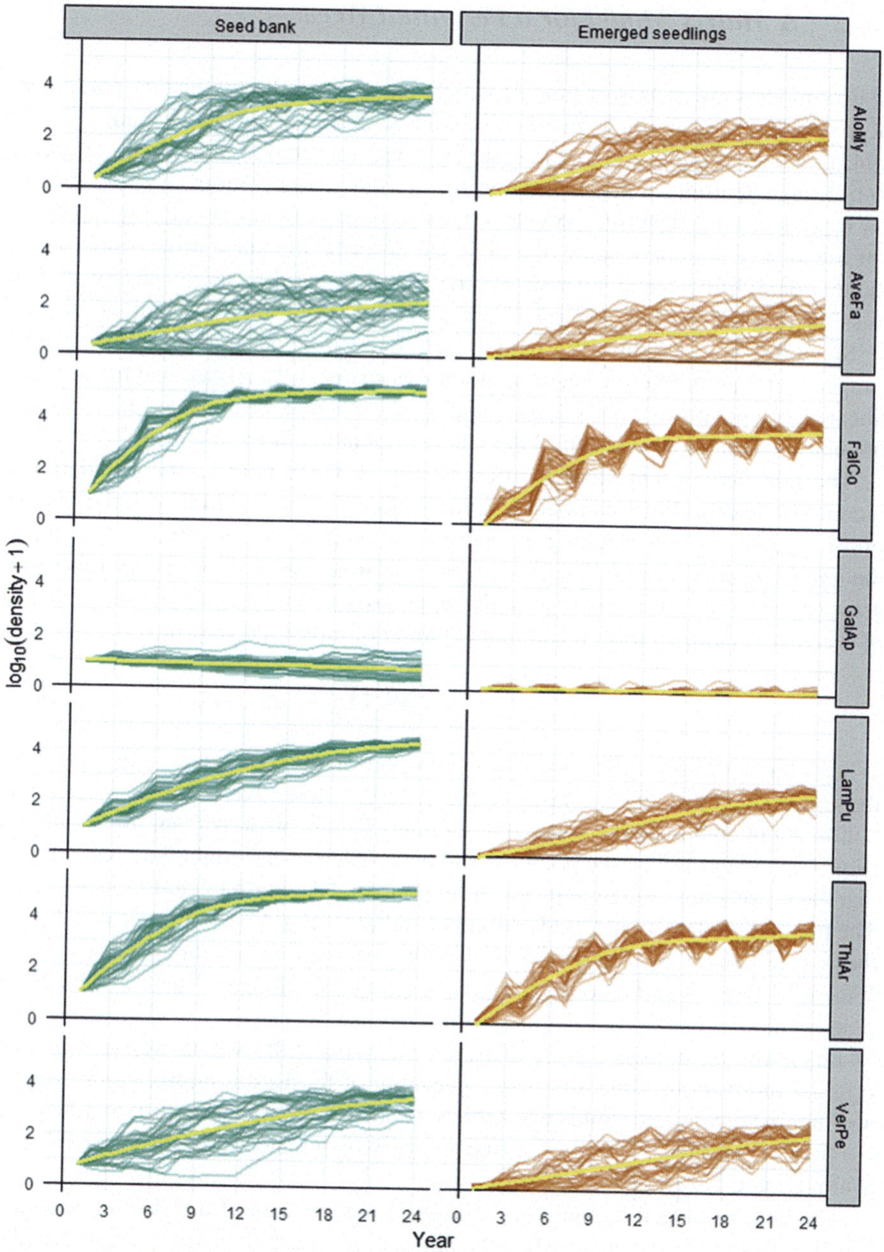


**Fig. 1.2** The result of 30 simulations of the crop rotation, maize-winter wheat-spring barley. Yellow curves show smoothed averages. For full species names, see Table 1.2. All populations started with ten seeds per m$^2$. Model formulated in Eqs. (1.5)–(1.8) and implemented in the dsswm-1-2.R script

### 1.6.2 A Matrix Model for a Perennial Weed

Matrix models are a class of models which summarises the life history parameters of a population in a single matrix (Table 1.3), a so-called Leslie matrix (Leslie 1945). In the columns, you find the fate over one time step of individuals according to life stage. Column sums <1 account for mortality, and column sums >1 account for reproduction. Likewise, rows show the origin of individuals entering the different life stages. Numbers below the diagonal describe life stage progression; above the diagonal, life stage regression; and on the diagonal, life stage conservation. In this concrete matrix, the seven stages are a mixture of life stage, age and size classes.

Since this is a deterministic model, only one run is necessary to explore what happens after an initial introduction of ten seeds (Fig. 1.3). Notice that the model is linear which means that it gives the same result, whether we consider the simulated population dynamics pertinent to the whole population or to, say, 1 $m^2$.

The two fields seem clearly different (Fig. 1.3). In field L, the population is increasing, approaching exponential growth after *c.* 10 years. In field J, the population is decreasing, approaching a negative exponential decline after *c.* 5 years. In theory, these matrix models will converge towards a state in which all life stages grow (or shrink) exponentially with the same growth rate, namely the *intrinsic rate of increase* ($r$) known from the classical model of unlimited growth:

$$N_t = N_0 \exp(rt) \tag{1.13}$$

When $r$ has stabilised, so has the relative proportion of the population in each stage; the *stable stage distribution* has been reached. It follows that when the stage distribution is not stable, then $r$ is not stable either. This is obvious from the simulation (Fig. 1.3); otherwise, all the points would have fallen on a straight line. Note though that the $y$-axis transformation bends the exponential decrease in field J towards zero. The population density in field L initially oscillates (Fig. 1.3), but the reason behind these oscillations is different than for the oscillations in the previous model (Fig. 1.2). Here, it is due to the unstable stage distribution, and there, it was due to crop rotation.

Leslie matrix models can be analysed mathematically which was part of their original motivation. Thus, the first eigenvalue of the Leslie matrix equals exp($r$), and the first eigenvector holds the stable stage distribution. For fields L and J, we get $r = 0.18\ y^{-1}$ and $r = -0.49\ y^{-1}$, respectively, which match the values arrived at by Werner and Caswell (1977).

The lines produced by these growth rates on a log scale (Eq. 1.13) are shown in Fig. 1.3, from the time about when the populations reach their stable stage distribution (again, the line is distorted on the approach towards zero for field J). The initial population size ($N_0$ in Eq. 1.13) was chosen to let the line pass through the average population size through years 10 to 20 for field L and through years 5–20 for field J. The stable stage distributions are computed in the dsswm-1-3.R script.

**Table 1.3** Leslie matrices for *Dipsacus sylvestris* (from Werner and Caswell (1977)) estimated for two fields, L and J

| | Seed0 | Seed1 | Seed2 | RosetteS | RosetteM | RosetteL | Flowering |
|---|---|---|---|---|---|---|---|
| ***Field L*** | | | | | | | |
| Seeds0 | 0 | 0 | 0 | 0 | 0 | 0 | 503 |
| Seeds1 | 0.43 | 0 | 0 | 0 | 0 | 0 | 0 |
| Seeds2 | 0 | 0.97 | 0 | 0 | 0 | 0 | 0 |
| RosettesS | 0.01 | 0.021 | 0.005 | 0 | 0 | 0 | 0 |
| RosettesM | 0.036 | 0.003 | 0 | 0.19 | 0.253 | 0 | 0 |
| RosettesL | 0 | 0 | 0 | 0.07 | 0.105 | 0.15 | 0 |
| Flowering | 0 | 0 | 0 | 0 | 0.002 | 0.517 | 0 |
| ***Field J*** | | | | | | | |
| Seeds0 | 0 | 0 | 0 | 0 | 0 | 0 | 476 |
| Seeds1 | 0.423 | 0 | 0 | 0 | 0 | 0 | 0 |
| Seeds2 | 0 | 0.987 | 0 | 0 | 0 | 0 | 0 |
| RosettesS | 0.024 | 0.009 | 0.006 | 0.007 | 0 | 0 | 0 |
| RosettesM | 0.044 | 0 | 0 | 0.05 | 0.158 | 0 | 0 |
| RosettesL | 0.001 | 0 | 0 | 0.002 | 0.008 | 0 | 0 |
| Flowering | 0 | 0 | 0 | 0 | 0 | 0.25 | 0 |

**Seeds of age 0, 1 or 2 years. Rosettes of size: small, medium or large. Diagonal cells greyed for easier reading**

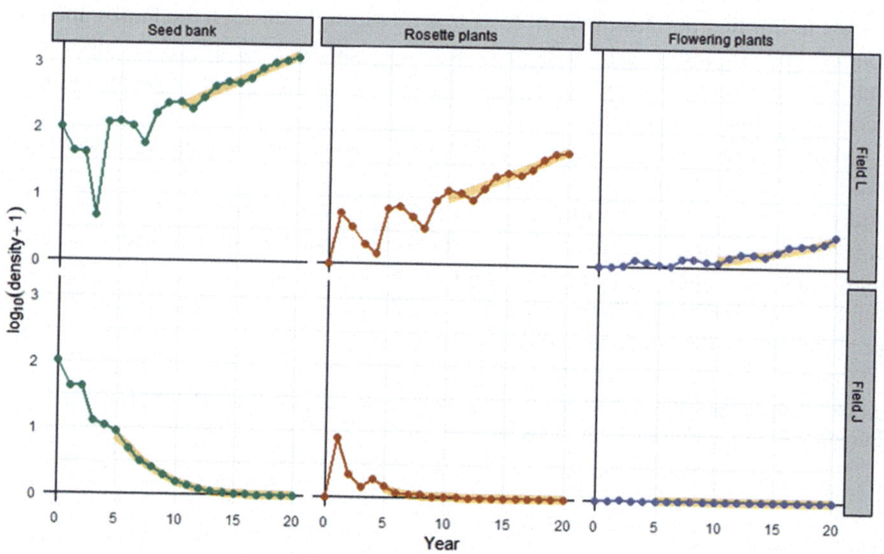


**Fig. 1.3** The result of a simulation starting with ten seeds of *Dipsacus sylvestris* running for 20 years based on Leslie matrices for field L and J. Seed bank numbers are the sum of all three age classes of seeds. Rosette plants are the sum of all three size classes of plants. Orange lines show the asymptotic population growth rate. Implemented in the dsswm-1-3.R script

The outcome of the model is interesting from a weed management point of view. The weed population in field J cannot maintain itself, so if by proper management, the conditions in field J could be mimicked in other fields, where the weed is a problem (such as in field L), then a solution would have been found. Werner and Caswell (1977) suggested that the high vegetation density in field J was outcompeting the weed.

Maybe the fields are representative of two stages in the natural succession going from young and susceptible to invasion (field L) to old and resistant to invasion (field J). The model tells us nothing about how quickly that succession may happen (another model would be needed). Still, the model could be used to predict the effect of control measures, such as grazing, which could possibly reduce the survival rate of the rosette stages and maybe the fecundity of flowering plants. It could also predict the fate of the population if all flowering plants were uprooted before seed set, simply by setting the value in the upper right corner of the Leslie matrix equal to zero.

## 1.7 Conclusions

DSS software provides a user interface to mathematical models that formulate the scientific knowledge relevant to the user's decision-making. The models are hidden for the user who operates them more or less unwittingly through the knobs and buttons of the DSS dashboard. Mathematical models can be as simple as the algorithms that define a data table query or as complicated as an agro-ecosystem or agro-economic simulation model.

Modellers should be prepared to collaborate within a multidisciplinary team during DSS construction. At the outset, (1) the whole team should define the domain addressed by the new DSS (Table 1.1); (2) the software engineers should choose a software development methodology; and (3) the modellers should design a model that will be amenable to validation and preferentially uncertainty analysis too. The final model should live up to scientific standards and be published scientifically, preferentially with full access to the open-source code.

**Acknowledgement** The author received funding from the Horizon 2020 Programme of the European Research Council (grant agreement 817617).

## References

Beck K (2002) Test driven development: by example. Addison-Wesley, Boston

CABI (2019) Invasive species compendium. CAB International, Wallingford

Conway GR (1984) Introduction. In: Conway GR (ed) Pest and pathogen control: strategic, tactical, and policy models. Wiley, New York, pp 1–11

Funtowicz SO, Ravetz JR (1993) Science for the post-normal age. Futures **25**:735–755

Holst N, Belete GF (2015) Domain-specific languages for ecological modelling. Eco Inform **27**:26–38

Holst N, Rasmussen IA, Bastiaans L (2007) Field weed population dynamics: a review of model approaches and applications. Weed Res **47**:1–14

Leslie PH (1945) On the use of matrices in certain population mathematics. Biometrika **35**:182–212

Martin RC (2006) Agile software development: principles, patterns, and practicese. Prentice Hall, Upper Saddle River, NJ

Nagel T (1974) What is it like to be a bat? Philos Rev **83**:435–450

Saltelli A, Ratto M, Andres T, Campolongo F, Cariboni J, Gatelli D, Saisana M, Tarantola S (2008) Global sensitivity analysis. Wiley, New York

Seemann M (2012) Dependency injection in.NET. Manning Publishing, Shelter Island, NY

Werner PA, Caswell H (1977) Population growth rates and age versus stage-distribution models for Teasel (*Dipsacus sylvestris* Huds.). Ecology **58**:1103–1111

Wittgenstein L (1922) Tractatus logico-philosophicus Harcourt. Brace and Co, New York

Zwerger P, Hurle K (1989) Studies on the relative importance of single life history parameters for weed infestation. Zeitschrift fur Pflanzenkrankheiten und Pflanzenschutz **96**:346–352

# Chapter 2
# Introduction to Decision Support Systems

José L. González-Andújar

**Abstract** Decision support systems (DSSs) are computer programs that, by using expert knowledge, simulation models and/or databases, are of assistance in the decision-making process as they offer management recommendations and/or options. The principal aim of a DSS is to improve the quality, speed and effectiveness of decisions. Since their beginnings in the 1960s, DSSs have been established as being an effective decision-making tool in different areas including agriculture. Weed science has not been immune to their influence, and since the end of the 1980s, a batch of DSSs have been developed towards the recognition and identification of seeds and seedlings, herbicide selection and the economic assessment of management strategies. Despite being powerful tools, DSSs have certain constraints and also a given resistance to their use. I hope that this chapter will serve to give a general insight into DSSs and their use in weed science, as well as to encourage the spreading of these systems in order to establish sustainable agriculture.

**Keywords** Decision-making · Information technology · Model · Data base · User interface · Agriculture · Herbicide · Agriculture · Computer-based systems

## 2.1 Introduction

Humans, as rational beings, make decisions between possible alternatives for solving conflicts, problems or proposals to be made in the future. This act of human intelligence requires a process of analysis and discernment called 'decision-making process'.

Ever since humans have existed on the earth, they have had to face making decisions, so it is difficult to imagine a field of a greater transcendence to humankind

J. L. González-Andújar (✉)
Department of Crop Protection, Institute for Sustainable Agriculture, Córdoba, Córdoba, Spain
e-mail: andujar@ias.csic.es


G. R. Chantre, J. L. González-Andújar (eds.), *Decision Support Systems for Weed Management*, https://doi.org/10.1007/978-3-030-44402-0_2

(Chen et al. 1999). The decision-making process has been inherently linked to human behaviour since ancient times. There is no doubt that the first human beings had to make decisions about the difficult tasks of feeding and protecting themselves in order to survive. Centuries later, as human settlements began to take shape due to the change towards the sedentary lifestyle, the activities became more elaborated as the groups grew and interacted with each other. Thus, the complexity of decision-making enormously increased.

In everyday life, we are constantly confronted with situations that we do not know how to resolve. Faced with a problem, it is necessary to make a decision (or not to make one). When doing so, people use their previous knowledge and experience, weighing up the consequences that each alternative will have, and then they choose that which appears to be sufficiently rational. The idea is to obtain the best result that, sometimes, involves maximizing a product to be obtained, or minimizing certain consequences of the action. In this process, it is also possible to predict future problems resulting from current decisions and even to foresee future ones. Uncertainty is also part of the decision-making process, and it is there that our intelligence is demonstrated.

The quality of decision-making is often influenced by the information available, so, frequently, the 'best decision' is not made due to the lack of necessary data, or due to the fuzziness or difficulty of the data process, or even due to lack of necessary time of analysis.

As civilization has evolved so have science and the art of making decisions (Coulson and Saunders 1987). Nowadays, computers have a great capacity for extracting precise, concise and relevant information through the use of databases and data managers or processors. Data processors perform the task of selecting, filtering and presenting information to the user in an easy-to-use way. Thus, what has been called 'a lack of information due to an excess of it' is prevented. Since computer technology burst into human society, diverse tools have been developed with the aim of helping people to resolve their oldest problem: decision-making. These tools are generically denominated decision support systems (DSSs).

## 2.2 Definition and History of Decision Support Systems

### *2.2.1 Definition*

It is no easy task to choose a definition of DSS among the many found in the literature. Actually, 'decision support system' is a highly generic term for which there are numerous definitions (Keen and Scott-Morton 1978; Bonczek et al. 1981; Schmoldt and Rauscher 1996; Sprague 1980, etc.). A DSS can be described as a computer system (software-, app- or weed-based products) that, through using knowledge from experts, simulation models and/or databases, helps in the making of decisions by providing management recommendations and/or options (Knight 1997). The main objective of a DSS is to improve the quality, speed and effectiveness of decisions made by users.

Different authors have envisioned this technology from different perspectives and have placed an emphasis on varied aspects, such as: (a) DSSs are computer-based systems; (b) their goal is to help in decision-making; (c) DSSs are applied to badly structured and complex problems that typically present themselves to users making tactical and/or strategic decisions; when it is a structural decision like, for instance, rejecting fruit of a smaller or larger diameter than that stipulated for a particular commercial category, or stopping the fruit conveyor belt when there are sufficient cases necessary for a shipment, the decision is so mechanical that no human action is needed; (d) DSSs are interactive (i.e. the user can communicate with the system); and finally (e) DSSs use data and models.

### 2.2.2 *Brief History*

It is difficult to pinpoint the exact date of the birth of the DSS. Keen and Scott-Morton (1978) mentioned that the concept of the DSS arose out of two research lines during the 1950s–1960s: (1) the theoretical studies of decision-making in the context of an organization, developed at the Carnegie Institute of Technology (USA), and (2) the technical studies on interactive computer systems, mainly performed at the Massachusetts Institute of Technology (MIT), (USA).

The DSS approach has evolved considerably since its advent. In principle, DSSs were oriented towards a series of highly specialized applications that, based on the knowledge of entrepreneurial sciences and statistics, assisted business management. They were later diversified to address other fields, incorporating ideas and knowledge of other scientific disciplines, among which operational research, simulation and artificial intelligence stand out (Gonzalez-Andujar 2003).

The introduction of information technology (IT) into the rural sphere was considered as a 'new revolution' in agriculture, similar to agricultural mechanization; thus, a great concern arose about the social impact that it might cause in contributing to the rural exodus (Zullo 1995). The objective of the initial IT-based programs elaborated for use in rural areas was to enable substitution of workers in repetitive tasks to free them for other more important activities. Those programs were geared to the solution of quite specific problems in rural properties on a local scale. Despite initial fears, no manpower exclusion was noted in those times, but the programs did free some workers to carry out more important jobs in their respective areas. They helped professionals to improve the reliability in their work, releasing them from activities that were reiterative and exhausting and incurred a risk of errors (Beraldo and Zullo Jr 1986).

Although in the early 1960s simulation models, which could be considered to be 'primitive' DSS, began to be employed in agriculture, the application of these systems in agricultural and agri-food contexts stopped by the 1980s and 1990s (Rubio 2002).

In the mid-1980s, scientific works made references to the usefulness of computers for performing environment samplings, for recording field data, for the development of models, and for database management (Teng and Rouse 1984). At that time, portable computers began to appear, and their potential for the organization of abundant information permitting a holistic approach to the agroecosystem was visualized. The models permitted the simulation of the performance of systems in different environments. The first ones were set up as games for the teaching of epidemiology and the management of plant diseases. That was the case of APPLESCAB at Michigan University (USA), an interactive simulator for the computer assigned to the teaching of the handling of mange in the apple tree, the most important fungal disease affecting that crop.

The DSSs in their original version offer numerous possibilities for agricultural development that can be specified in the following points. First, they constitute an ideal tool for the transfer of technology, permitting the handling, interpretation and transmission of scientific information from research centres (universities, private-sector researchers) to extension services and agricultural entrepreneurs (Gonzalez-Andujar 1995). Second, through the information modelling process, they point out gaps in knowledge and their importance, informing researchers on where they should direct their investigation efforts. Third, they are of assistance in decision-making using abundant and complex information. Finally, they combine suitable conditions for being used as educational tools, since they are interactive, graphic and affordable, allowing the user to situate in different scenarios, facilitating the understanding of the key aspects determining decision-making.

Nowadays, most farmers in industrialized countries use computers for aiding decisions. In underdeveloped ones, their dissemination has been more restricted, but it is rapidly increasing. Despite the fact that the use of computers was originally concentrated by farmers with a greater economic power owning large-sized properties (Francisco and Pino 2002), nowadays, the popularization of electronic networks like Internet and the social networks is providing new possibilities in the dissemination of DSS information.

### *2.2.3 Structure*

Three fundamental components of DSS architecture are (Marakas 1999) (Fig. 2.1):

1. The database (or knowledge base).
2. The model (i.e. the decision context and user criteria).
3. The user interface

In order to offer appropriate assistance in decision-making, DSS require abundant and precise information (Bajwa and Kogan 2001). The value of any DSS is proportional to the extent and quality of its knowledge base. The models are the programs representing the decision context. The interface with the user is the set of components employed by people to communicate with the computer. The quality of

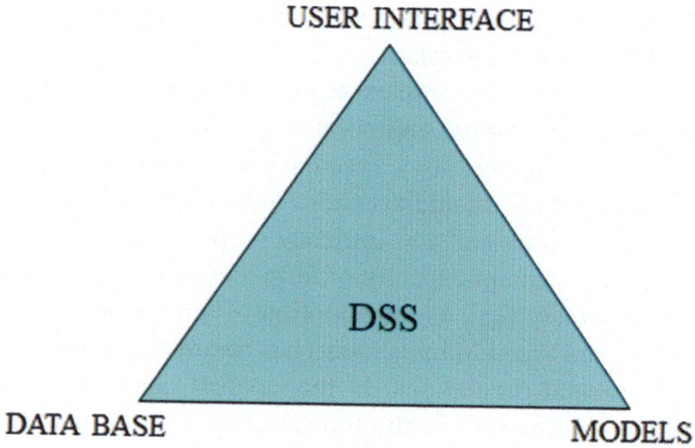


**Fig. 2.1** General structure of a decision support system

the user interface could be one of the reasons that lead to the success (or failure) of the system, since, if it is not appropriate, the user will refuse to use the system even when it has been elaborated in the best way possible.

### *2.2.4 Characteristics*

**The following characteristics of DSS can be addressed**

- *Interactivity*: provide dynamic and interactive options that permit the user to see and interpret the answers easily.
- *Easy to use*: be as simple and intuitive as possible even for unexperienced users.
- *Integration between systems*: permit access to information stored in global databases.
- *Decision types*: provide assistance for structured or non-structured decisions; the former are those which can be resolved by the tool without needing any human help (algorithms); the latter are decisions for which it is not possible to design an algorithm.
- *Adequate information for each user*: each user will only have the information that they require and that is related to their profiles.

## 2.3 Applications of DSS in Agriculture

The agricultural sector has been characterized as being constituted by a large amount of production units spread over wide geographical areas. This sector has traditionally been relegated as far as its access to information and to services is concerned mainly in Third World countries. Its production is intended for the consumer markets in large cities that are distinguished by their constant evolution. This means that the agricultural sector has to be permanently kept up-to-date in order to be able to meet the new demands of these markets.

Agricultural production is a complex activity that requires abundant technical knowledge in areas like economy, engineering, production (horticulture, fruit and flower growing, etc.), phytopathology, weed science, entomology, post-harvesting and marketing, among others. In addition, also the complexity and changeability of agricultural legal regulations and the intricacy of production systems sometimes rapidly exceed the management capacity of farmers and technicians. Such inherent complexity has partly increased, as the adoption of more sustainable management practices (e.g. Integrated Weed Management) has become an actual demand by both society and governmental regulations. In this context, agricultural extension systems have to play a leading role in the transference of high-quality technological knowledge to farmers and other related sector agents.

Another prominent characteristic of agriculture is the high degree of uncertainty existing when making decisions. This uncertainty could be associated to several factors such as lack of sufficient data and incomplete knowledge of biological systems and to specific features of biological processes. Although the number of variables intervening in agricultural processes is relatively small, their behaviour and relationships with each other are aleatory and imprecise (Rubio 2002). Thus, farmers are averse to taking risks, and they react by simplifying the production system and deliberately using external inputs with the aim of reducing the risks. One example of this is the excessive application of pesticides in order to diminish the risk of losses from pests (diseases, weeds or insects).

The increase in complexity in primary production due to the demands of the world agri-food market, the need to employ sustainable production practices and an increase in information and information flow have led to the development of DSS (Bonczek et al. 1981). Numerous DSSs have been set up for different purposes for agricultural sciences and have been triggered by the tremendous progress made in IT aimed at increasing the efficiency and effectiveness of the decisions made by farmers (Lentz 1998; Divya and Sreekumar 2014). These systems have been created in different agronomic areas such as plant production, animal production, plant protection, management and planning, agri-food production, environment and forestry production (Carrascal and Pau 1992; Gonzalez-Andujar and Recio 1996; Manos et al. 2004). An example of these is Decision Support System for Agrotechnology Transfer (DSSAT) to estimate production, resource use and risks associated with different crop production practices. DSSAT contains crop-soil simulation models; databases for weather, soil and crops; and strategy evaluation programs integrated with a user-friendly interface on computers and comprises crop simulation models for over 42 crops (Hoogenboom et al. 2019). An example of practical application of DSSAT is presented in Chap. 15.

In crop protection, one of the first DSSs was PLANT/ds developed in 1983 by Michalski et al. (1983), which diagnosed diseases in soybean in Illinois (USA) and which served as a reference for the creation of other systems, especially for diagnoses.

The exchange of information by electronic means is regenerating the role of the extension services to provide information, training and assistance in decision-making to agricultural producers. Nowadays, a digital photograph of a plant can be sent to an agricultural assessor in order to obtain a presumptive diagnosis and suitable advice or treatment recommendations. These services are rapidly accepted and highly appreciated by the producers.

## 2.4 Decision Support Systems in Weed Science

In this section, it is not intended to make an exhaustive review of DSSs developed in weed science but, rather, to give an overview of their history. In the following chapters, some examples of specific applications will be shown.

DSSs began to be used in weed science at the end of the 1980s, beginning of the 1990s, the main beneficiaries of their employment being three areas:

### *2.4.1 Recognition and Identification of Seedlings and Seeds*

Some systems were oriented towards identifying weed seeds (Naidu et al. 2015), but they were mainly focused on the identification of weeds in their seedling state, when they are easier to control. Their identification in that state is much harder than when they reach the adult stage. The difficulty found by non-botanists for weed detection in using classic tools like handbooks (too technical) led to the creation of new computer-assisted identification tools.

The first models were generally based on the transference to the computer of already available dichotomous keys, with a certain refinement and presented in an accessible manner (Lonchamp et al. 1991; Pascual 1994).

Although they are usually of some help, those programs are far from act in the way in which the experts resolve this taxonomical problem; by highlighting a few relevant characteristics, they arrive at determining the species, without recovering all the elements of the dichotomous keys. One advance was the incorporation of specialist knowledge and graphic aids, permitting a greater flexibility and speed in identification (Gonzalez-Andujar et al. 1990; Ballegaard and Haas 1990). SIMCE (Gonzalez-Andujar et al. 2006), a DSS developed to identify weed seedlings in cereals (Fig. 2.2) by means of the combination of a text, expert knowledge and photographs, allows the identification of 41 species commonly found in cereal crops in Spain. The inclusion of specialist knowledge enabled heuristic information to be included in the identification process. This system was validated for its educational capacity with a group of students with no knowledge of identifying weeds. Aided by the SIMCE, they were able to correctly identify 70% of the species evaluated.

Another system with similar characteristics is WEED-ONE, created for the visual recognition of weed seedlings (Olmo and Recasens 1995). It consists of a

**Fig. 2.2** Snapshot of the start page of the SIMCE DSS

collection of photographs, a multimedia database and a self-evaluation module. It contains approximately a 100 species and has been used for educational purposes.

Recognition and identification systems have gone on evolving, with the growth of technology, from computers to smartphones. An example of this is WEEDSCOUT, an application developed by Bayer that recognizes weeds by means of images taken with the telephone. The system possesses a large database of weed images. With over 30,000 images registered, the application relies on a collaborative update system, which is compiled through the use of photos taken by users to enhance algorithms.

### 2.4.2 *Herbicide Selection*

A second group of DSSs is formed by those systems that select the treatments based on the effectiveness of herbicides (Linker et al. 1990; Sonderskov et al. 2014). First approaches were designed by computerizing weed guides used by the agricultural extension services. Regardless of how their databases were organized, all of them selected the most suitable method based on herbicide efficacies.

Another step forward was the selection of herbicides based on their cost-benefit (Mortensen and Coble 1991). Within this group, some examples like CHEX (Bolte et al. 1988), HERB (Bennet et al. 2003; Wilkerson et al. 1991) and SOYHERB (Renner and Black 1991) can be cited.

In HERB, the determination of herbicide efficacy depends on the growth stages of weeds (three classes) and soil moisture (two levels). As such, six efficacy levels are given for each herbicide and weed species. SOYHERB, developed by Michigan University (USA), gave pre- and post-emergence guidance for weed control in soybean. The pre-emergence advice was based on weed infestation estimates.

### 2.4.3 DSS with Models in Their Structure

The last group is formed by DSSs that include models inside their structure. The most common are bioeconomic models, which permit management strategy selection to be based on a cost-benefit analysis (Wilkerson et al. 1991; Nordblom et al. 2003; Lindsay et al. 2017; Lacoste and Powles 2015). Bioeconomic models are integrated by a population dynamics model and/or a weed-crop competition and an economic model (Fig. 2.3).

SEMAGI was created by Castro-Tendero and García-Torres (1995) to assess the potential yield reduction from weeds and parasitic plants in sunflower and to ascertain a suitable herbicide selection. It combines databases related to herbicides, weeds and their interaction. It comprises a total of 34 species of weeds and 26 herbicides. The system processes and selects the herbicide(s) considering the limitations of the effectiveness data of the herbicide and of a weed competition model. In addition, SEMAGI provides an economic study of any herbicide treatment selected or introduced by the user, based on the cost of the herbicide, the increase in the yield expected from the weed control treatment and the sale price of the sunflower.

DSSAVENA is a system that permits the economic evaluation of different types of strategies to control wild oats (*Avena sterilis*) in winter wheat in the central area of Spain (Gonzalez-Andujar et al. 2011). It employs agronomic (infestation density, previous crop, etc.) and economic data (inflation rate, price of herbicide, etc.) (Fig. 2.4) (for a more extensive description, see Chap. 14).

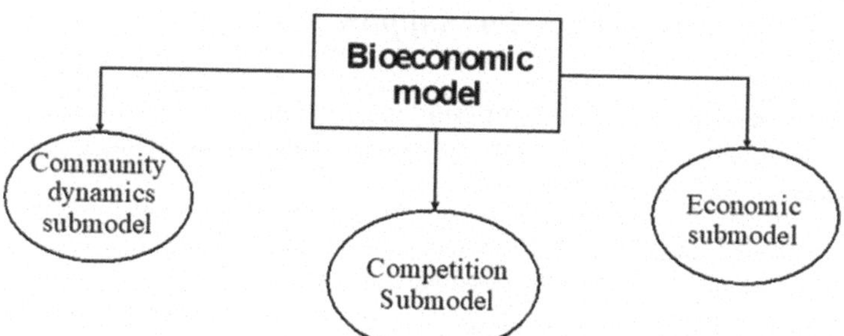


**Fig. 2.3** Bioeconomic model structure

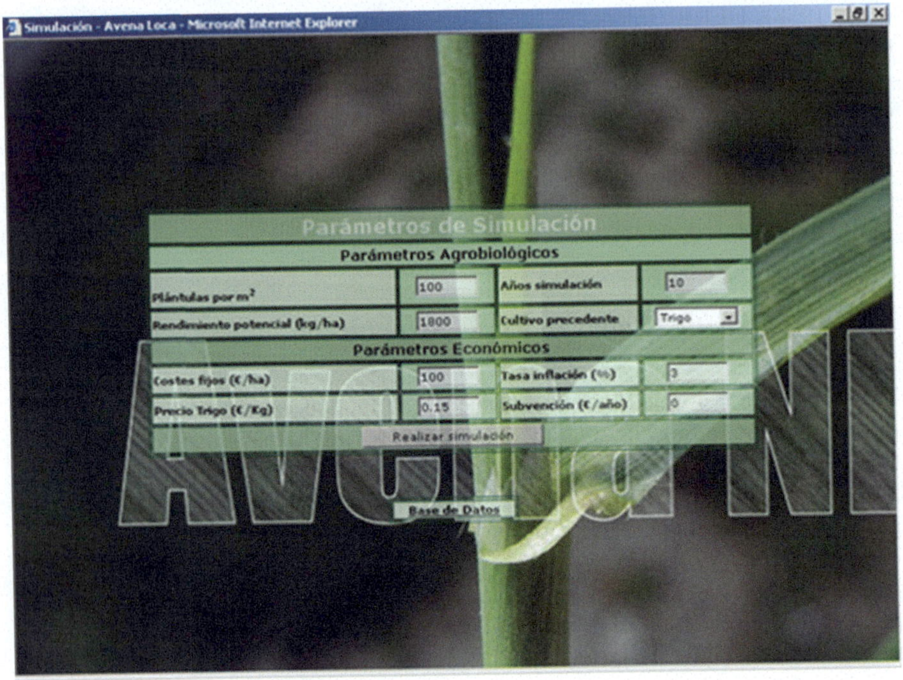


**Fig. 2.4** Data entry screen of DSSAVENA

Other DSSs are those based on weed emergence models that help the users to optimize the use of weed seedling control measures. For instance, RotPiña is a DSS developed to predict the emergence of *Rottboellia cochinchinensis* (Leon et al. 2015) in pineapple in Costa Rica. It employs meteorological information for predicting both timing and quantity of cumulative percentage of emergence (Fig. 2.5). For a detailed description of weed emergence tools, the reader is referred to Chap. 5.

### 2.4.4 Constraints of Decision Support Systems

Being too dependent on a decision support system and depositing an unusual amount of confidence in it is not a healthy sign. There are many uncertainties associated with the DSS (Parker 2004), like:

**An Excess of Information** A DSS can sometimes end up with an overload of information. Since it analyses all the aspects of a problem, it leaves the user with the dilemma of what to take into consideration and what not to; it is not necessary to use

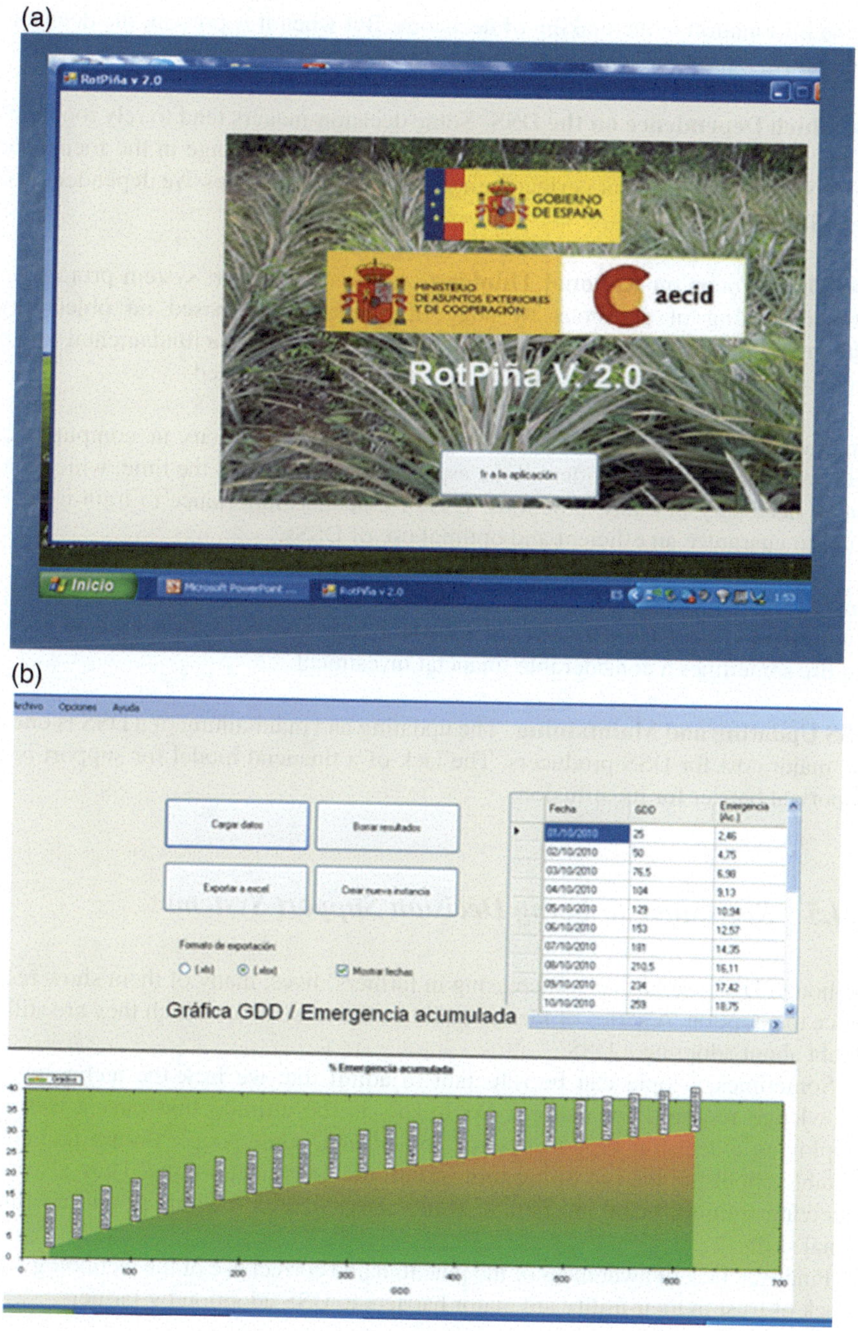


**Fig. 2.5** Start (**a**) and output (**b**) screens in the DSS RotPiña

all the information in the making of decisions. But when it is present, the decision-maker finds it hard to ignore the information that is not a priority.

**Too Much Dependence on the DSS** Some decision-makers tend to rely too much on the computerized decision-making. There is clearly a change in the focus, and decision-makers might not perfect their skills due to their excessive dependence on the DSS.

**Too Much Focus on Rational Thinking** A decision support system promotes a rational making of decisions by suggesting alternatives based on objectivity. Although limited rationality or constrained irrationality plays a fundamental role in decision-making, subjectivity cannot and should not be shunned.

**An Excessive Focus in Decision-Making** Clearly, the focus in computerized decision-making is to consider all the aspects of a problem all the time, which may not be necessary in many situations. It is of essential importance to train users in order to guarantee an efficient and optimal use of DSSs.

**Cost of Development** The cost of decision-making is reduced once a decision support system is installed. However, the development and implementation of a DSS require sometimes a considerable financial investment.

**DSS Updating and Maintaining** The updating and maintaining of a DSS is one of the major cost for DSS producers. The lack of a financial model for support is an important barrier for its diffusion.

### *2.4.5 Resistance to Using Decision Support Systems*

Although DSS use has been increasing in farmers' lives, many of them show resistance to adoption. A series of factors could be the reasons for which they are still in doubt about adopting a DSS.

Sometimes, people can be reluctant to admit that we lack the technological knowledge required for using a DSS. This is the attitude that puts a user off employing a decision support system. In addition, some people do not feel comfortable about the idea of doing things with the latest technology. They also fear receiving training or taking part in workshops directed towards providing functional skills.

Finally, a lack of awareness of the potential and correct use of the technology and a lack of trust in their utility are major barriers to DSS adoption by farmers.

## References


Bajwa WI, Kogan M (2001) Informática y apoyo de decisiones de MIP con base en internet. In: Radcliffe EB, Hutchison WD (eds) Radcliffe's IPM world textbook. University of Minnesota, St. Paul, MN

Ballegaard T, Haas H (1990) WEEDEX—an expert system identification of weed seedlings. In: Skov S, Ballegaard T, Mikkelsen SO, Moller MF, Rasmussen LK (eds) Expert systems in agricultural research. Pub Udgivet af Statens Planteavisforg, Denmark, pp 27–31

Bennet AC, Proce AJ, Sturgill MC et al (2003) HADSS, Pocket HERB, and WebHADSS: decision aids for field crops. Weed Technol 17:412–420

Beraldo AL, Zullo J Jr (1986) Sistema Topográfico Computacional, 38ª Reunião Anual da SBPC

Bolte JP, Moss ND, Burrill L, Appleby A (1988) A herbicide recommendation expert system. Abstract. American Society of Agronomy, p 66

Bonczek RH, Holsapple CW, Whinston A (1981) Foundations of decision support systems. Academic Press, New York

Carrascal MJ, Pau LF (1992) A survey of expert systems in agriculture and food processing. AI Applications 6: 27–49

Castro-Tendero AJ, García-Torres L (1995) Semagi—an expert system for weed-control decision-making in sunflowers. Crop Prot 14:543–548

Chen WC, Hong TP, Jeng R (1999) A framework of decision support systems for use on the world wide web. J Netw Comput Appl 22:1–17

Coulson RN, Saunders MC (1987) Computer-assisted decision-making as applied to entomology. Annu Rev Entomol 32:415–437

Divya J, Sreekumar K (2014) A survey on expert systems in agriculture. Int J Comput Sci Inform Technol 5:7861–7864

Francisco VLFS, Pino FA (2002) Farm computer usage in Sao Paulo state, Brazil. Rev Bras Agroinform 4:81–89

Gonzalez-Andujar JL (1995). Modelling the effects of climate change and climatic variability on crops at the site scale: effects on cereal weeds. En: Climate change and Agriculture in Europe: Assessment of Impacts and Adaptations (Ed. Harrison, PA, Butterfield RE & Downing, TE) Research report nº 9, Environmental Change Unit. University of Oxford.

Gonzalez-Andujar JL (2003) Los sistemas de soporte a la decisión en protección de cultivos. Su utilización en Malherbologia. Phytoma España 146:37–40

Gonzalez-Andujar JL , Fernandez-Quintanilla C, Izquierdo J, Urbano JM (2006) SIMCE: an expert system for seedling weed identification in cereals. Computers & Electronics in Agriculture 54:115–123

Gonzalez-Andujar JL, Recio B (1996) Aplicación de los Sistemas Expertos en Agricultura. MAPA-Mundiprensa, Madrid

Gonzalez-Andujar JL, Rodriguez J, Navarrete L (1990) Development of a prototype expert system (SIEXMAL) for identification of weeds in cereals. In: Proceedings of the European Weed Research Society, integrated weed management in cereals, pp 429–434

Gonzalez-Andujar JL, Fernandez-Quintanilla C, Bastida F, Calvo R, Izquierdo J, Lezaun JA (2011) Assessment of a decision support system for chemical control of annual ryegrass (*Lolium rigidum*) in winter cereals. Weed Res 51:304–309

Hoogenboom G, Porter CH, Shelia V, Boote KJ, Singh U, White JW, Hunt LA, Ogoshi R, Lizaso JI, Koo J, Asseng S, Singels A, Moreno LP, Jones JW (2019) Decision support system for Agrotechnology transfer (DSSAT) version 4.7.5 (https://DSSAT.net). DSSAT Foundation, Gainesville, FL

Keen PGW, Scott-Morton MS (1978) Decision support systems: an organizational perspective. Addison-Wesley, Reading, MA

Knight JD (1997) The role of support systems in integrated crop protection. Agric Ecosyst Environ 64:157–163

Lacoste M, Powles S (2015) RIM: anatomy of a weed management decision support system for adaptation. Weed Sci 63:676–689

Lentz W (1998) Models applications in horticulture: a review. Sci Hortic 74:151–174

Leon RG, Izquierdo J, Gonzalez-Andujar JL (2015) Characterization and modelling of itchgrass (*Rottboellia cochinchinensis*) biphasic seedling emergence patterns in the tropics. Weed Sci 63:623–630

Lindsay K, Popp M, Norsworthy J, Bagavathiannan M, Powles S, Lacoste M (2017) PAM: decision support for long-term palmer Amaranth (Amaranthus palmeri) control. Weed Technol 31:915–927

Linker MM, York AC, Wilthite DR (1990) Weeds—a system for developing a computer based herbicide recommendation program. Weed Technol 4:380–385

Lonchamp JP, Barralis G, Gasquez J, Janzein P, Kerguelen M, Le Klerk J, Maillet J (1991) Malherb, logiciel de reconnaissance des mauvaises herbes des cultures: approche botanique. Weed Res 31:237–245

Manos BD, Ciani A, Bournaris T, Vassiliadou I, Papathanasiou J (2004) A taxonomy survey of decision support systems in agriculture. Agric Econ Rev Greek Assoc Agric Econ 5:1–15

Marakas GM (1999) Decision support systems in the twenty-first century. Prentice Hall, Upper Saddle River, NJ

Michalski RS, Davis JH, Bisht VS, Sinclair JB (1983) A computer-based advisory system for diagnosing soybean diseases in Illinois. Plant Dis 67(4):459–463

Mortensen DA, Coble HD (1991) Two approaches to weed control decision-aid software. Weed Technol 5:445–452

Naidu VSGR, Ravisanker H, Dhagat S, Kamalvanshi V, Sharma AR (2015) Knowledge based system for weed seed identification. Indian J Weed Sci 47(2):197–200

Nordblom TL, Jones RE, Medd RW (2003) Economics of factor adjusted herbicide doses: a simulation analysis of best efficacy targeting strategies (BETS). Agr Syst 76:863–882

Olmo JJ, Recasens J (1995) Weed-one: Sistema multimedia de enseñanza asistida por ordenador para el aprendizaje en el reconocimiento visual de plántulas de malas hierbas. In: Actas Congreso de la Sociedad Española de Malherbología, pp 109–117

Parker CG (2004) Decision support tools: barriers to updake and use. Asp Appl Biol 72:31–41

Pascual GM (1994) Development of an expert system for the identification and control of weeds in wheat, triticale, barley and oat crops. Comput Electron Agric 10:117–134

Renner KA, Black JR (1991) SOYHERB: a computer program for soybean herbicide decision making. Agron J 83:921–925

Rubio FN (2002) Paradigma de planificación avanzada empleando técnicas de propagación de restricciones: aplicación al sector agrícola/alimentario. Ph.D. Thesis. Politecnic University of Madrid

Schmoldt DL, Rauscher HM (1996) Building knowledge-based systems for natural resources management. Chapman & Hill, New York, 386 pp

Sonderskov M, Kudsk P, Mathiassen S, Bøjer OM, Rydahl P (2014) Decision support system for optimized herbicide dose in spring barley. Weed Technol 28:19–27

Sprague R (1980) A framework for the development of decision support systems. MIS Q 4:1–25

Teng PS, Rouse DI (1984) Understanding computers: applications in plant pathology. Plant Dis 68:539–543

Wilkerson GG, Modena SA, Coble HD (1991) HERB: decision model for postemergence weed control in soybeans. Agron J 83:413–417

Zullo J Jr (1995) A Utilização da Informática na Agropecuária. In: Agrosoft 95. Anais, Juiz de Fora

# Chapter 3
# Optimization in Decision Support Systems

**Aníbal M. Blanco**

**Abstract** Two important components of decision support systems (DSSs) are a model that describes the behaviour of the system under study along a certain period of time and an optimization algorithm that searches for one or several 'good' options within the space of admissible human interventions. Both elements received a great deal of attention in recent decades, mostly motivated by industrial and management applications. There are also many examples of decision support projects in agronomy, in particular, in weed control for crop protection. Since the bioeconomic models arising in Integrated Weed Management (IWM) studies are nonlinear, mixed integer and large scale, they are quite difficult to optimize. In this chapter, some basic elements of optimization are reviewed, with special emphasis in practical issues (modelling, programming), which are less covered than theoretical topics in the open literature.

**Keywords** DSS · Optimization · LP · NLP · MINLP · Mathematical programming · Metaheuristics

## 3.1 Introduction

Optimization has to do with identifying a good option among many … hopefully the best. It is a natural activity carried out by human beings during their decision-making processes in most aspects of their lives. If the situation is too complex to be solved by intuition or mental enumeration (too many options or too intricate search space), the optimization process can benefit from a mathematical model that hopefully contains all the alternatives in a compact fashion. Optimal solutions can be extracted from mathematical models by theoretical considerations or, more practically, by numerical algorithms.

A. M. Blanco (✉)
Planta Piloto de Ingeniería Química, PLAPIQUI (Universidad Nacional del Sur-CONICET), Bahía Blanca, Argentina
e-mail: ablanco@plapiqui.edu.ar


G. R. Chantre, J. L. González-Andújar (eds.), *Decision Support Systems for Weed Management*, https://doi.org/10.1007/978-3-030-44402-0_3

In particular, agronomical production environments are indeed complex systems. For example, Doole and Pannell (2008) estimate in $10^{119}$ the number of possible alternatives to operate with six possible rotation scenarios including five crops and 12 different weed control strategies for a 20-year period in the Western Australian wheat belt. The underlying model is known by its acronym RIM (Ryegrass Integrated Management) which was designed as a decision support system (DSS) with emphasis in herbicide resistance (a detailed description of RIM and its adaptations are presented in Chap. 12). Clearly, such a search space cannot be efficiently explored by enumeration without the aid of some algorithmic procedure.

Mathematical models and numerical optimization tools are the core components of modern DSS that can help managers to automate their decision-making processes. This chapter will focus on the optimization part as mathematical modelling is thoroughly addressed along the volume.

Due to its paramount importance to all fields of science and engineering, numerical optimization had a tremendous progress in the last decades and is still one of the most active and challenging fields within applied mathematics and computers science. Literally, hundreds of books, thousands of scientific papers and millions of lines of programming code have been written on the topic generating an overwhelming amount of theoretical results, practical tools and applications.

Although rooted in mathematics and computer science, the development of numerical optimization was definitely boosted by its application in areas such as Operations Research and Management Science. A particularly important landmark in the evolution of the field was the emergence in the early 1960s of the Process Systems Engineering (PSE), which is the application of systematic computer-based methods to process engineering. This field was pioneered by chemical engineers, in particular by Prof. Roger Sargent (Imperial College, UK) who is considered the father of the discipline. The link between chemical processes and numerical optimization becomes evident when considering that chemical plants typically operate in a continuous fashion for about 8000 hours per year. Therefore, even a modest improvement in the adopted performance criterion, typically economic benefit, has a tremendous impact when integrated along time.

The developments rapidly disseminated across all technical disciplines at a research level all around the world and were even adopted by industrial and commercial sectors in countries of technological leadership. For example, the family of commercial software AspenTech[1] allowed the implementation of Real-Time Optimization (RTO) of large petrochemical plants for more than two decades already.

Currently, most engineering careers of all disciplines worldwide include at least some elements of numerical optimization in their curricula. Many advanced courses in all branches of optimization are also offered in most post-graduate programs of science and engineering.

[1] https://www.aspentech.com/

This chapter intends to briefly present some topics on numerical optimization, which hopefully will be of interest to weed scientists and agronomists engaged in automated DSS. In the first place, a review of the literature on optimal model-based weed management is presented. Although such review does not pretend to be exhaustive, it will certainly provide some insight in recent developments on the field.

Regarding the abundant material on numerical optimization, such a broad discipline distinguishes 'theoretical' and 'practical' areas. Theoretical optimization is proficiently and didactically covered in a huge amount of textbooks, and many readers might be somewhat familiar with such contents. Therefore, only a panoramic view will be presented here. Emphasis will be done instead on 'practical' optimization, which is the main source of both frustration and delight of the 'hands-on optimizer'. Although there is also a lot of material on the topic, it is much less covered in the open bibliography, the reason being that practical optimization is heavily dependent on the specific adopted programming software or modelling platform. An overview of the different modelling philosophies will be presented together with specific tools to implement them, both commercial and non-commercial.

In all cases, relevant references will be provided. Such bibliography will be non-exhaustive and, of course, unavoidably biased since they reflect my personal background and preferences. However, they undoubtedly constitute a good starting point for further research.

## 3.2 Optimization-Based Agronomical DSS: A Review

This section concentrates on a small number of references that present mathematical model-based weed control studies that explicitly adopt some type of optimization approach. The review is limited to weed management on specific crop production systems. This is indeed a quite narrow scope within the agronomical DSS literature. In fact, it is even limited within the computer-based weed management literature (see Chap. 2). On one hand, it should be highlighted that DSS research based on weed control was in a large extent driven by simulation studies rather than by optimization efforts. Additionally, many studies address optimal control of weed infestations in ecosystems different than industrial crop settings, in particular to study invasive species. Finally, a great deal of research has been also devoted to weed control with emphasis in their spatial distribution. There is also a large body of research on optimization methods in pest management, which is also beyond the scope of this section. Finally, it should also be recognized that although the review intends to be exhaustive within its limited scope, the literature is so vast that it is quite likely that some valuable contributions will be missing. The reviewed optimization-based studies for weed control are detailed in Table 3.1.

A pioneer work by Fisher and Lee (1981) proposed an optimization approach to solve a crop rotation problem considering wild oats (*Avena fatua/Avena ludoviciana*) infestation in cereal crops of the wheat belt of New South Wales (Australia). Besides crop-weed competition, the effect of crown rot disease was also included.

**Table 3.1** Optimization-based studies in weed control

| Reference | Weed/crop | Country | Optimization | Control | Scope | | Objective | | Resistance |
|---|---|---|---|---|---|---|---|---|---|
| | | | | | Operational | Tactical | Economic | Environmental | |
| Fisher and Lee (1981) | Wild oats/wheat | Australia | DP | Chemical | | X | X | | |
| Taylor and Burt (1984) | Wild oats/wheat | USA | Stochastic DP | Chemical | | X | X | | |
| Pandey and Medd (1991) | Wild oats/wheat | Australia | Stochastic DP | Chemical/cultural | | X | X | | |
| Sells (1995) | Wild oats/ wheat-barley | UK | Stochastic DP | Chemical/cultural | | X | X | | |
| Gorddard et al. (1995) | Ryegrass/wheat | Australia | NLP | Chemical/ non-chemical | | X | X | | X |
| Wu (2001) | Foxtail-cocklebur/corn | USA | Optimal Control theory | Chemical | | X | X | | |
| Neeser et al. (2004) | Several/several | USA | Ranking | Chemical | X | | X | X | |
| Doole and Pannell (2008) | Ryegrass/several | Australia | Compressed annealing | Chemical/mechanical/ cultural | | X | X | | X |
| Chalak-Haghighi et al. (2008) | Californian thistle/pasture | New Zealand | DP | Chemical/mechanical/ cultural/biological | | X | X | | |
| Benjamin et al. (2009) | Several/several | UK | Stochastic DP | Chemical/mechanical/ cultural | | X | X | | |
| Lodovichi et al. (2013) | Wild oat/wheat | Argentina | MINLP | Chemical | X | | X | X | |
| Martinez et al. (2018) | Teosinte/corn | Spain | NLP | Manual/cultural | | X | X | | |

*DP* dynamic programming, *NLP* nonlinear programming, *MINLP* mixed-integer nonlinear programming

The approach seeks to maximize the net present value of the revenue stream over a 10-year planning horizon. The main optimization decision is the crop-fallow sequence out of the following options: (1) wheat (no control), (2) wheat + herbicide, (3) sorghum, (4) winter-fallow and (5) summer-fallow. Dynamic Programming (DP) was adopted as optimization framework.

Taylor and Burt (1984) presented a model similar to Fisher and Lee's that includes stochastic elements and allows the use of a pre-emergent herbicide for wild oat control within the decisions set. The model was parameterized for the spring-wheat production area of north central Montana (USA), and it was solved using stochastic DP optimization.

Pandey and Medd (1991) worked on wheat cropping systems of southern Australia. A multi-period bioeconomic model was developed and solved with a stochastic DP approach. In this case, the decision alternatives are the different doses of a selective graminicide (diclofop-methyl) for wild oat control ranging from non-use (no-control) to the maximum permitted dose.

In Sells (1995), a mathematical model is applied to a typical (250 ha) heavy land cereal farm and also optimized with the stochastic DP methodology. Optimization options are (1) four cereals (winter and spring wheat and barley), (2) timing of planting of each variety, (3) possible break crop (oilseed rape) and (4) five chemical options for wild oat control.

Gorddard et al. (1995) is a pioneer example of the application of nonlinear programming in IWM. The objective was to maximize the net present value over a 30-year horizon in wheat crops of Western Australia under annual ryegrass (*Lolium rigidum*). It was one of the first attempts to deal with diclofop-methyl (FOP) resistance evolution through a combination of chemical and non-chemical tactics.

Wu (2001) used results from optimal control theory to analytically find the amount (lb/acre) of herbicide to be applied each year in a 5-year planning horizon in order to maximize the present value of profits. The approach was illustrated with foxtail (*Setaria pumila*) and cocklebur (*Xanthium strumarium*) control using atrazine in corn fields of Iowa (USA).

WeedSOFT (Neeser et al. 2004) is a DSS based on a bioeconomic model that estimates crop yield loss from a multispecies weed complex and evaluates the net return of the system under different herbicide treatments. The software was developed by the University of Nebraska to address weed control studies in cotton, soybeans, sugar beet, corn and wheat in various agricultural areas of USA. The package includes an extensive database of pre- and post-emergence treatments considering crop rotational and environmental restrictions. A querying procedure ranks (in descending order) among potential eligible treatments according to final maximum yield (or net gain) values allowing for further analysis by the decision-maker. The optimization algorithm performs an exhaustive enumeration of the possible alternatives. WeedSOFT also includes rules to disallow treatments that do not meet environmental guidelines.

As previously mentioned, *Ryegrass Integrated Management* (RIM) is a deterministic bioeconomic simulation model that describes *L. rigidum* population dynamics across a 20-year horizon. It was originally developed by Pannell et al. (2004) for the central wheat belt of Western Australia, and it has been adapted to several other agroecosystems around the world. Doole and Pannell (2008) optimized the RIM model using a technique called compressed annealing. Annual ryegrass control options comprise 40 treatments, including chemical and non-chemical alternatives. Five crops (wheat, lupins, barley, clover, and serradella (*Ornithopus* spp.)) are combined in six possible rotation schemes. Such an amount of options along a 20-year horizon expands a search space of about $10^{119}$ possible intervention strategies, which cannot be explored by exhaustive enumeration. Good, near optimal solutions are therefore sought by compressed annealing at a reasonable computational cost.

Chalak-Haghighi et al. (2008) developed a bioeconomic model to study possible integrated management strategies to deal with Californian thistle (*Cirsium arvense*) in pastures of New Zealand. The model is solved with DP for a 40-year time horizon. Control options include the use of two hormonal herbicides (MCPA, MCPB), mechanical and cultural methods, biological methods such as cattle grazing, mycoherbicides (*Sclerotinia sclerotiorum*) and an insect biocontrol agent (weevil, *Apion onopordi*).

A stochastic DP approach was applied to maximize gross margin over the rotation using the Weed Manager DSS (Benjamin et al. 2009). The system is based on a population dynamic model parameterized for 12 common annual weed species in the United Kingdom. Rotations are defined by the user out of nine possible crops. The optimization variables are cultivation type (non-tillage, ploughing, rotary), sowing time (early, mid, late) and herbicide control efficiency in each crop (low, moderate and high).

In Lodovichi et al. (2013), mathematical programming was adopted as optimization approach to solve an operational-planning herbicide-based weed control problem, formulated as a mixed-integer nonlinear model. Differently to most previous approaches which have a strategic scope of several years, this model covered a calendar year with a daily discretization of the timeline. The adopted objective function was the economic benefit of the activity which explicitly considered the environmental impact as an external cost. The model was applied to a winter wheat/wild oat (*Avena fatua*) system of the semiarid temperate region of Argentina. Discrete optimization variables were the optimal time of chemical intervention (day of the planning horizon) within a range of eight non-selective (glyphosate and paraquat) and selective graminicides (FOPs, DEMs and DINs).

Martinez et al. (2018) proposed an optimization model to study optimal control strategies of the invasive teosinte (*Zea mays* ssp.) in corn growing settings of northeastern Spain. As no selective herbicide exists for teosinte control in corn, chemical interventions are unfeasible. The proposed bioeconomic model allows the prediction of the impact of seven control strategies (no control, manual control, false seed bed, and four rotations: barley-sunflower, pea-sunflower, alfalfa, wheat-alfalfa) on the net benefit along a 15-year horizon. Model analysis is applied from both single farmer economic level and from the farmer's community viewpoint, considering also public expenditures from the government, accounting activities such as divulgation, research, surveying, monitoring, etc. The resulting nonlinear model is solved with mathematical programming.

As can be observed in Table 3.1, the majority of such research has been performed on developed countries. Regarding the optimization approach, Dynamic Programming largely dominates the solution framework of the underlying bioeconomic models. DP is in fact the first adopted methodology to address the resulting multi-period problems, and only recently other methodologies such as mathematical programming and stochastic optimization arose within IWM DSS. It should also be highlighted that only a few studies explicitly consider the environmental impact or take into account the probability of resistance evolution associated to chemical control. Besides the specifics, it is clear that optimization has been reputed as a valid tool to address decision-making in crop/weed management scenarios at a research level for almost 40 years.

## 3.3 Elements of Optimization

From a formal point of view, a quite general formulation of an optimization problem is presented in Eq. (3.1). The main elements of such formula are:

**x:** Continuous variables
$\mathbf{x}^{lo}$, $\mathbf{x}^{up}$: Lower and upper bounds on continuous variables (data)
**y**: Binary variables
**Θ**: Parameters (data)
**h**(.): Equality constraints
**g**(.): Inequality constraints
OF(.): Objective function

$$
\begin{aligned}
&\min_{\mathbf{x},\mathbf{y}} \mathrm{OF}(\mathbf{x},\mathbf{y},\Theta) \\
&\text{s.t.} \\
&\mathbf{h}(\mathbf{x},\mathbf{y},\Theta) = \mathbf{0} \\
&\mathbf{g}(\mathbf{x},\mathbf{y},\Theta) \le \mathbf{0} \\
&\mathbf{x}^{lo} \le \mathbf{x} \le \mathbf{x}^{up} \\
&\mathbf{y} \in \{0,1\}
\end{aligned}
\tag{3.1}
$$

Bold letter indicates a vector, meaning that there may be many variables, parameters and constraints in our optimization model. The formulation is general in the sense that there are both continuous and binary variables, and the constraints admit any relation among variables, in particular nonlinear relations. This formulation is known as a Mixed-Integer Nonlinear Problem (MINLP). OF is the objective function, the index adopted to evaluate the performance of the system under study. Equality constraints **h**(.) establish the relations among variables. Inequality constraints **g**(.) establish conditions that should not be violated. Constraints and bounds define the so called 'feasible region'. Of course, we are interested in finding the combination of variables **x** and **y** within the feasible region that optimizes (in this case minimizes) our OF.

An important idea in optimization is that of 'degrees of freedom' (DOF). DOF is the number of continuous variables that can be independently manipulated in the search of the optimum solution. DOF is calculated as the number of continuous variables minus the number of equality constraints. If DOF equals zero, the solution(s) of the optimization problem is 'simply' given by the solution of the system of equality constraints $\mathbf{h}(\mathbf{x}, \mathbf{y}, \boldsymbol{\Theta}) = \mathbf{0}$. If DOF is equal or greater than one, there is probably some specific combination of variables that is better than the rest.

Continuous variables have very straightforward meaning in models and usually represent quantities that can be measured, such as plant densities and crop yields. Binary variables can have several sources. A binary variable can be used to model a decision (performing or not a particular action on the system, e.g. applying a specific herbicide in certain moment of the planning horizon). Binary variables are also used to model integer variables (an integer number can be modelled through a combination of binaries). Binary variables can also be used to model discontinuous functions such as saturations, absolute values and minimum/maximum among quantities.

## 3.4 Complexity of Optimization Problems

MINLPs represent lots of realistic systems. Most planning and scheduling problems in all sort of activities (including IWM) admit an MINLP formulation. Unfortunately, they are quite difficult to solve. There exist two main sources of complexity: nonlinear and combinatorial. On one hand, nonlinear functions can introduce non-convexities, which usually implies that the feasible region is complex (narrow, intricate). Non-convexities may also introduce local optima, meaning that there exist many feasible solutions that are in a certain region better than its neighbours. However, identifying the 'global best solution' among the possible 'locally best solutions' can be a task of considerable difficulty.

On the other hand, the presence of binary variables imposes integrality constraints (the variable can be only 0 or 1). Since binary variables are (selectively) fixed by many algorithms along the search process, they introduce a 'combinatorial complexity' which has to do with the problem of enumerating a large number of possibilities. The number of possible combinations for a number $n$ of binary variables is $2^n$, which is huge even for a modest $n$ (e.g. $2^{20}$=1,048,576).

If not all the elements are present in the formulation, several simplifications are possible. For example, if there are no binary variables and the relationship among variables is linear, Eq. (3.1) reduces to a linear problem (LP). If still there are no binaries but at least some variables are nonlinearly related, the problem is nonlinear (NLP). For example, in Fig. 3.1, a nonlinear objective function with one maximum and one minimum, subject to box constraints (bounds on variables $x_1$ and $x_2$) is shown.

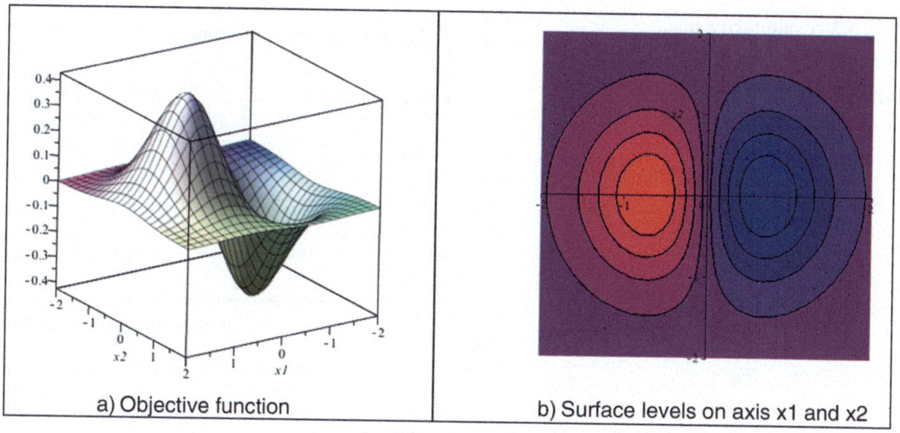


**Fig. 3.1** Nonlinear function subject to box constraints. (**a**) Objective function. (**b**) Surface levels on axes $x_1$ and $x_2$

If there are binary variables but OF and constraints are linear, the problem is Mixed Integer Linear (MILP). If all the complexity is allowed (nonlinear functions and binary variables), we recover our original MINLP. Of course, LPs are the easiest to solve, followed perhaps by MILPs. MINLPs are usually harder than plain NLPs.

As might be evident, there exists an additional source of complexity: scale. Large-scale problems are usually harder to solve than small ones. Scale can be represented on a hypothetical coordinate through an index made up of some combination of the number of variables, constraints, degrees of freedom and nonlinearities. In general, the larger this hypothetical index, the hardest the optimization problem. In Fig. 3.2, the optimization universe is represented on a *sui generis* complexity space. It is likely that our optimization problem of any kind will lay in some point of this space.

Before going a step further into optimization algorithms, it is necessary to mention an important result derived from the so called no-free lunch theorem of optimization (Ho and Pepyne 2002). According to such theorem, 'universal optimizers are impossible'. In other words, 'there is no strategy that outperforms all others on all problems'. Therefore, despite the numerous efforts in developing general optimization solvers, they are expected to efficiently perform only on a sub-set of the universe of optimization problems.

By exploiting the special characteristics of each of the different problems, numerical algorithms (solvers) have been developed to solve them more or less efficiently. In particular, if the variables are linearly related (LPs and MILPs), the problems are significantly simpler, and very powerful solvers exist to address very large instances efficiently. Conversely, nonlinear models do not have a particular structure that can be easily exploited, becoming far more difficult to solve. However, there exist some algorithms capable to identify local optima of large nonlinear models and also some global optimization solvers that can find the global solution if certain conditions are satisfied. The amount and diversity of algorithms are so vast

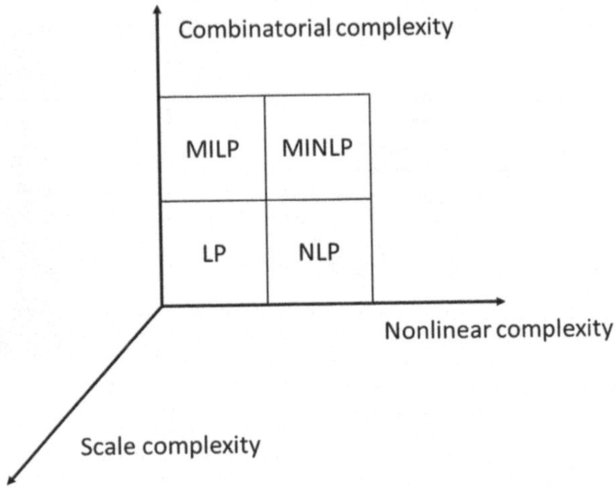


**Fig. 3.2** Complexity space of optimization problems

that only a brief reference will be done in this chapter, based on a broad typical classification that recognizes deterministic and stochastic solvers.

## 3.5 Deterministic Versus Stochastic Algorithms

Deterministic algorithms deal with binary variables using the so called 'branch and bound' techniques, which are 'intelligent' strategies to explore combinatorial spaces. Linear problems are addressed through simplex and interior point algorithms. Nonlinear solvers make strong use of function derivatives. These algorithms have been under development for several decades. They are mathematically sophisticated and work very efficiently in a large amount of science and engineering problems allowing the identification of exact solutions. For example, Fig. 3.3 illustrates a possible path followed by a deterministic algorithm to find the minimum of the function of Fig. 3.1 from a given starting point. The provided initial point is sequentially improved by the algorithm in the steepest descent direction (provided by the derivatives of the function) until optimality conditions are met.

There are lots of excellent textbooks on these topics. Some of my favourites are Floudas (1995), Nocedal and Wright (2006), Tawarmalani and Sahinidis (2002) and Biegler (2010). These methods are very successful but can be extremely time and memory consuming for real-world problems (large dimension, multimodal). In these cases, decomposition techniques are usually required to solve such instances in reasonable computation times (Conejo et al. 2006). Another weakness is that since they proceed by improving an initial solution, sometimes, they can be trapped in local optima (or end in an unfeasible point) rather than achieving the global solution.

The other large family of algorithms is the stochastic solvers, also known as metaheuristics. Metaheuristics are designed to solve in an approximate fashion a

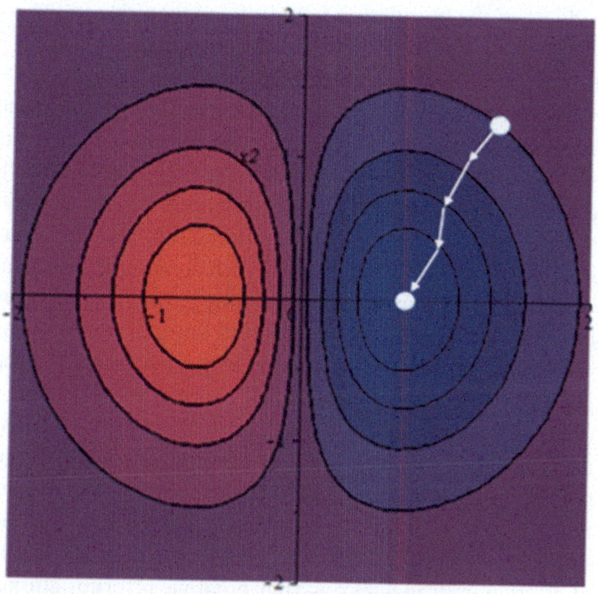

**Fig. 3.3** Hypothetical path followed by a deterministic algorithm to find the global minimum of an objective function

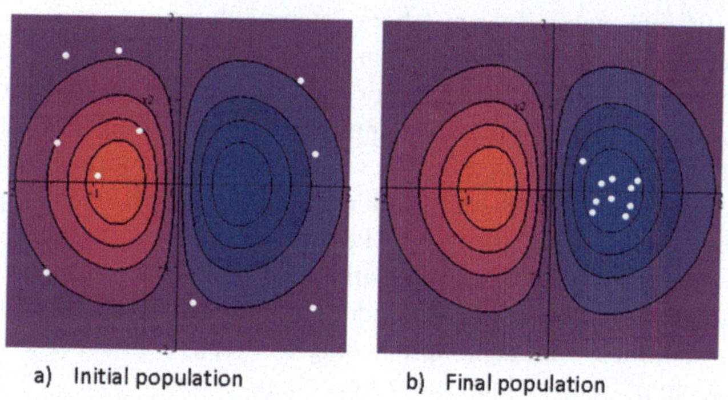


**Fig. 3.4** A population-based metaheuristic. (**a**) Initial population. (**b**) Final population

wide range of difficult optimization problems without specific adaptions to each one. Broadly speaking, these techniques operate by improvement of a population of points in the search space according to a set of rules usually inspired in natural or biological processes. A very important feature of these type of algorithms is the stochastic components. A proper balance between exploration of the feasible region and exploitation of the most promising points is required for efficient performance of a metaheuristic. Genetic algorithms and particle swarm optimization are among the most popular alternatives.

In Fig. 3.4, the initial and final states of a population-based metaheuristic are illustrated. First, the initial population is randomly generated within the box. As the algorithm proceeds, information from each individual and random components are used to decide the next position of every member of the swarm. Finally, the individuals of the population hopefully concentrate in the region of the global solution. The interested reader is referred to Boussaïd et al. (2013) for a complete survey on the subject. The main feature of these techniques is that they basically use function values of objective and constraints in the search procedure, without resorting to derivative calculation. Although this avoids a lot of mathematical complexity and sometimes precludes the entrapment in local optima, it is also the source of one main weakness: the difficulty of dealing efficiently with equality constraints. The reader is referred to Coello Coello (2002) and Crawford et al. (2017) for comprehensive reviews on how to deal with constraints in metaheuristic optimization.

A word on Dynamic Programming. As previously mentioned in the review, DP has been largely adopted to address optimal decision-making in weed management scenarios. DP is an optimization approach that does not fit exactly within the categories provided by the previous classification. Rather, it is an optimization technique based on the decomposition of original problem into a series of small, easy-to-solve sub-problems. Therefore, it is well suited for models that present a recursive/sequential/multi-stage structure. Since many industrial and agricultural management scenarios present such structure, DP has been one of the most popular options in these disciplines for many decades (Shoemaker 1981; Carter 1998). One of its major drawbacks is that creative reformulations of the models are usually required to apply this technique effectively. Additionally, the problem may become computationally intractable if the number of states is large.

Some Final Comments on Formulation (3.1). Although quite general, formulation (3.1) does not explicitly account for many important optimization problem variants, frequently raising in practical applications. For example, most real world systems are multi-objective (i.e. there are several different objective functions to be optimized, ideally in a simultaneous fashion). Moreover, such OF are typically conflicting, meaning that each objective can be improved only at the expense of worsening the others. Multi-objective optimization has received a great deal of attention in literature. Interestingly, many methodologies to deal with several objectives involve reformulations which end up in a single objective model (e.g. a composite function made up of the weighted sum of the single objectives).

Another important sub-class of optimization problems involves dynamic models. In these cases, it is required to optimize some performance index at a particular point in the timeline, or to ensure that a specific variable never violates some safety bound along its temporal evolution. To do so, it is necessary to solve an ordinary differential equation (ODE) system. One possible approach is to transform the ODE system into an algebraic equation system by approximating the differential terms through appropriate difference equations. This way, the dynamic optimization problem can be translated into our well-known MINLP, typically of a very large scale (Biegler 2010).

Other type of problems arises when our data ($\boldsymbol{\Theta}$, $\mathbf{x}^{lo}$, $\mathbf{x}^{up}$) are imperfectly known, meaning that they are uncertain, described, for example, through bounds or by some statistical distribution (mean and deviation). In this case, techniques have been developed which, for example, discretizes the parameter space and translates therefore the uncertainty into a very large MINLP. An overview on recent advances on the topic is provided in Grossmann et al. (2016).

Therefore, although Eq. (3.1) might not straightforwardly represent every problem under study, it is much likely that any problem can be reformulated as a large-scale MINLP. In particular, most bioeconomic models for IWM are nonlinear since many agronomical and biological variables are nonlinearly related (e.g., Cousens's hyperbolic yield loss function). These models may also involve discrete decisions (e.g., to apply or not to apply a given control action) in a specific moment of the planning horizon. Moreover, depending on the scope, planning horizons usually cover from several years to several decades. The discretization of the timeline may also consider several periods within each year to represent stages of agronomic significance. Such discretization of the timeline over such extended horizons expands very large models. Therefore, optimization models to assist in decision-making in Integrated Weed Management are, in principle, large-scale MINLPs. Moreover, the problem is necessarily multi-objective since at least two objectives of outstanding importance can be identified: economic and environmental. Finally, uncertainty is of course present along the whole system. Biological and agronomical variables/parameters of crops and weeds are difficult to measure and usually require long-term field, greenhouse and laboratory experiments to be estimated. Weather variables such as daily temperatures and particularly precipitations are very difficult to forecast even within short-term periods.

## 3.6 Practical Optimization

Practical optimization has to do with putting hands on optimization problems. Basic theoretical topics can be efficiently covered in one or two courses. However, solving a specific optimization problem with a computer requires (1) skills on mathematical modelling and (2) mastering some programming language or development platform. Usually, the required modelling skills and the adopted development software are quite interrelated. Although the acquisition of such abilities can of course be tutored, a self-educated experience-based process is usually unavoidable. Most solvers perform well close to the origin of the complexity space of Fig. 3.2. The further we move in any direction, the more sophisticated software is needed, and the more specialized modelling skills are required.

Modelling skills basically imply the wise application of a specific set of techniques, notions and concepts to formulate the problem under study in order to solve it as efficiently as possible. Good modelling is somewhat an art, besides a technical discipline. A specific set of modelling abilities are needed to address Mixed-Integer Linear Problems and another set to address nonlinear problems.

It should be also mentioned that both types of problems are in some extent interchangeable. Specifically, binary variables can be reformulated in a continuous fashion through a rather standard reformulation: (1) each variable is relaxed between its bounds ($0 \leq y \leq 1$), and (2) an additional nonlinear constraint is introduced per variable to force integrality ($y(1 - y) = 0$). Then, an MILP or MINLP can be transformed into a plain NLP. This way, the combinatorial complexity is avoided at the expense of increasing the nonlinear and scale complexities. On the other hand, a nonlinear function can be approximated by linear segments with a large degree of accuracy through piecewise linearization. In this technique, the linear segments are organized with the aid of binary variables and additional sets of linear constraints. Then, an NLP or MINLP can be transformed through piecewise linearization into an MILP. In other words, the nonlinear complexity can be overridden by increasing the combinatorial and scale complexities.

Modelling techniques are widely spread over many sources for both nonlinear and mixed-integer problems. Many can be drawn directly from textbooks. For example, the modelling of logic constraints and logic inference, a very important part in developing integer models, is well developed in Biegler Lorenz et al. (1997). Other concepts can be obtained from the documentation of the optimization platforms/solvers themselves. For example, the section 'Hints on good model formulation' of the CONOPT Solver manual is a very good compilation of techniques to develop nonlinear models. Finally, excellent material can be found on the Internet. For example, the online material by consultant Erwin Kalvelagen[2] is highly recommended.

Regarding integer models, the wise use of binary variables to represent on/off situations is vital. With the aid of big $M$ constraints, lots of different logics can be modelled. Just to provide some insight in the type of required modelling skills, consider the situation where a continuous variable $x$ can only take a specific value, say, $a$, if some particular situation occurs represented by binary variable $y$ taking value 1. Equation (3.2) represents a big $M$ formulation. Parameter $M$ is a large constant, much larger than $a$. If $y = 0$, variable $x$ is relaxed: $(a - M) \leq x \leq (a + M)$, meaning that it can take any value within its bounds. If $y = 1$, big $M$ terms vanish; therefore, $a \leq x \leq a$, and then $x = a$. Interestingly, Eq. (3.2) is just a linear set of equations.

$$x \leq a + M.(1 - y) \quad \text{and} \quad x \geq a - M.(1 - y) \tag{3.2}$$

In turn, good nonlinear modelling typically requires the provision of very good bounds and starting points on continuous variables. Since derivatives are involved, good scaling of variables and equations is also crucial for good performance. Other good modelling practices involve reformulations. For example, $x_1 = x_2/x_3$ is much better posed for numerical optimization purposes programmed as $x_1.x_3 = x_2$.

## 3.7 Development Software

Regarding development software, an advantage is the large amount of available options, both commercial and non-commercial. Commercial optimization software requires the purchase of an appropriate license. Typically, optimization software companies offer academic licenses for teaching and research institutions at quite

[2] http://yetanothermathprogrammingconsultant.blogspot.com/

affordable prices. Licenses for companies, on the other hand, are much more expensive. Of course, commercial optimization platforms undergo permanent development and count with very friendly modelling interfaces and very powerful solvers. Moreover, they provide excellent documentation and support. There are many commercial optimization software platforms. Some current popular choices are GAMS[3], MATLAB[4] optimization toolbox and Excel/Solver[5], among many others.

Non-commercial optimization software is freely available and is, in many cases, open source, meaning that the code is open for inspection and modification. However, usually a great deal of programming skills are required to put them to work. Although there is a huge amount of non-commercial optimization projects, most usually have low rates of development and limited maintenance and support. Very interesting options are APMonitor[6], SCIP Optimization Suite[7], SciPy/Optimize[8], Pyomo[9] and Excel/OpenSolver[10], among many others.

Independently of the adopted platform/software, the user will be required to provide, in some specific format, the involved functions: objective and constraints. If the adopted algorithms are of the deterministic type, first order derivatives (and eventually second order derivatives) will be also required. Some platforms, however, provide automatic differentiation options (usually approximate). In particular, the GAMS platform provides exact automatic differentiation. If the adopted solver is of the stochastic type, only the functions will be required. However, since metaheuristics use to perform poorly in highly equality-constrained models, a convenient practice is to work in the 'reduced space', meaning to eliminate as much as possible equations from the model by explicitly resolving some of the variables from previously calculated variables.

An alternative to adopting optimization platforms/solvers (commercial or noncommercial) is to develop from scratch on a certain programming language both the underlying model and the optimization engine. This alternative overlooks the power of tools with many years of development in favour of projects without third-party components which usually behave as 'black boxes' and might have significant costs.

One element in favour of this strategy is that due to the 'no free lunch' results on optimization, counting with the most sophisticated and expensive general purpose algorithms does not necessarily guarantee to solve easily problems located far from the origin of the complexity space (Fig. 3.2). In these cases, a great deal of additional programming (reformulations, decompositions) might be required to achieve satisfactory results even with the most friendly development platforms and competitive solvers. Another argument in favour of this approach is that, from a practical

---

[3] https://www.gams.com/

[4] https://www.mathworks.com/products/optimization.html

[5] https://www.solver.com/products-overview

[6] http://apmonitor.com/

[7] https://scip.zib.de/

[8] https://docs.scipy.org/doc/scipy/reference/tutorial/optimize.html

[9] http://www.pyomo.org/

[10] https://opensolver.org/

point of view, suboptimal solutions obtained with less sophisticated algorithms can be good enough in most uncertain systems with extensive planning horizons.

The major drawback of this approach is that a more interdisciplinary development team is required since specialized programmers and professionals with a strong background on simulation/optimization are needed, other than agronomic engineers and biologists who might not have a specific background on applied mathematics and software development.

Regarding programming languages, there are also several options for from-scratch developments, each with strengths and weaknesses. In fact, 'the best' programming language is a hot topic among professional programmers. Two very popular current options are C and Python. Language C is, maybe, the strongest general purpose programming platform. For example, the Linux kernel and many other programming languages (such as Python) are programmed in C. C is a 'low-level' language which allows a close interaction with the hardware and produces very fast and resource-efficient programs. Since it has been widely used in the last decades, any development in C can benefit from lots of freely available code. Python[11], on the other hand, is a 'high-level' language with a very expressive syntax, optimal for quick code development. Python implementations are therefore less efficient and fast compared with those of C, for example, although they can be accelerated with several techniques. There is also a large amount of scientific and engineering codes over there which can indeed contribute to Python developments.

As can be concluded from the previous paragraphs, there are lots of options to implement the optimization module within decision support projects. However, as mentioned before, nonlinear and mixed-integer models are difficult to optimize (Fig. 3.2), which means that finding good suboptimal or even feasible solutions may be a hard task for any algorithm. In other words, an algorithm may take a lot of clock time (hours, days) to find a satisfactory solution for our problem, even though modern computer processors are very fast devices. In these situations, some optimization algorithms can benefit from parallelization, which basically implies performing calculations simultaneously instead of sequentially, whenever possible. In fact, modern computers include multi-core (several cores) processors allowing a certain degree of parallelism in many applications.

Relatively recently, the computer power at a desk level has been boosted by the availability of general purpose graphics processing units (GPUs), which are rather cheap devices developed originally for the gaming industry. These GPUs are massive parallel processors (hundreds of cores) which can be programmed to implement accelerated versions of numerical algorithms.

In particular, NVIDIA GPUs can be programmed with the CUDA[12] programming model, a specific set of instructions, resembling C language, to exploit at a very low level the hardware of the device. The potential of such devices is also

[11] https://www.python.org/

[12] https://developer.nvidia.com/cuda-zone

available from other programming languages, for example, to Python through Numba[13] tools.

These programming options add up to the already huge box of optimization tools and should be considered when initiating a new DSS effort or an upgrade of existing ones.

## 3.8 Conclusions

Optimization projects for decision support of real systems are complex and extensive efforts. The best option (commercial, non-commercial, deterministic, stochastic, etc.) depends heavily on each particular application. Each developer team usually has a favourite set of tools and modelling/programming skills. In particular, bioeconomic models for decision-making are typically designed in research labs, but they should be ideally developed in conjunction with farmers and agricultural advisors. In order to share the code, there should not be licensing constraints. Non-commercial options or widely spread software such as Excel spreadsheets might be preferable if budgets will be a limitation sometime along the project. It should be considered that modelling for decision support projects usually demands several years and is really never completely finished. A flexible platform that ensures a sustainable development in time should be adopted. Finally, website applications are very effective channels to interact with broader audiences, although their develpment requires the involvement of additional software specialists.

Alternative approaches to optimization in decision-making support are, of course, valid. For example, the RIM model has recently undergone some re-development, and interesting conclusions were drawn by their developers (Lacoste and Powles 2016). Specifically, they preferred to retain the 'what-if' approach (scenario simulation) rather than to implement an optimization engine to automate the analysis and generate ranking recommendations. According to Lacoste and Powles (2016), the simulation approach better contributes to both practical and educational aims by allowing a more direct and trustworthy interaction and feedback between the model and the decision-makers' abilities and experience. Such considerations, rooted in a practical vision of decision-making in agriculture, are indeed valid and reasonable. It is our belief, however, that the gap between this type of research methodologies and its technological application at a farm level will eventually decrease with the advent of (even) cheaper, easily accessible large computation power, (even) increased adoption of electronics at a field level and a larger involvement of highly skilled professionals (even PhDs) in the daily decision-making process within the farm.

---

[13] http://numba.pydata.org/

## References

Benjamin LR, Milne AE, Parsons DJ, Cussans J, Lutman PJW (2009) Using stochastic dynamic programming to support weed management decisions over a rotation. Weed Res 49:207–216

Biegler LT (2010) Nonlinear programming: concepts, algorithms, and applications to chemical Processes, MOS-SIAM Series on Optimization

Biegler Lorenz T, Grossmann IE, Westerberg AW (1997) Systematic methods of chemical process design. Prentice Hall—Technology & Engineering, Upper Saddle River, NJ

Boussaïd I, Lepagnot J, Siarry P (2013) A survey on optimization metaheuristics. Inform Sci 237:82–117

Carter R (1998) Pipeline optimization: dynamic programming after 30 years. PSIG Annual Meeting, 28–30 October, Denver, Colorado

Chalak-Haghighi M, Van Ierland EC, Bourdôt GW, Leathwick D (2008) Management strategies for an invasive weed: a dynamic programming approach for Californian Thistle in New Zealand. N Z J Agric Res 51(4):409–424

Coello Coello CA (2002) Theoretical and numerical constraint-handling techniques used with evolutionary algorithms: a survey of the state of the art. Comput Methods Appl Mech Eng 191:1245–1287

Conejo AJ, Castillo E, Minguez R, Garcia-Bertrand R (2006) Decomposition techniques in mathematical programming: engineering and science applications. Springer Science & Business Media, Berlin

Crawford B, Soto R, Astorga G, García J, Castro C, Paredes F (2017) Putting continuous metaheuristics to work in binary search spaces. Complexity 19:8404231. https://doi.org/10.1155/2017/8404231

Doole GJ, Pannell DJ (2008) Optimisation of a large, constrained simulation model using compressed annealing. J Agric Econ 59(1):188–206

Fisher BS, Lee RR (1981) A dynamic programming approach to the economic control of weed and disease infestations in wheat. Rev Mktg Agr Econ 49:175–187

Floudas C (1995) Nonlinear and mixed-integer optimization. fundamentals and applications. Oxford University Press, Oxford

Gorddard RJ, Pannell DJ, Hertzler G (1995) An optimal control model for integrated weed management under herbicide resistance. Aust J Agric Econ 39(1):71–87

Grossmann IE, Apap RM, Calfa BA, García-Herreros P, Zhang Q (2016) Recent advances in mathematical programming techniques for optimization of process systems under uncertainty. Comp Chem Eng 91:3–14

Ho YC, Pepyne DL (2002) Simple explanation of the no-free-lunch theorem and its implications. J Optim Theor Appl 115(3):549–570

Lacoste M, Powles S (2016) Beyond modelling: considering user-centred and post-development aspects to ensure the success of a decision support system. Comput Electron Agric 121:260–268

Lodovichi MV, Blanco AM, Chantre GR, Bandoni JA, Sabbatini MR, López R, Vigna M, Gigón R (2013) Operational planning of herbicide-based weed management. Agr Syst 121:117–129

Martinez Y, Cirujeda A, Gomez MI, Mari AI, Pardo G (2018) Bioeconomic model for optimal control of the invasive weed Zea Mays Subspp. (Teosinte) in Spain. Agr Syst 165:116–127

Neeser CJ, Dille A, Krishnan G, Mortensen DA, Rawlinson JT, Martin AR, Bills LB (2004) Weedsoft®: a weed management decision support system. Weed Sci 52(1):115–122

Nocedal J, Wright S (2006) Numerical optimization; Springer series in operations research. Springer, New York

Pandey S, Medd RW (1991) A stochastic dynamic programming framework for weed control decision making: an application to Avena Fatua L. Agric Econ 6:115–128

Pannell DJ, Stewart V, Bennett A, Monjardino M, Schmid C, Powles SB (2004) RIM: A bioeconomic model for integrated weed management of Lolium rigidum in Western Australia. Agric Syst 79: 305–325

Sells JE (1995) Optimising weed management using stochastic dynamic programming to take account of uncertain herbicide performance. Agr Syst 48:271–296
Shoemaker CA (1981) Applications of dynamic programming and other optimization methods in pest management. IEEE Trans Autom Control 26:1125–1132
Tawarmalani M, Sahinidis N (2002) Convexification and global optimization in continuous and mixed-integer nonlinear programming: theory, algorithms, software, and applications. Springer Science, Berlin
Taylor CR, Burt OR (1984) Near-optimal management strategies for controlling wild oats in spring wheat. Am J Agric Econ 66:50–60
Wu J (2001) Optimal weed control under static and dynamic decision rules. Agric Econ 25:119–130

# Part II
# Bio-Ecological and Site-Specific Based Models

# Chapter 4
# Modelling Weed Seedbank Dormancy and Germination

**Diego Batlla, Cristian Malavert, Rocío Belén Fernández Farnocchia, and Roberto Benech-Arnold**

**Abstract** Weeds are usually more vulnerable to control practices at the seedling stage or at early stages of their growth. Therefore, developing models to predict the timing and extent of weed emergence is useful to assist farmers and agronomist to time pre- and post-emergence control practices to increase their efficacy. However, many important weeds forming persistence seedbanks in agricultural fields present dormancy. In those species, the number of established seedlings is strongly related to the dormancy level of the seedbank, and the timing of seedling emergence depends on the seasonal variation in seedbank dormancy level. Therefore, if we pretend to predict timing and extent of seedling emergence, we should include the regulation of the seedbank dormancy level in our predictive models. In this chapter, we present a conceptual framework to understand how dormancy and germination of weed seedbanks are regulated by the environment. This framework is based on the distinction between those factors that regulate seasonal changes in the seedbank dormancy level (i.e. temperature in interaction with seed moisture content) and those factors that terminate dormancy (i.e. light and alternating temperatures). Changes in the seedbank dormancy level are related to changes in the range of environmental conditions permissive for seed germination, as, for example, the thermal range permissive for germination which is defined by the lower and the higher limit temperatures. Seeds germinate when environmental conditions are within the permissive range, for example, seeds begging to accumulate thermal time towards germination once soil temperature overlaps the permissive thermal range. We present examples of how these concepts can be used to establish functional relationships between dormancy and germination regulating factors (i.e. temperature) and changes in seedbank population dormancy level and germination dynamics in order

D. Batlla (✉)
Facultad de Agronomía, Cátedra de Cerealicultura, IFEVA (UBA/CONICET), Buenos Aires, Argentina
e-mail: batlla@agro.uba.ar

C. Malavert · R. B. F. Farnocchia · R. Benech-Arnold
Facultad de Agronomía, Cátedra de Cultivos Industriales, IFEVA (UBA/CONICET), Buenos Aires, Argentina
e-mail: malavert@agro.uba.ar; rbfernan@agro.uba.ar; benech@agro.uba.ar


G. R. Chantre, J. L. González-Andújar (eds.), *Decision Support Systems for Weed Management*, https://doi.org/10.1007/978-3-030-44402-0_4

to develop mechanistic models to predict the timing and extent of weed seedling emergence in the field.

**Keywords** Dormancy · Predictive models · Seed-bank · Soil temperature · Soil water content · Thermal time · Weed emergence

## 4.1 Introduction

To maximize the success of weed management strategies, we should optimize the effect of control practices to avoid crop yield losses due to weed competition in the short term and to maintain low weed population levels in the long term. To achieve these goals, it is essential to understand the biological and ecological bases of the weediness process and to determine those stages in the life cycle of weeds that are critical in its regulation. Once those critical stages have been determined, the development of models allowing the prediction of their occurrence in time and space becomes essential for assisting farmers and agronomist in weed management decisions (Ogg and Dawson 1984; Buhler et al. 1997).

Weed emergence is usually a critical stage for the application of weed control practices because young plants and weed seedlings are more vulnerable in that stage (Radosevich et al. 1997). Consequently, control methods are more effective when weeds are controlled soon after they emerge (Fenner 1987; Batlla and Benech-Arnold 2007). The possibility of predicting the timing and extent of weed emergence from soil seedbanks is therefore of paramount importance to increase the effectiveness of control practices (Grundy et al. 2000).

The weed seedbank is considered the primary source of weed infestations in crop fields (Buhler 1999; Grundy and Mead 2000). Although buried seeds can be non-dormant, most weed seed populations composing persistence seedbanks usually present dormancy. In weed species showing dormancy, the number of established seedlings is strongly related to the dormancy level of the seedbank, and the timing of seedling emergence depends on the seasonal variation in seedbank dormancy level (Benech-Arnold et al. 2000). Therefore, if we pretend to predict timing and extent of seedling emergence from weed seedbanks, we should consider dormancy in our predictive models (Batlla and Benech-Arnold 2010).

Weed emergence can be divided into different sub-processes which are regulated by different factors (Fig. 4.1): (A) the passage of seeds from a dormant to a non-dormant state, and vice versa, (B) the germination process and (C) pre-emergence growth (Vleeshouwers and Kropff 2000). Although factors regulating pre-emergence growth can occasionally affect timing and extent of seedling emergence, under most agricultural situations, emergence can be adequately predicted just taking into account factors affecting dormancy and germination; therefore, this chapter will

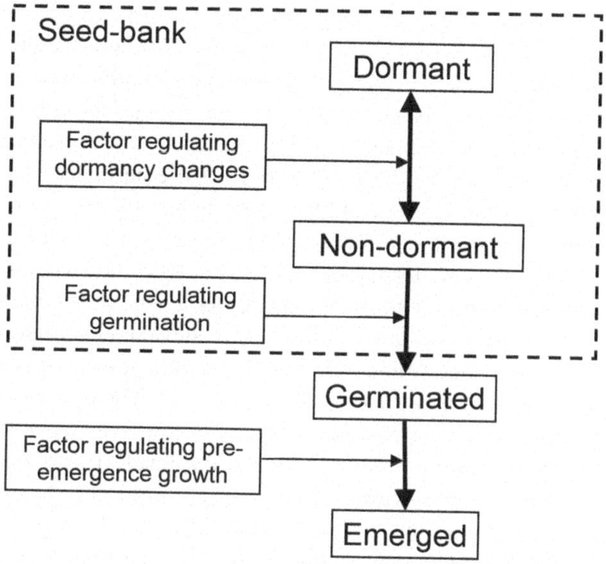


**Fig. 4.1** Schematic representation of sub-processes and factors regulating seedling emergence from weed seedbanks

focus on these two latter processes. To predict seedling emergence, we should (1) know which are the 'key' environmental factors regulating seed dormancy and germination in a certain ecological scenario, (2) have a clear notion of how those key environmental factors affect both processes and (3) establish 'functional relationships' between regulating factors and the target processes. In this chapter, we propose a conceptual framework of how dormancy and germination of weed seedbanks are regulated by environmental factors. Based on that framework, we show examples of how these concepts can be used to predict the timing and extent of weed seedling emergence in the field.

## 4.2 Environmental Regulation of Dormancy in Weed Seedbanks

Dormancy can be defined as an internal impedance of the seed that prevents germination under moisture, thermal and gaseous conditions that, otherwise, would result suitable for germination (Egley 1986; Benech-Arnold et al. 2000). The nature of this impedance could be physiological (i.e. hormonal), morphological or merely physical (Nikolaeva 1967), and based on this, seed dormancy can be classified into five different classes (see Baskin and Baskin (2004) for the classification). However, within those classes, physiological dormancy is possibly the most common type of dormancy in seeds of major agricultural weeds in temperate climates. So, this chapter is mostly referred to this type of dormancy.

Dormancy can be also classified into primary dormancy and secondary dormancy (Karssen 1982). While primary dormancy refers to the dormant state seeds present at dispersal, secondary dormancy results from the re-induction of dormancy in dispersed seeds that had been previously released from primary dormancy or had attained a low dormancy level (Fig. 4.2). Usually, primary dormancy level decreases with time after dispersal determining the emergence of a proportion of the seedbank when environmental conditions are favourable for germination (Fig. 4.2). The fraction of the seedbank that did not emerge, either because the seeds did not achieve a sufficiently low dormancy level or because the environmental conditions were not favourable for germination, may enter a state of secondary dormancy (Fig. 4.2).

Exit from dormancy followed by subsequent re-inductions into secondary dormancy may determine indefinite seasonal cyclic changes in the seedbank dormancy level (Baskin and Baskin 1988) (Fig. 4.2). The dynamics of these cyclic changes in relation to season changes throughout the year depends on the species' life cycle. For many summer annual species, the level of dormancy of the seedbank usually decreases during winter, determining a minimum level of dormancy at the beginning of spring, and increases again at the end of spring-beginning of the summer (Fig. 4.2). This dynamic ensures that the seasonal 'emergence window' takes place timely during spring allowing the resultant plants to place their reproductive phase during summer avoiding frost damage. Conversely, species displaying an annual autumn-winter life cycle generally show the reverse dormancy pattern (i.e. the dormancy level decreases during summer and increases during winter).

The environmental factors that affect dormancy of buried seeds can be divided into two classes, those that regulate the level of dormancy and those that terminate dormancy (Benech-Arnold et al. 2000).

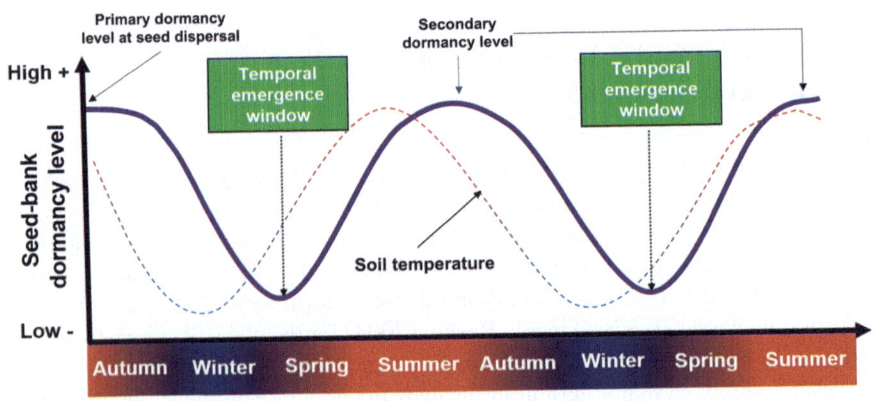


**Fig. 4.2** Schematic representation of cyclic seasonal changes in seedbank dormancy level for a summer annual weed. Solid line indicates seedbank dormancy level, and dotted line indicates soil temperature (Adapted from Batlla and Benech-Arnold 2007)

### 4.2.1 Factors That Regulate the Level of Dormancy

Seasonal changes in the seedbank dormancy level are regulated by environmental factors which ensure emergence occurrence at the 'right' season. Temperature has been pointed out as the main factor regulating changes in dormancy level in seeds of many weed species (Baskin and Baskin 1985; Batlla and Benech-Arnold 2010). The way in which temperature affects seed dormancy level can be different depending on the species' life cycle. In the case of summer annuals, low winter temperatures decrease dormancy level determining minimum dormancy at the entrance of spring, while high summer temperatures increase dormancy level leading to the entrance into secondary dormancy (Fig. 4.2). This type of behaviour has been observed in many spring emergence species, such as *Chenopodium album* L., *Sisymbrium officinale* L., *Polygonum persicaria* L. (Bouwmeester 1990), *Polygonum aviculare* L. (Kruk and Benech-Arnold 1998) and *Ambrosia artemisiifolia* L. (Baskin and Baskin 1980). In contrast, high summer temperatures determine dormancy release in winter annual species, while low winter temperatures might induce secondary dormancy (Karssen 1982). Dormancy release due to high summer temperatures was observed in many winter annual species, such as *Avena fatua* (Baskin and Baskin 1998), *Lolium rigidum* (Steadman et al. 2003), *Bromus tectorum* (Christensen et al. 1996), *Buglossoides arvensis* (Chantre et al. 2009) and *Cynara cardunculus* (Huarte et al. 2018) among others. This effect of temperature is modulated by the seed moisture content, which in turn depends on the soil water content. For example, in many summer annuals, dormancy release occurs when imbibed seeds perceive low temperatures during winter (i.e. stratification). Conversely, seeds from winter annual species are released from dormancy by high temperatures during the summer, normally at low seed moisture content (i.e. dry after-ripening) (Karssen 1982; Probert 1992; Bair et al. 2006). Beyond these general patterns of response, there are species in which, depending on their seed moisture content, low and high temperatures can both provoke dormancy release, although at difference rates (e.g. *Arabidopsis thaliana*; Finch-Savage et al. 2007).

### 4.2.2 Factors That Terminate Dormancy

In many weed species, once a minimum dormancy has been attained, dormancy needs to be terminated or the last impediments must be removed, for germination to proceed. The termination of dormancy requires the action of specific environmental cues which are different from those that regulate changes in dormancy level providing information at the 'seasonal level'. On the contrary, from an ecological point of view, these dormancy-terminating factors indicate if the 'place' is safe enough for germination and establishment (Finch-Savage and Leubner-Metzger 2006). Among these factors, the most common under field conditions are light and alternating temperatures, although there are many others factors that can elicit dormancy termination under specific conditions, such as nitrates, ethylene and carbon dioxide (Benech-Arnold et al. 2000).

Light responses are of paramount importance for the functioning of weed seedbanks (Batlla and Benech-Arnold 2014). Seeds perceive the light stimulus through photoreceptors, mainly those belonging to the phytochrome family. Phytochromes have two interconvertible forms: Pfr (considered the active form for germination) that presents its maximum absorption around 730 nm (FR light) and Pr that presents its maximum absorption around 660 nm (R light) (Borthwick et al. 1954). Phytochromes are synthesized in the seed in the Pr form, and the proportion that is converted to the active form (Pfr) depends on the light environment (spectral composition and irradiance) to which the seeds are exposed. For example, exposures to a light environment characterized by a high R/FR ratio results in a high proportion of the molecule being in its active form (Pfr) which will trigger a cascade of events ending in dormancy termination. Phytochrome action in the LFR (low-fluence response) mode provides the seeds with information regarding the eventual presence of established neighbours which would impair the establishment of the new seedling, a mechanism known as gap-sensing (Pons 1992). In contrast, phytochrome action in the VLFR (very low-fluence response) mode provides the seeds with other kind of information, usually related to soil disturbance in agricultural environments being an opportunity for seedling establishment (Batlla and Benech-Arnold 2014).

Many seeds require the stimulus of temperature fluctuations for dormancy termination. The stimulus can be exerted through different attributes like the thermal amplitude of the cycle, the cumulative effect of stimulating cycles (i.e. number of cycles) or the upper temperature of the cycle (Totterdell and Roberts 1980). As in the case of phytochromes acting in the LFR or in the VLFR mode, the requirement of temperature fluctuations to terminate the dormant state and further elicit germination has been related to the possibility of detecting depth of burial, soil disturbance and vegetation gaps (Benech-Arnold et al. 2000).

### 4.2.3 Seedbank Dormancy Level and Its Relationship with the Range of Environmental Conditions Permissive for Germination

Seasonal changes in seed dormancy level driven by temperature are related to changes in the range of environmental conditions under which germination can occur (Vegis 1964; Vleeshouwers et al. 1995). For example, as dormancy is relieved, the range of temperatures permissive for germination gradually widens until it is maximal, while as dormancy is induced, the range of temperatures narrows until germination is no longer possible at any temperature and full dormancy is reached (Karssen 1982; Bouwmeester and Karssen 1992) (Fig. 4.3). This range is usually determined by two threshold limit temperatures: (1) the lower-limit temperature ($T_l$; below which dormancy is expressed) and (2) the higher-limit temperature ($T_h$; above

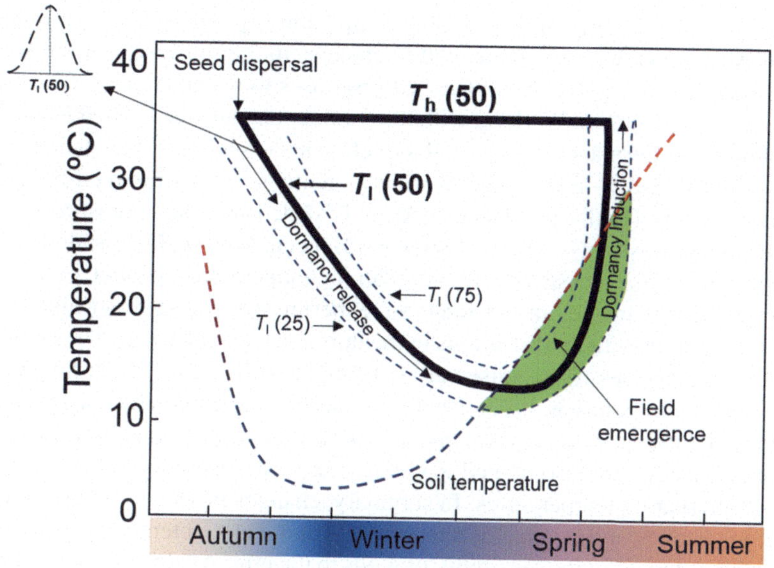


**Fig. 4.3** Schematic representation of seasonal changes in the permissive thermal range for seed germination and its relation with soil temperature dynamics for a summer annual weed. Solid black lines indicate the mean lower ($T_l(50)$) and mean higher ($T_h(50)$) limit temperatures of the permissive thermal range for seed germination. Dashed black lines indicate $T_l$ for the 25 and 75% seed population fractions. Dashed red-blue lines indicate the soil temperature. The green zone represents the moment when germination occurs once the soil temperature enters in the permissive thermal range for seed germination. Black arrows indicate the lowering and increase in $T_l$ during dormancy release and induction, respectively. The bell-shaped dashed curve indicates that $T_l$ is assumed to be normally distributed within the seed population (originally from Probert 2000, adapted from Malavert et al. 2017)

which dormancy is expressed) which are assumed to be normally distributed within the seed population showing a median, $T_l(50)$ and $T_h(50)$, and their corresponding standard deviations, $\sigma_{Tl}$ and $\sigma_{Th}$ (Washitani 1987; Batlla and Benech-Arnold 2015). In summer annuals, changes in dormancy level are mainly the result of the decrease or the increase in $T_l$, for dormancy release and induction, respectively (Fig. 4.3), while in winter annuals, these changes are mainly driven through the increase and the decrease in $T_h$ for dormancy release and induction, respectively.

In species that do not require the effect of dormancy-terminating factors (i.e. light and/or alternating temperatures), the broadening of the thermal permissive range for germination, as a result of dormancy release (as has been referred above), allows the germination of a given seedbank fraction when the soil temperature enters this range (Fig. 4.3). If the latter does not occur, or if water conditions are insufficient for germination, the latter process will be inhibited, and the population might be induced into secondary dormancy depending on the prevailing thermal environment, thus restarting the cycle (Probert 1992).

In addition to changes in the range of temperatures permissive for germination, dormancy modifications are associated to changes in the sensitivity of the seeds to many other factors, as, for example, seed responses to water availability, light and alternating temperatures. Working with the summer annual weed *P. aviculare*, it has been shown that dormancy release and induction are associated to an increase and a decrease in the width of the range of water potentials (*Ψ*) permissive for germination, respectively (Batlla and Benech-Arnold 2004; Batlla and Agostinelli 2017), changes in the sensitivity to light (Batlla and Benech-Arnold 2005; Malavert 2017) and changes in the sensitivity to alternating temperatures (Batlla et al. 2003; Malavert 2017). Changes in the range of *Ψ* permissive for seed germination and sensitivity to light with modifications in the dormancy level have also been reported in other weed species (Christensen et al. 1996; Hawkins et al. 2017; Scopel et al. 1991; Derkx and Karssen 1993, 1994). Similarly, studies conducted by Benech-Arnold et al. (1990a, b) in *Sorghum halepense* L. showed that, as in the case of light, dormancy changes are also related to changes in seed sensitivity to the stimulatory effect of alternating temperatures. In summary, changes in seed dormancy level are related to changes in the range of temperatures and water potentials permissive for seed germination, as in the sensitivity of seeds to the effect of dormancy-terminating factors (i.e. light and alternating temperatures).

### 4.2.4 Conceptual Model

On the basis of the above discussed concepts, we propose a conceptual model (Fig. 4.4) in which seeds with a high level of dormancy can germinate in a narrow range of temperatures and water potentials and present a low sensitivity to light and/or to alternating temperatures, or they can directly not germinate under any environmental conditions, showing full dormancy. On the contrary, seeds showing a low dormancy level can germinate in a wide range of thermal and water potentials and present a high sensitivity to light and/or to alternating temperatures. Changes in dormancy level (from high to low and vice versa) are regulated by soil temperature and modulated by seed moisture content which in turn depends on soil water content.

Germination and consequent emergence in the field will take place when the prevailing environmental conditions coincide with those required for germination (the latter depends on the level of dormancy of the seedbank). It is important to note that the effect of temperature and its modulation by seed water content depends on the species' life cycle. In summer annuals, the transition from a high to a low dormancy level is mainly determined by the exposure of partially imbibed seeds to low temperatures and the transition from a low to a high dormancy level by exposure to high temperatures. In winter annuals, the transition from high to low dormancy is determined by the exposure of partially dry seeds to high temperatures and the transition from a low to a high dormancy level by exposure to low temperatures. Finally, in species that do not require light or alternating temperatures to terminate

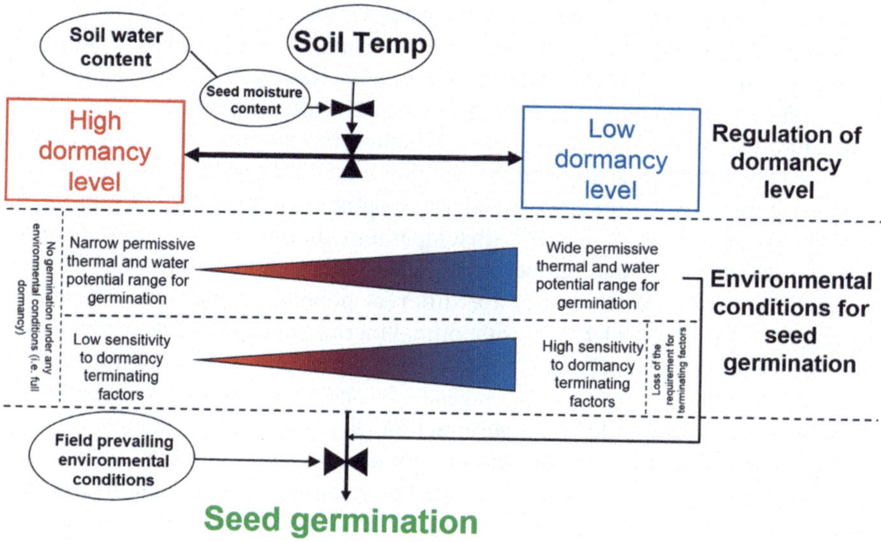


**Fig. 4.4** Flowchart representing the most relevant environmental factors regulating dormancy level and changes in the range of environmental conditions for seed germination in soil seedbanks (originally from Benech-Arnold et al. 2000, adapted from Batlla and Benech-Arnold 2010)

dormancy, germination only depends on soil temperature and water potential to be within the germination permissive range. In species requiring light or alternating temperatures, a low dormancy level could determine not only an increase in the sensitivity of the seeds to the effect of such factors but even the loss of this requirement in a fraction of the population (i.e. germination in the dark or under constant temperatures).

## 4.3 Temperature and Water Availability Effects on Seed Germination

To predict germination in non-dormant species, or in those seedbank fractions in which dormancy is not expressed because prevailing environmental conditions are within those permissive for germination, it is only required to establish functional relationships between germination and its modulating factors. When oxygen is not limiting, the main environmental factors controlling germination are temperature and water availability.

In the thermal range within which germination is possible, temperature acts only modulating the rate of germination (Bewley et al. 2013; Washitani 1985; Batlla and Benech-Arnold 2015). The germination rate, defined as the inverse of the time required to reach a certain germination percentage at a given incubation temperature, could be described as a bilinear regression characterized by three cardinal temperatures (Fig. 4.5), although alternative approaches for relating germination rate to temperature have been developed as well (Hardegree 2006). As observed in Fig. 4.5, the germination rate increases linearly with temperature from a base temperature ($T_b$; that temperature below which germination rate becomes zero), considered common for all seeds in the population, until it reaches a maximum at a temperature that is consequently defined as the optimum temperature ($T_o$). From $T_o$, germination rate decreases linearly with temperature until it becomes zero at a maximum temperature ($T_m$; that temperature above which germination rate becomes zero), which is considered to vary for different population fractions. The range between $T_b$ and $T_o$ is considered the suboptimal thermal range for seed germination, while that between $T_o$ and $T_m$ is considered the supra-optimal thermal range. The inverse of each linear slope in both ranges (sub- and supra-optimal) is regarded as the thermal time required for seed germination ($\vartheta_T$), with units being degree days (°Cd) or degree hours (°Ch). The thermal time for seed germination in the sub-optimal thermal range varies within the seed population (i.e. a normal distribution defined by the median, $\vartheta_T(50)$, and its standard deviation, $\sigma_{\vartheta T}$), so different fractions of the population (10%, 25%, 75%, etc.) will need to accumulate different values of $\vartheta_T$ to germinate as the relationship between temperature and germination rate for each fraction has a different slope. Instead, in the supra-optimal range, the different seed fractions often show the same slopes (parallel lines) resulting in different intercepts on the *x*-axis (i.e. different $T_m$; Covell et al. 1986; Ellis et al. 1986); however, common $T_m$ values for all seeds in a population have been reported as well (Garcia-Huidobro et al. 1982; Hardegree 2006).

Above-mentioned parameters ($T_b$, $T_o$, $T_m$ and $\vartheta_T$) can be used to describe, or eventually predict, the dynamics of germination of a seed population as a function of time and temperature using a thermal-time modelling approach (Washitani 1987;

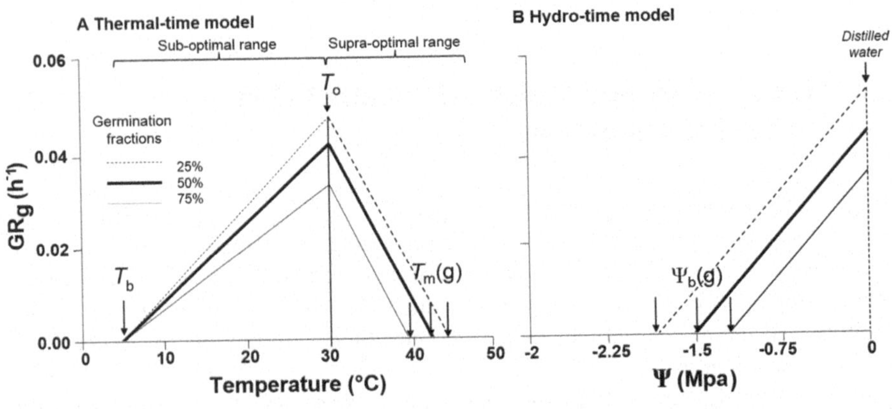


**Fig. 4.5** (**a**) Schematic representation of the relationship between germination rates ($GR_g = 1/t_g$) and temperature at the suboptimal and the supra-optimal thermal range for 25, 50 and 75% of a seed population. (**b**) Relationship between $GR_g$ and water potential for 25, 50 and 75% of a seed population

Benech-Arnold and Sánchez 1995). For example, in the case of the suboptimal thermal range for seed germination (between $T_b$ and $T_o$), if we know the values of $T_b$ and $\vartheta_T(50)$ of a particular weed species, we can predict the time required to achieve 50% of germination using the following equation:

$$\vartheta_T(50) = (T_s - T_b)t(50) \tag{4.1}$$

where $T_s$ is the soil temperature and $t(50)$ is time for germination of the 50% of the seed population. The time for germination of other seed fractions can also be calculated knowing their corresponding $\vartheta_T$. The thermal-time approach was successfully used to analyse and model the germination and the emergence of different species as a function of time and temperature (Garcia-Huidobro et al. 1982; Covell et al. 1986; Leguizamon et al. 2009). A similar model can be used to predict germination as a function of time and temperature in the supra-optimal thermal range (between $T_o$ and $T_m$); for details, see Bewley et al. (2013).

Another factor that has a marked influence on germination is water availability. Variations in soil water content can affect both germination rate and the seed fraction capable to germinate. Similarly, to the thermal-time model, seeds require a certain amount of hydro-time ($\vartheta_H$) to germinate, and that amount of $\vartheta_H$ (in MPa day or MPa h) is accumulated above a threshold value called base water potential ($\Psi_b$; the minimum water potential at which germination will occur for a given seed). However, there are differences between models. In the suboptimal thermal range of the thermal-time model, seeds accumulate $\vartheta_T$ above a common $T_b$ for the entire seed population, and the $\vartheta_T$ required for germination is different for each fraction of the population. Instead, in the hydro-time model, the $\Psi_b$ above which seeds accumulate hydro-time is distributed in the population (i.e. a normal distribution defined by the median, $\Psi_b(50)$, and its standard deviation, $\sigma_{\Psi b}$), while the amount of $\vartheta_H$ required for germination is equal for all the seeds in the population (i.e. similar slopes of the relationship between germination rate and $\Psi$ for different seed population fractions) (Fig. 4.5b). The response of the seeds to water availability can be analysed in a similar way to that commented before for the analysis of temperature effects on seed germination. So, in analogy to the thermal-time model, seed germination response to water availability can be described or predicted using the hydro-time model equation:

$$\vartheta_H = (\Psi_s - \Psi_b(50))t(50) \tag{4.2}$$

where $\Psi_s$ is soil water potential in MPa, $\Psi_b(50)$ is the average water potential of the seed population in MPa and $t(50)$ is the time required for the germination of 50% of the seed population.

An integrative model, called the hydrothermal-time model, is based on the two models explained above (the thermal-time, Eq. (4.1), and the hydro-time, Eq. (4.2)). The hydrothermal-time model allows describing the effect of temperature and water potential on seed germination together (Gummerson 1986; Bradford 1995, 2002).

Thus, for the suboptimal range, germination dynamics can be described using the following equation:

$$\vartheta_{\mathrm{HT}} = \left(\Psi_{\mathrm{s}} - \Psi_{\mathrm{b}}(50)\right)\left(T_{\mathrm{s}} - T_{\mathrm{b}}\right)t(50) \tag{4.3}$$

where $\vartheta_{\mathrm{HT}}$ is the hydrothermal-time (with units being °C MPa d or °C MPa h) required for seed germination. The hydrothermal-time model can be also applied to the supra-optimal thermal range (for details, see Bewley et al. 2013).

## 4.4 The Use of Dormancy and Germination Models to Predict Timing and Extent of Emergence from Weed Seedbanks

As commented previously, to model seedbank dormancy changes, we should establish functional relationships between key regulating factors and the rates of dormancy alleviation and induction. Because soil temperature is the most important environmental factor controlling annual dormancy cycles of buried seeds in most temperate zones, relationships should be established between temperature and some parameters able to characterize the seedbank dormancy level. As commented before, one way to characterize seedbank dormancy level is through the range of temperatures permissive for seed germination, which can be quantified through their limit temperatures, $T_{\mathrm{l}}$ and $T_{\mathrm{h}}$ (Fig. 4.3). These limit temperatures can be estimated from the final fraction of germinated seeds within a temperature range assuming that the fraction of germinated seeds in relation to temperature can be predicted based on the normal distribution of these two limit temperatures (see Washitani (1987) and Batlla and Benech-Arnold (2015) for details). With the purpose of establishing relationships between time, soil temperature and seed dormancy level, Batlla and Benech-Arnold (2003) characterized *P. aviculare* seed dormancy loss for seeds stratified at different temperatures through changes in the range of temperatures permissive for germination as a consequence of changes in the mean lower limit temperature of the range ($T_{\mathrm{l}}$ (50)) (Fig. 4.6a). *P. aviculare* is a cosmopolitan summer annual weed that invades winter crops and forms persistence seedbanks (Costea and Tardif 2005). As many other summer annuals, *P. aviculare* imbibed seeds loss dormancy when exposed to low winter temperatures and go into secondary dormancy when exposed to high summer temperatures (Batlla et al. 2009). To quantify the effects of stratification time and temperature on seed population dormancy level (assessed through changes in $T_{\mathrm{l}}$ (50)), authors used a thermal-time equation:

$$S_{\mathrm{tt}} = \mathrm{days}\left(T_{\mathrm{c}} - T_{\mathrm{s}}\right) \tag{4.4}$$

where $S_{\mathrm{tt}}$ is stratification thermal-time units (°Cd), $T_{\mathrm{c}}$ is the dormancy release 'ceiling' temperature (°C) (the temperature at, or over, which dormancy release does not

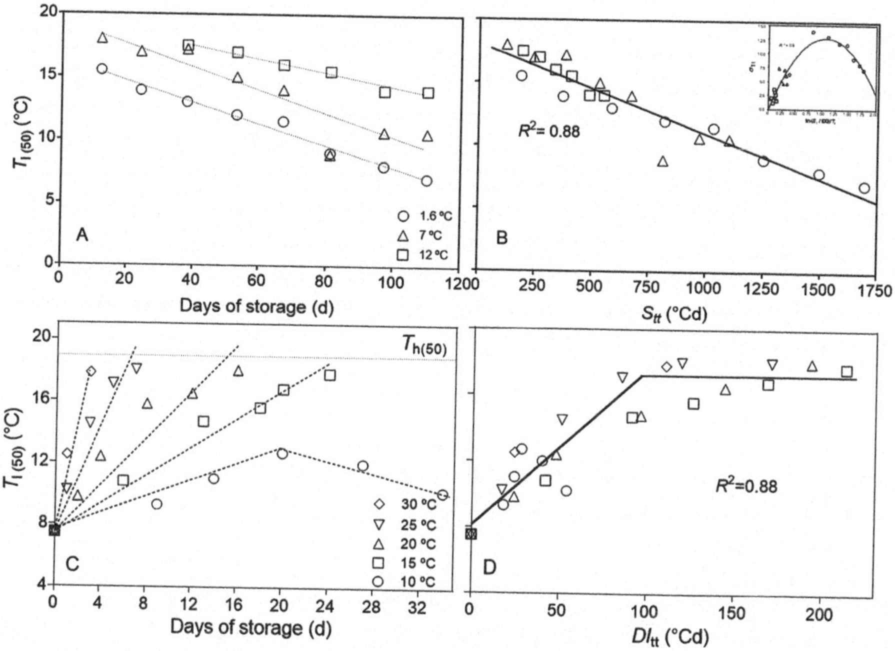


**Fig. 4.6** Changes in the mean lower limit temperature ($T_l(50)$) for *P. aviculare* seeds during dormancy release and induction. (**a**) Changes in $T_l(50)$ during dormancy release for seeds stored at 1.6, 7 and 12 °C, plotted against days of storage and (**b**) against stratification thermal time ($S_{tt}$; Eq. 4.4). The dotted lines in (**a**) were fitted linear equations for each storage temperature with $R^2$ values of 0.98 (1.6 °C), 0.84 (7 °C) and 0.96 (12 °C). The fitted line in (**b**) ($T_l(50) = -0.007\ S_{tt} + 18.07$; Eq. 4.5) is the result of repeated regression analysis to obtain the threshold 'ceiling' temperature ($T_c$) with the best fit according to Eq. 4.4. Inset in (**b**) is estimated values of standard deviation of the lower limit temperature ($\sigma_{Tl}$) for *P. aviculare* seeds stored at 1.6, 7 and 12 °C plotted against the $\ln(S_{tt}/100)/T_s$, where $S_{tt}$ is the stratification thermal time (Eq. 4.4) and $T_s$ is the daily mean storage or soil temperature. The line was fitted according to equation $\sigma_{Tl} = -11.28\ (\ln (S_{tt}/100)/T_s)^2 + 23.91\ (\ln (S_{tt}/100)/T_s)$ with an $R^2$ of 0.9. (**c**) Changes in $T_l(50)$ during dormancy induction for seeds stored at 10, 15, 20, 25 and 30 °C plotted against days of storage and (**d**) against dormancy induction thermal time ($DI_{tt}$). The dashed lines in (**c**) were fitted by linear equations for each storage temperature with $R^2$ values of 0.96 (10 °C), 0.99 (15 °C), 0.87 (20 °C), 0.89 (25 °C) and 0.96 (30 °C), while the dotted straight line indicates the mean higher limit temperature for seed germination of the seed population ($T_h$ (50)). The fitted bilinear line in (**d**) is the result of repeated regression analysis to obtain the threshold 'dormancy induction temperature' ($T_{uDI}$) with the best fit according to equation $T_l\ (50) = 0.12\ DI_{tt} + 7.5$, if $DI_{tt} \geq 96.5$ °Cd $T_l\ (50) = 18$ °C (Eq. 4.7) (Figures **a** and **b** adapted from Batlla and Benech-Arnold 2003; figures **c** and **d** adapted from Malavert et al. 2017)

occur) and $T_s$ is the daily mean storage or soil temperature (°C). The optimal 'ceiling' temperature for dormancy loss in *P. aviculare* seeds was 17 °C (according to Batlla and Benech-Arnold 2003).

This thermal-time approach is similar to that usually used to relate other biological processes (i.e. crop phenology) to time and temperature. However, in contrast to common thermal-time models in which degree days are accumulated over a base temperature, $S_{tt}$ accumulates degree days below a ceiling threshold temperature below which dormancy loss occurs. Using this thermal-time equation, the lower the stratification temperature, the more thermal-time units are accumulated and the lower the seed dormancy level (i.e. lower $T_l(50)$). Similar thermal-time approaches were used successfully by other authors to quantify time and temperature effects on seed dormancy status (Pritchard et al. 1996; Bauer et al. 1998; Steadman and Pritchard 2003). The possibility of quantifying temperature effects using a thermal-time approach allows the prediction of the dormancy level of a seed population exposed to the variable soil field thermal environment. For example, in *P. aviculare* seeds $T_l(50)$ could be predicted as a negative linear function of accumulated $S_{tt}$ using the following function (Fig. 4.6b):

$$T_l(50) = -0.007S_{tt} + T_{l(hd)}(50) \tag{4.5}$$

where $T_{l(hd)}(50)$ is the initial $T_l(50)$ of the population (for recently dispersed or for seeds showing a high dormancy level) which was determined to be 18 °C by Batlla and Benech-Arnold (2003).

Equations (4.4) and (4.5) can be used to predict the time at which 50% of the seedbank population would emerge (i.e. when soil temperature surpasses $T_l(50)$; see Fig. 4.3). However, weed seedbank populations are usually large, and emergence of even a small fraction (e.g. 10%) can result in a high seedling density. Consequently, it is important to predict even low levels of emergence. To achieve this goal, we should know the $T_l$ for different fractions of the seedbank population (e.g. $T_l$ (10), $T_l$ (20), etc.); in other words, we should know the distribution of $T_l$ within the seed population. Assuming a normal distribution of $T_l$, Batlla and Benech-Arnold (2003) developed an additional function to predict changes in the standard deviation of $T_l$ ($\sigma_{Tl}$) in relation to accumulated $S_{tt}$ units and the daily mean storage or soil temperature (see inset in Fig. 4.6b). This population threshold model can be used to predict the $T_l$ for different fractions of the seed population (i.e. the distribution of dormancy levels within the population) allowing for the estimation of the fraction of the seedbank that would germinate under a given thermal environment.

However, to predict the seasonal pattern of weed emergence, we should not only be able to predict dormancy loss but also dormancy induction. The effect of temperature on *P. aviculare* dormancy induction assessed through changes in $T_l$ was recently quantified by Malavert et al. (2017) (Fig. 4.6c). Obtained results showed that, as for many other summer annuals weeds, the higher the temperature, the higher the rate of induction into secondary dormancy (i.e. higher rate of $T_l(50)$ increase). In order to relate time and temperature to changes in seed dormancy level characterized by $T_l(50)$, authors used a positive thermal-time index in which temperature was accumulated over a base temperature for dormancy induction to proceed:

$$DI_{tt} = \text{days}(T_s - T_{uDI}) \tag{4.6}$$

where $DI_{tt}$ is dormancy induction thermal time (°Cd) and $T_{uDI}$ is the threshold temperature for induction into secondary dormancy (7.9 °C according to Malavert et al. (2017), temperature at or below which dormancy induction does not occur). $T_l(50)$ could be predicted through a bilinear function that depends on $DI_{tt}$ accumulation:

$$T_l(50) = 0.12DI_{tt} + T_{l(Id)}(50) \quad \text{if} \quad DI_{tt} \geq 96.5^{\circ}\text{Cd} \;\; T_l(50) = 18^{\circ}\text{C} \qquad (4.7)$$

where $T_{l\,(Id)}(50)$ is the initial $T_l(50)$ of the seed population, that before induction into secondary dormancy (i.e. after a period of stratification or for seeds showing a low dormancy level). Authors found that $\sigma_{Tl}$ did not change during dormancy induction, so they assumed a fixed value of 2.7 °C.

Assuming that just $T_l$ changes with changes in dormancy level, while $T_h$ remains constant (with $T_h(50)$ = 18 °C, $\sigma_{Th}$ = 0.1 °C), both models can be used simultaneously to predict how the thermal range permissive for seed germination changes as a consequence of changes in $T_l$ during dormancy release and induction in relation to soil temperature. An example of how both models work together can be observed in Fig. 4.7. In summary, both models, $S_{tt}$ (operating at soil temperatures below 17 °C; Eq. (4.4)) and $DI_{tt}$ (operating at soil temperatures above 7.9 °C; Eq. (4.6)), accumulate °Cd simultaneously according to soil temperature on a daily basis once autumn mean soil temperature is below 17 °C (inset in Fig. 4.7a) (the model assumes that seeds dispersed in autumn are fully dormant and that their dormancy level does not change until soil temperature is below the threshold temperature for dormancy release, $T_c$ in Eq. (4.4)). From this date onwards, dormancy loss prevails because low autumn and winter temperatures determine the accumulation of more $S_{tt}$ than $DI_{tt}$, so the decrease in $T_l$ is predicted using Eq. (4.5) based on accumulated $S_{tt}$ (Fig. 4.7a). However, during spring, rising temperatures establish a higher accumulation of $Di_{tt}$ than $S_{tt}$ because $DI_{tt}$ accumulates °Cd over 7.9 °C and $S_{tt}$ below 17 °C. The date accumulated $DI_{tt}$ surpasses accumulated $S_{tt}$, dormancy induction starts and the increase in $T_l$ is predicted using Eq. (4.7) based on accumulated $Di_{tt}$ (Fig. 4.7a). When $T_l(50)$ equals $T_h(50)$, full dormancy is reached. The model would re-start next autumn when soil temperature values get below 17 °C (i.e. $T_c$, Eq. 4.4). Germination of a certain fraction of the seedbank takes place when mean soil temperature surpasses the $T_l$ for that fraction (i.e. soil temperatures overlap the permissive thermal range for germination of that fraction; Fig. 4.7b). Therefore, assuming that the fraction of germinated seeds in relation to temperature can be described based on the normal distribution of $T_l$ and $T_h$ (Batlla and Benech-Arnold 2015), it can be predicted using the following function:

$$GF(T) = (\Phi((T - T_l(50))/\sigma_{Tl}) - (1 - \Phi((T - T_h(50))/\sigma_{Th}) \qquad (4.8)$$

where $GF(T)$ is the fraction of seeds germinating at temperature $T$ and $\Phi$ is the normal probability integral. This allows an acceptable description of the temporal window of seedling emergence in the field and the fraction of the seedbank able to emerge (Fig. 4.7b).

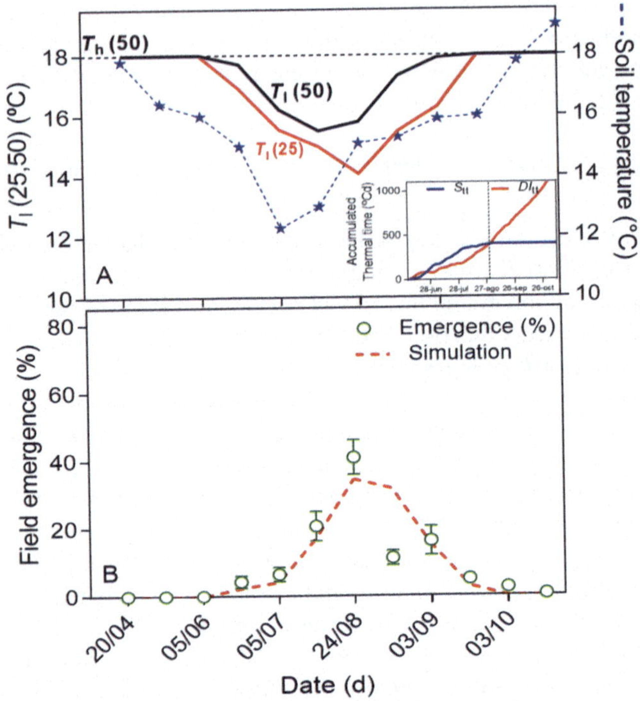


**Fig. 4.7** Evaluation of models for dormancy release (Eqs. 4.4 and 4.5 and inset in Fig. 4.6b) and dormancy induction (Eqs. 4.6 and 4.7) under field conditions. (**a**) Simulated changes in the lower limit temperature ($T_l$) for *P. aviculare* seeds during burial in the field. The solid black and red lines represent the lower limit temperature for 50% ($T_l(50)$) and 25% ($T_l(25)$) of the seed population, respectively. The dashed black line represents the higher limit temperature for 50% ($T_h(50)$) of the seed population. The dashed blue line indicates the soil temperature at 5 cm and stars the date seeds were exhumed from the soil. Inset graph in (**a**) shows the accumulation of stratification thermal time ($S_{tt}$, blue line) and dormancy induction thermal time ($DI_{tt}$, red line) according to soil temperature using Eqs. (4.4) and (4.6), respectively. When $S_{tt} > DI_{tt}$, $T_l$ values were predicted using Eq. (4.5) and that of the inset in Fig. 4.6b, when $DI_{tt} > S_{tt}$, $T_l$ values were predicted using Eq. (4.7) and assuming a fixed value of 2.7 for the standard deviation of $T_l$. (**b**) Recorded and simulated field emergence percentages for buried *P. aviculare* seeds

Similar approaches were developed to predict weed seed dormancy loss through dry after-ripening in winter annual weeds. For example, Chantre et al. (2009) developed a model to predict dormancy loss as a consequence of changes in the mean maximum limit temperature for seed germination ($T_m(50)$) in relation to temperature for *Buglossoides arvensis*, a common winter annual weed in the south area of the semiarid pampean region of Argentina. Changes in $T_m(50)$ during dormancy loss can be predicted through a quadratic equation in relation to the accumulation of after-ripening thermal-time units (°Cd) above a base temperature of −6 °C for the after-ripening process to occur. The model also accounts for a decrease in the thermal-time required for seed germination (i.e. an increase in germination rate) as a consequence of dormancy loss.

Above examples used changes in the thermal range permissive for seed germination as a proxy of changes in the dormancy status of the seed population. However, as commented before, changes in dormancy level are also related to changes in the base water potential for seed germination ($\Psi_b$), sensitivity to light and alternating temperatures. So, any of this responses can be used as a proxy of the dormancy level of the seed population. For example, shifts in $\Psi_b$ can be used as a proxy of changes in germination behaviour in relation to variations in seed population dormancy level (Bradford 2002); a decrease in $\Psi_b$ during dormancy loss determines an increase in germination percentage and germination rate, while an increase in $\Psi_b$ determines a decrease in germination percentage and rate as seeds go into secondary dormancy. Bauer et al. (1998) and Christensen et al. (1996) successfully used changes in $\Psi_b$ as a proxy of changes in seedbank dormancy status to model seed dormancy loss in the annual invasive grass *Bromus tectorum* L. These authors accurately predicted the increase in germination percentage and rate during dormancy loss through a progressive decrease in $\Psi_b$ of the seed population. Similar to previous explained models, the decrease in $\Psi_b$ can be predicted through negative linear or exponential functions based on the accumulation of thermal-time units above a base temperature for the after-ripened process to occur (Chantre et al. 2010). Recently, using a similar approach, the model was extended to include induction into secondary dormancy through an increase in $\Psi_b$ due to the exposition of seeds to low temperature under moderate water stress (Hawkins et al. 2017). Changes in $\Psi_b$ were also used as a proxy of seed dormancy level for modelling dormancy loss through stratification in *P. aviculare* seeds by Batlla and Benech-Arnold (2004). The authors found that $\Psi_b(50)$ became more negative as seeds were released from dormancy while other hydro-time parameters, such as $\sigma_{\Psi b}$ and $\vartheta_H$, did not change significantly during the dormancy loss process. Changes in $\Psi_b(50)$ in relation to soil temperature were predicted using a negative exponential equation according to the accumulation of $S_{tt}$ units (Eq. 4.4). As commented for *B. tectorum*, the model was recently extended to include dormancy induction due to the effect of high temperatures (Batlla and Agostinelli 2017). However, in this case, the decrease in seed germination due to induction into secondary dormancy was a consequence of a progressive decrease in $\Psi_b(50)$, together with changes in other hydro-time parameters ($\sigma_{\Psi b}$ and $\vartheta_H$). Using above-mentioned models, changes in the germination response of seedbanks to temperature and water potential can be simulated as a function of soil temperature.

As mentioned before, dormancy-terminating factors could affect the emergence pattern of many weed species under field conditions. For example, light is often an important factor controlling weed seedbank emergence under conventional tillage systems. Batlla et al. (2003) and Batlla and Benech-Arnold (2005) developed models to simulate changes in the sensitivity of buried seeds to the stimuli of alternating temperatures and light, respectively, as a function of stratification temperature using the previously presented stratification thermal-time index ($S_{tt}$; Eq. 4.4). For example, the light sensitivity model allows the simulation of the progressive increase in buried seed sensitivity to light during winter and early spring showing how different fractions of the seedbank acquire an extreme sensitivity to light presenting VLFR type responses. Due to the fact that the acquisition of a VLFR would permit buried seeds to germinate in response to the light flash perceived during soil disturbance (Casal and Sánchez 1998), this model could be used to predict the proportion of the seedbank able to germinate in response to tillage operations after the accumulation of a certain amount of $S_{tt}$ during winter (Batlla and Benech-Arnold 2014). Recently, complementary models able to predict how sensitivity to light and alternating temperatures decreases during dormancy induction based on the accumulation of $Di_{tt}$ units were developed (Malavert 2017). These models could be coupled to previously described models to predict changes in the sensitivity of seedbanks to dormancy-terminating factors.

As pointed out before, soil water content can also affect seedbank dormancy changes through its effects on seed hydration level. However, the effect of soil water content or seed moisture on seed dormancy status under natural environments has been scarcely studied (Hu et al. 2018), and there are just few models including this factor as a modulator of seed dormancy level.

One nice example is the model developed by Bair et al. (2006) for *B. tectorum*. These authors quantified the effect of solutions with different water potentials on the rate of dormancy loss in *B. tectorum* seeds through its effects on the rate of decrease in $\Psi_b(50)$. These results were further related to the water potential that seeds can experience when buried in the soil, showing that soil water potential effects on seed dormancy loss rate can be divided into four range zones: (1) one in which seeds are too dry for after-ripening to occur (below −375 MPa), (2) an intermediate range in which the dormancy loss rate increases linearly with an increase in soil water potential (between −375 and − 150 MPa), (3) a third range within which the dormancy loss rate just depends on prevailing soil temperature and is not affected by soil water status (between −150 and − 40 MPa) and (4) a fourth range in which seeds are too wet for after-ripening to occur (above −40 MPa). Bair et al. (2006) showed that including the effect of soil water potential as a dormancy regulating factor in previous models that were driven just by soil temperature generally improved predictions of dormancy loss under dry soil conditions.

Recently, Malavert (2017) quantified the effect of seed moisture content on the rate of dormancy release and induction in *P. aviculare* seeds. Dormancy changes were null for a seed moisture content below 15% (a similar threshold value was recently reported for *Chenopodium album* L. by Hu et al. 2018), while above this threshold, the rate of dormancy release increases up to 31% (above 31%, the rate of dormancy loss depended only on temperature). The inclusion of the effect of seed moisture content on dormancy changes when the soil water content in the seed zone establishes a seed moisture content below 31% improved the prediction of seedling emergence in relation to predictions made using only temperature as a driver of dormancy changes.

Although dormancy is a common feature of many weed seedbank populations, there are some weed species in which their sexual and/or asexual propagules show a very low dormancy level soon after dispersal or show no dormancy at all. In those species, emergence can be predicted just based on thermal-time and hydrothermal-time models previously explained (Eqs. 4.1 and 4.3). These models can be used as tools to predict the period in which these weeds are more susceptible to the application of chemical control practices improving significantly their efficacy (Ghersa et al. 1990). A throughout description and analysis of empirical-based models is presented in the next chapter.

## 4.5 Conclusions

Models which are able to predict timing and extent of seedling emergence forming soil seedbanks in agricultural fields can be useful tools to assist farmers and agronomist to increase the efficacy of weed control methods. Knowing the temporal pattern of weed emergence not only allows us to apply post-emergence control practices when a high proportion of the plants had already emerged but are still in the seedling stage but also to time the application of pre-emergent control tactics to affect a higher proportion of emerged seedling and avoid weed escapes from control practices. In this chapter, we show the way in which a conceptual framework of how the environment regulates seedbank dormancy can be used to develop mechanistic models able to predict seedling emergence patterns in the field. Although this type of models requires a clear conceptualization about the functioning of the system, they have the advantage that once they are developed for a certain weed, for example, a summer annual weed requiring stratification for dormancy release, they can be re-parametrize for other species showing a similar response to the environment. This raises the possibility of developing models for weed dormancy and/or germinative syndromes (group of species that shows similar dormancy and/or germinative response to environmental factors), which can then be re-parametrized for different weed species within each syndrome (Duarte et al. 2015). On the other hand, these mechanistic models can be used to forecast the behaviour of weed seedbanks in changing climatic environments and are flexible enough as to introduce the effects of other factors affecting seed dormancy level and field emergence, as, for example, the effect of the maternal environment (Fernández Farnocchia et al. 2019). Finally, although in this chapter we present models dealing with seeds showing physiological dormancy, models accounting for the environmental regulation of dormancy in weed seeds showing other types of dormancy, such as physical (Gama-Arachchige et al. 2013) or combinational type (physical/physiological) dormancy (Renzi et al. 2016, 2018) have been developed.

## References

Bair NB, Meyer SE, Allen PS (2006) A hydrothermal after-ripening time model for seed dormancy loss in *Bromus tectorum* L. Seed Sci Res 16:17–28

Baskin JM, Baskin CC (1980) Ecophysiology of secondary dormancy in seeds of *Ambrosia artemisiifolia*. Ecology 61:475–480
Baskin JM, Baskin CC (1985) The annual dormancy cycle in buried weed seeds: a continuum. BioScience 35:492–498
Baskin CC, Baskin JM (1988) Germination ecophysiology of herbaceous plant species in temperate region. Am J Bot 75:286–305
Baskin CC, Baskin JM (1998) Seed dormancy and germination. In: Ecology, biogeography and evolution. Academic Press, San Diego (EEUU)
Baskin JM, Baskin CC (2004) A classification system for seed dormancy. Seed Sci Res 14:1–16
Batlla D, Agostinelli A (2017) Thermal regulation of secondary dormancy induction in *Polygonum aviculare* seeds: a quantitative analysis using the hydrotime model. Seed Sci Res 27:231–242
Batlla D, Benech-Arnold RL (2003) A quantitative analysis of dormancy loss dynamics in *Polygonum aviculare* L. seeds. Development of a thermal time model based on changes in seed population thermal parameters. Seed Sci Res 13:55–68
Batlla D, Benech-Arnold RL (2004) Seed dormancy loss assessed by changes in *Polygonum aviculare* L. population hydrotime parameters. Development of a predictive model. Seed Sci Res 14:277–286
Batlla D, Benech-Arnold RL (2005) Changes in the light sensitivity of buried *Polygonum aviculare* seeds in relation to cold-induced dormancy loss: development of a predictive model. New Phytol 165:445–452
Batlla D, Benech-Arnold RL (2007) Predicting changes in dormancy level in weed seed soil banks: implications for weed management. Crop Prot 26:189–197
Batlla D, Benech-Arnold RL (2010) Predicting changes in dormancy level in natural seed soil banks. Plant Mol Biol 73:3–13
Batlla D, Benech-Arnold RL (2014) Weed seed germination and the light environment. Implications for weed management. Weed Biol Manag 14:77–87
Batlla D, Benech-Arnold RL (2015) A framework for the interpretation of temperature effects on dormancy and germination in seed populations showing dormancy. Seed Sci Res 25:147–148
Batlla D, Verges V, Benech-Arnold RL (2003) A quantitative analysis of seed responses to cycle-doses of fluctuating temperatures in relation to dormancy level. Development of a thermal-time model for *Polygonum aviculare* L. seeds. Seed Sci Res 13:197–207
Batlla D, Grundy A, Dent KC, Clay HA, Finch-Savage WE (2009) A quantitative analysis of temperature-dependent dormancy changes in *Polygonum aviculare* seeds. Weed Res 49:428–438
Bauer MC, Meyer SE, Allen PS (1998) A simulation model to predict seed dormancy loss in the field for *Bromus tectorum* L. J Exp Bot 49:1235–1244
Benech-Arnold RL, Sánchez RA (1995) Modelling weed seed germination. In: Kigel J, Gallili G (eds) Seed development and germination. Marcel Dekker, New York, pp 545–566
Benech-Arnold RL, Ghersa CM, Sanchez RA, Insausti P (1990a) A mathematical model to predict *Sorghum halepense* (L.) Pers. seedling emergence in relation to soil temperature. Weed Res 30:91–99
Benech-Arnold RL, Ghersa CM, Sánchez RA, Insausti P (1990b) Temperature effects on dormancy release and germination rate in *Sorghum halepense* (L.) Pers seeds: a quantitative analysis. Weed Res 30:81–89
Benech-Arnold RL, Sánchez RA, Forcella F, Kruk BC, Ghersa CM (2000) Environmental control of dormancy in weed seed soil banks. Field Crop Res 67:105–122
Bewley JD, Bradford K, Hilhorst H, Nonogaki H (2013) Seeds: physiology of development, germination and dormancy, 3rd edn. Springer, New York
Borthwick HA, Hendricks SB, Toole EH, Toole VK (1954) Action of light on lettuce seed germination. Bot Gaz 115:205–225
Bouwmeester HJ (1990) The effect of environmental conditions on the seasonal dormancy pattern and germination of weed seeds. PhD thesis. Agricultural University, Wageningen, The Netherlands. p 157

Bouwmeester HJ, Karssen CM (1992) The dual role of temperature in the regulation of the seasonal changes in dormancy and germination of seeds of *Polygonum persicaria* L. Oecologia 90:88–94

Bradford KJ (1995) Water relations in seed germination. In: Kigel J, Galili A (eds) Seed development and germination. Marcel Dekker, New York, pp 351–396

Bradford KJ (2002) Applications of hydrothermal time to quantifying and modeling seed germination and dormancy. Weed Sci 50:248–260

Buhler DD (1999) Expanding the context of weed management. J Crop Prod 2:1–7

Buhler DD, Hartzler RG, Forcella F, Gunsolus J (1997) Relative emergence sequence for weeds of corn and soybeans. Iowa State University Publication on Pest Management, File 9 (SA-11), Ames, IA

Casal JJ, Sánchez RA (1998) Phytochromes and seed germination. Seed Sci Res 8:317–329

Chantre G, Batlla D, Sabbatini M, Orioli G (2009) Germination parameterization and development of an after-ripening thermal-time model for primary dormancy release of *Lithospermum arvense* seeds. Ann Bot 103:1291–1301

Chantre GR, Sabbatini MR, Orioli GA (2010) An after-ripening thermal-time model for *Lithospermum arvense* seeds based on changes in population hydrotime parameters. Weed Res 50:218–227

Christensen M, Meyer S, Allen PS (1996) A hydrothermal time model of seed after-ripening in *Bromus tectorum* L. Seed Sci Res 6:147–153

Costea M, Tardif FJ (2005) The biology of Canadian weeds. *Polygonum aviculare* L. Can J Plant Sci 85:481–506

Covell S, Ellis RH, Roberts EH, Summerfield RJ (1986) The influence of temperature on seed germination rate in grain legumes. 1. A comparison of chickpea, lentil, soybean and cowpea at constant temperatures. J Exp Bot 37:705–715

Derkx MPM, Karssen CM (1993) Changing sensitivity to light and nitrate but not to gibberellins regulates seasonal dormancy patterns in *Sisymbrium officinale* seeds. Plant Cell Environ 16:469–479

Derkx MPM, Karssen CM (1994) Are seasonal dormancy patterns in *Arabidopsis thaliana* regulated by changes in seed sensitivity to light, nitrate and gibberellin? Ann Bot 73:129–136

Duarte VAC, Batlla D, Ghersa CM y Ferraro DO (2015) Cuando la clave es integrar: introducción al desarrollo de grupos de comportamiento de dormición de malezas en cultivos agrícolas. Agron Ambiente 35(2): 153–169. FA-UBA, Buenos Aires, Argentina

Egley G (1986) Stimulation of weed seed germination in soil. Rev Weed Sci 2:67–89

Ellis RH, Covell S, Roberts EH, Summerfield RJ (1986) The influence of temperature on seed germination rate in grain legumes. II. Interspecific variation in chickpea (*Cicer arietinum* L.) at constant temperature. J Exp Bot 37:1503–1515

Fenner M (1987) Seedlings. New Phytol 106:35–47

Fernández Farnocchia RB, Benech-Arnold RL, Batlla D (2019) Regulation of seed dormancy by the maternal environment is instrumental for maximizing plant fitness in *Polygonum aviculare*. J Exp Bot 70(18):4793–4806. https://doi.org/10.1093/jxb/erz269

Finch-Savage WE, Leubner-Metzger G (2006) Seed dormancy and the control of germination. New Phytol 171:505–523

Finch-Savage WE, Cadman CSC, Toorop PE, Lynn JR, Hilhorst HWM (2007) Seed dormancy release in Arabidopsis cvi by dry after-ripening, low temperature, nitrate and light shows common quantitative patterns of gene expression directed by environmentally specific sensing. Plant J 51:60–78

Gama-Arachchige NS, Baskin JM, Geneve RL, Baskin CC (2013) Quantitative analysis of the thermal requirements for stepwise physical dormancy-break in seeds of the winter annual *Geranium carolinianum* (Geraniaceae). Ann Bot 111:849–858

Garcia-Huidobro J, Monteith JL, Squire GR (1982) Time, temperature and germination of pearl millet (*Pennisetum typhoides* S & H). I. Constant temperature. J Exp Bot 33:288–296

Ghersa CM, Satorre EH, Van Esso ML, Pataro A, Elizagaray R (1990) The use of thermal calendar models to improve the efficiency of herbicide applications in *Sorghum halepense* (L.) Pers. Weed Res 30(3):153–160

Grundy AC, Mead A (2000) Modelling weed emergence as a function of meteorological records. Weed Sci 48:594–603

Grundy AC, Phelps K, Reader RJ, Burston S (2000) Modelling the germination of *Stellaria media* using the concept of hydrothermal time. New Phytol 148:433–444

Gummerson RJ (1986) The effect of constant temperatures and osmotic potentials on the germination of sugar beet. J Exp Bot 37:729–741

Hardegree SP (2006) Predicting germination response to temperature. I. Cardinal-temperature models and sub-population specific regression. Ann Bot 97:1115–1125

Hawkins K, Allen P, Meyer S (2017) Secondary dormancy induction and release in *Bromus tectorum* seeds: the role of temperature, water potential and hydrothermal time. Seed Sci Res 27(1):12–25

Hu XW, Ding XY, Baskin CC, Wang YR (2018) Effect of soil moisture during stratification on dormancy release in seeds of five common weed species. Weed Res 58(3):210–220

Huarte HR, Borlandelli F, Varisco D, Batlla D (2018) Understanding dormancy breakage and germination ecology of *Cynara cardunculus* (Asteraceae). Weed Res 58:450–462

Karssen CM (1982) Seasonal patterns of dormancy in weed seeds. In: Khan AA (ed) The physiology and biochemistry of seed development, dormancy and germination. Elsevier, Amsterdam, pp 243–270

Kruk BC, Benech-Arnold RL (1998) Functional and quantitative analysis of seed thermal responses in prostate knotweed (*Polygonum aviculare*) and common purslane (*Portulaca oleracea*). Weed Sci 46:83–90

Leguizamon ES, Rodriguez N, Rainero H, Perez M, Perez L, Zorza E, Fernandez-Quintanilla C (2009) Modelling the emergence pattern of six summer annual weed grasses under no tillage systems in Argentina. Weed Res 49:98–106

Malavert C (2017) Regulación ambiental de los cambios en el nivel de dormición durante la salida de la dormición primaria y la entrada en dormición secundaria en semillas de *Polygonum aviculare* L.: el papel de la temperatura y la disponibilidad hídrica del suelo [Tesis Doctoral]. Facultad de Agronomía, Universidad de Buenos Aires, Argentina, 2017

Malavert C, Batlla D, Benech-Arnold RL (2017) Temperature-dependent regulation of induction into secondary dormancy of *Polygonum aviculare* L. seeds: a quantitative analysis. Ecol Model 352:128–138

Nikolaeva MG (1967) Physiology of deep dormancy in seeds. National Science Foundation, Washington, DC, p 219

Ogg A, Dawson J (1984) Time of emergence of eight weed species. Weed Sci 32(3):327–335

Pons TL (1992) Seed responses to light. In: Fenner M (ed) Seeds. The ecology of regeneration in plant niches. C.A.B. International, Wallingford, pp 259–284

Pritchard HW, Tompsett PB, Manger KR (1996) Development of a thermal time model for the quantification of dormancy loss in *Aesculus hippocastanum* seeds. Seed Sci Res 6:127–135

Probert RJ (1992) The role of temperature in germination ecophysiology. In: Fenner M (ed) Seeds. The ecology of regeneration in plant communities. C.A.B. International, Wallingford, pp 285–325

Radosevich SR, Holt J, Ghersa CM (1997) Weed ecology: implications for management. Wiley, New York

Renzi JP, Chantre GR, Cantamutto MA (2016) Effect of water availability and seed source on physical dormancy break of *Vicia villosa* ssp. *villosa*. Seed Sci Res 26(3):254–263

Renzi JP, Chantre GR, Cantamutto MA (2018) *Vicia villosa* ssp. *villosa* Roth field emergence model in a semiarid agroecosystem. Grass Forage Sci 73(1):146–158

Scopel AL, Ballaré CL, Sánchez RA (1991) Induction of extreme light sensitivity in buried weed seeds and its role in the perception of soil cultivation. Plant Cell Environ 14:501–508

Steadman KJ, Pritchard HW (2003) Germination of *Aesculus hippocastanum* seeds following cold-induced dormancy loss can be described in relation to a temperature-dependent reduction in base temperature (Tb) and thermal time. New Phytol 161:415–425
Steadman KJ, Crawford AD, Gallagher RS (2003) Dormancy release in *Lolium rigidum* seeds is a function of thermal after-ripening time and seed water content. Funct Plant Biol 30:345–352
Totterdell S, Roberts EH (1980) Characteristics of alternating temperatures which stimulate loss of dormancy in seeds of *Rumex obtusifolius* L. and *Rumex crispus* L. Plant Cell Environ 3:3–12
Vegis A (1964) Dormancy in higher plants. Annu Rev Plant Physiol 15:185–224
Vleeshouwers LM, Kropff MJ (2000) Modelling field emergence patterns in arable weeds. New Phytol 148:445–457
Vleeshouwers LM, Bouwmeester HJ, Karssen CM (1995) Redefining seed dormancy: an attempt to integrate physiology and ecology. J Ecol 83:1031–1037
Washitani I (1985) Germination-rate dependency on temperature of *Geranium carolinianum* seeds. J Exp Bot 36:330–337
Washitani I (1987) A convenient screening test system and a model for thermal germination responses of wild plant seeds: behaviour of model and real seed in the system. Plant Cell Environ 10:587–598

# Chapter 5
# Weed Emergence Models

**Aritz Royo-Esnal, Joel Torra, and Guillermo R. Chantre**

**Abstract** Weed emergence models are practical tools that aim to describe the dynamics of emergence in the field. Such models can be conceptualized from two main perspectives: a reductionist/mechanistic approach and an empirical modelling viewpoint. While the former provides a close description of the basic ecophysiological processes underlying weed emergence (i.e. seed dormancy, germination and pre-emergence growth), they usually require a large amount of difficult to estimate species-specific parameters, as well as sometimes unavailable or missing experimental data for model development/calibration/validation. Conversely, the latter aims to describe the emergence process as *a whole* by seeking a general mathematical description of field emergence data as a function of field environmental variables, mainly temperature and precipitation. As reviewed in the literature, most emergence models have been developed using nonlinear regression (NLR) techniques. NLR sigmoidal type models which are based on cumulative thermal or hydrothermal time have become the most popular approach as they are easy to develop and use. However, some statistical and bioecological limitations arise, for example, the lack of independence between samplings, censored data, need for threshold thermal/hydric parameter estimation and determination of 'moment zero' for thermal/hydrothermal-time accumulation, among other factors, which can lead to inaccurate descriptions of the emergence process. New approaches based on soft computing techniques (SCT) have recently been proposed as alternative models to tackle some of the previously mentioned limitations. In this chapter, we focus on empirical weed emergence models with special emphasis in NLR models, highlighting some of the main advantages, as well as the statistical and biological limitations that could affect their predictive accuracy. We briefly discuss new SCT-based

A. Royo-Esnal (✉) · J. Torra
Grup de Malherbologia i Ecologia Vegetal, Department of HBJ, Agrotecnio, ETSEA, Universitat de Lleida, Lleida, Spain
e-mail: aritz.royo@udl.cat; joel.torra@udl.cat

G. R. Chantre
Departamento de Agronomía, Universidad Nacional del Sur - CERZOS/CONICET, Bahía Blanca, Buenos Aires, Argentina
e-mail: gchantre@criba.edu.ar


G. R. Chantre, J. L. González-Andújar (eds.), *Decision Support Systems for Weed Management*, https://doi.org/10.1007/978-3-030-44402-0_5

approaches, such as artificial neural networks which have recently been used for weed emergence modelling.

**Keywords** Empirical modelling · Field emergence data · Non-linear regression · Hydrothermal-time · Soft computing · Artificial neural networks · Uncertainty

## 5.1 Introduction

### *5.1.1 Weed Management: Field Emergence Dynamics*

Weeds are the major risk for crop production (Oerke 2006). Field management strategies always aim to minimize the impact of weeds, directly by applying specific weed control practices (chemical or mechanical) or indirectly by cultural management methods (e.g. crop selection, soil management and fertilization, sowing dates). The latter also determine the effectiveness of weed control (Cardina et al. 2007). In this regard, two important reasons can define the necessity for a precise characterization of weed species emergence dynamics in the field: (1) primarily, the relative moment of emergence determines the probability of individuals' successful recruitment within the population (Forcella et al. 2000); (2) secondly, early growth stages of the population show the highest susceptibility to control interventions and both intra- and interspecific competition (Menalled and Schonbeck 2011). Thus, knowing both time and magnitude of field emergence is a key aspect of weed management.

As described in the previous chapter, mechanistic models are valuable approximations as they provide a close description of the basic ecophysiological processes underlying weed emergence (i.e. seed dormancy, germination and pre-emergence growth) (Fig. 5.1). They require a large amount of species-specific parameters which can be difficult to estimate, and sometimes, the complexity of the model conspires against the level of parsimony desired for practical decision-making (Grundy 2003). Reductionist approaches are rather scarce in the literature (e.g. Vleeshouwers and Kropff 2000; Gardarin et al. 2012; Renzi et al. 2018) as they can be very time-consuming to develop/calibrate/validate. Ideally, they should be part of more integrative frameworks including demographic (Gardarin et al. 2012; D'Amico et al. 2018; Renzi et al. 2019), economic (Pannell et al. 2004; Beltran et al. 2012) and environmental (Lodovichi et al. 2013; Lammoglia et al. 2017) elements of weed management systems.

In this chapter, we focus on empirical weed emergence approaches with special emphasis on sigmoidal type nonlinear regression (NLR) models, highlighting some of their main advantages and also the statistical and biological limitations that could affect their predictive accuracy. We briefly discuss new soft computing techniques

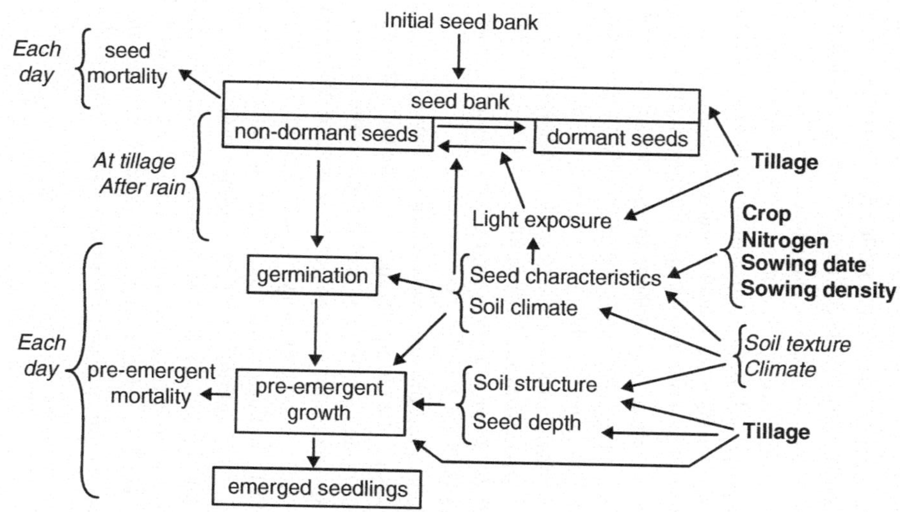


**Fig. 5.1** A schematic model of the weed emergence process, from the seedbank to seedling emergence, including intervening biological, environmental (in italic) and crop management (in bold) factors. The arrows show the direction of the effect (modified from Colbach et al. 2006)

(SCT), such as artificial neural networks, which have recently been used for weed emergence modelling.

### 5.1.2 A Summary of Modelling Efforts: A Literature Review

Empirical NLR approaches have been extensively used for weed emergence modelling (Gonzalez-Andujar et al. 2016a, b). Among them, Weibull, Logistic and Gompertz are by far the most conspicuous (Tables 5.1 and 5.2).

Models for 58 dicotyledonous (Table 5.1) and 37 monocotyledonous weed species (Table 5.2) belonging to 24 botanical families, 22 of them dicotyledonous and two monocotyledonous, were developed. For some of these species, only germination models have been proposed, but because of their field applicability described by the corresponding authors, they have also been included.

As observed in Tables 5.1 and 5.2, the number of empirical models developed for each weed species denotes their importance. In this sense, *Avena fatua* could be considered the most concerning weed, as substantial efforts have been dedicated by the scientific modelling community (Table 5.2). Other species, such as *Chenopodium album*, *Amaranthus retroflexus*, *Ambrosia trifida*, *Echinochloa crus-galli*, *Sorghum halepense*, *Ambrosia artemisiifolia*, *Xanthium strumarium*, *Thlaspi arvense*, *Abutilon theophrasti*, *Polygonum aviculare* and *Digitaria sanguinalis*, also received considerable attention (Tables 5.1 and 5.2).

Among NLR models, Weibull and its variations have largely been used for parameterization, but others like Gompertz and Logistic have also successfully been applied. The least used models have been probit regression, Boltzmann, Chapman and Hill functions, Gaussian, Linear, General-logistic and Wang and Engel functions (Tables 5.1 and 5.2). As explanatory variables, both thermal-time (TT) and hydrothermal-time (HTT) indices have been used to integrate the effects of soil temperature and soil water potential on the emergence process. Also, TT and HTT are at least at some extent influenced by the cropping system. For example, models for weed species from summer irrigated crops have been successfully parameterized using TT, while HTT has been useful in describing autumn-winter annual weeds that occur in cereal crops and other rain-fed winter crops (Tables 5.1 and 5.2).

## 5.2 Empirical NLR Models

### 5.2.1 Basics

Empirical NLR models are tools for predicting both timing and quantity of cumulative percentage using environmental variables, such as temperature, soil moisture and, more recently, light (Royo-Esnal et al. 2015a). They are based on the thermal

**Table 5.1** Dicotyledonous families and species (with their common names) for which germination (Germ) and emergence (Emerg) models considering empirical and mechanistic approaches have been developed, based on either thermal time (TT) of hydrothermal time (HTT)

| Family | Species | Common name | Germ | Emerg | TT | HTT | Model | References |
|---|---|---|---|---|---|---|---|---|
| Amaranthaceae | *Amaranthus blitoides* | Prostrate pigweed | | X | X | | Logistic/Gompertz/GA | Haj Seyed Hadi and Gonzalez-Andujar (2009) |
| | *Amaranthus powellii* | Green pigweed | X | | X | X | Weibull? | Oryokot et al. (1997) |
| | *Amaranthus retroflexus* | Redroot pigweed | | X | | X | Weibull | Werle et al. (2014b) |
| | | | X | X | X | X | Logistic/Gompertz | Masin et al. (2010) |
| | | | | X | X | | Logistic/Gompertz/GA | Haj Seyed Hadi and Gonzalez-Andujar (2009) |
| | | | X | | X | X | Weibull? | Oryokot et al. (1997) |
| | *Amaranthus tuberculatus* | Common waterhemp | | X | | X | Weibull | Werle et al. (2014b) |
| | *Bassia scoparia* | Bassia | | X | | X | Weibull | Werle et al. (2014b) |
| | *Beta vulgaris* | Weed beet | X | X | X | X | Weibull/Sigmoidal | Sester et al. (2007) |
| | *Chenopodium album* | Common lambsquarter | | X | | X | Weibull | Werle et al. (2014b) |
| | | | X | X | X | X | Logistic/Gompertz | Masin et al. (2010) |
| | | | | X | X | | Weibull | Leblanc et al. (2003, 2004) |
| | | | | X | | | Logistic | Grundy et al. (2003b) |
| | | | | X | | X | Weibull | Roman et al. (1999) |
| | | | X | X | X | | Gompertz | Vleeshouwers and Kropff (2000) |
| Apocynaceae | *Apocynum androsaemifolium* | Spreading dogbane | | X | X | | Weibull | Wu et al. (2013) |
| Compositae | *Ageratum conyzoides* | Tropic ageratum | | X | | X | Weibull | Ekeleme et al. (2005) |
| | *Ambrosia artemisiifolia* | Common ragweed | | | X | X | Weibull | Barnes et al. (2017) |
| | | | | X | | X | Weibull | Werle et al. (2014b) |
| | | | X | | | X | Weibull | Shrestha et al. (1999) |

(continued)

**Table 5.1** (continued)

| Family | Species | Common name | Germ | Emerg | TT | HTT | Model | References |
|---|---|---|---|---|---|---|---|---|
| | *Ambrosia trífida* | Giant ragweed | | X | | X | Weibull | Goplen et al. (2018) |
| | | | | X | | X | Weibull | Werle et al. (2014b) |
| | | | | X | | X | Weibull | Davis et al. (2013) |
| | | | | X | | X | Weibull | Schutte et al. (2008) |
| | *Centaurea diluta* | North African knapweed | | X | X | X | Logistic/Gompertz/ Weibull | Sousa-Ortega et al. (2020b) |
| | *Erigeron bonariensis** | Hairy fleabane | X | | X | | Tb, To, Tc estim | Valencia-Gredilla et al. (2020) |
| | | | X | X | | X | Probit/Gompertz | Zambrano-Navea et al. (2013) |
| | *Helianthus annuus* | Common sunflower | | X | | X | Weibull | Werle et al. (2014b) |
| | | | | X | X | X | Weibull | Clay et al. (2014) |
| | *Senecio vulgaris* | Common groundsel | X | | X | | Tb estim | Masin et al. (2017) |
| | | | | X | X | X | Weibull | McGiffen et al. (2008) |
| | *Sonchus oleraceus* | Annual sowthistle | X | | X | X | Sigmoidal | Ali et al. (2020) |
| | | | X | | X | | Tb estim | Masin et al. (2017) |
| | | | | X | X | | Gompertz | Dorado et al. (2009) |
| | *Taraxacum officinale* | Dandelion | X | | X | | Tb estim | Masin et al. (2017) |
| | | | | X | | X | Weibull | Werle et al. (2014a) |
| | *Xanthium strumarium* | Common cocklebur | | X | | X | Weibull | Werle et al. (2014b) |
| | | | | X | X | | Weibull/Gompertz | Dorado et al. (2009) |
| | | | | X | | X | Weibull | Norsworthy and Oliveira (2007) |
| Boraginaceae | *Buglossoides arvensis* | Field gromwell | X | | X | X | Probit-regression | Chantre et al. (2010) |
| | | | X | | X | | Normal dist. | Chantre et al.(2009) |
| Brassicaceae | *Brassica napus* | Volunteer canola | | X | X | | Weibull | Soltani et al. (2018) |
| | *Camelina microcarpa* | Small-seeded false flax | | X | | X | Weibull | Royo-Esnal et al. (2015c) |

| | | | | | | | | |
|---|---|---|---|---|---|---|---|---|
| | *Camelina sativa* | False flax | | X | X | | Hill | Allen et al. (2014) |
| | *Capsella bursa-pastoris* | Shepherd's purse | | X | | X | Weibull | Werle et al. (2014a) |
| | | | | X | X | | Gompertz | Hill et al. (2014) |
| | *Descurainia pinnata* | Pinnate tansymustard | | X | | X | Weibull | Werle et al. (2014a) |
| | *Descurainia sophia* | Flixweed | | | X | | Weibull/Gompertz | Aboutalebian et al. (2017) |
| | *Lepidium virginicum* | Virginia pepperweed | | X | | X | Weibull | Werle et al. (2014a) |
| | *Neslia paniculata* | Ball mustard | X | X | X | X | Boltzmann/Weibull | Royo-Esnal et al. (2019) |
| | *Rapistrum rugosum* | Annual bastardcabbage | X | | X | X | Sigmoidal | Ali et al. (2020) |
| | *Thlaspi arvense* | Field pennycress | | X | | X | Weibull | Royo-Esnal et al. (2015a) |
| | | | | X | | X | Weibull | Werle et al. (2014a) |
| | | | | X | X | | Gompertz | Hill et al. (2014) |
| Caryophyllaceae | *Drymaria cordata* | West Indian chickweed | X | | X | X | Probit-regression | Cardoso and Pereira (2008) |
| | *Spergula arvensis** | | X | X | X | | Gompertz | Vleeshouwers and Kropff (2000) |
| | *Stellaria media* | Chickweed | | X | X | | Gompertz | Hill et al. (2014) |
| | | | | X | | | Logistic | Grundy et al. (2003b) |
| Convolvulaceae | *Cuscuta campestris* | Field dodder | X | | | | Weibull | Goldwasser et al. (2016) |
| | *Ipomoea hederacea* | Ivyleaf morningglory | | X | | X | Weibull | Werle et al. (2014b) |
| | *Ipomoea purpurea* | Common morningglory | | X | X | | Logistic/Gompertz/GA | Haj Seyed Hadi and Gonzalez-Andujar (2009) |
| Cucurbitaceae | *Sicyos angulatus* | Burcucumber | | X | | X | Weibull | Werle et al. (2014b) |

(continued)

**Table 5.1** (continued)

| Family | Species | Common name | Germ | Emerg | TT | HTT | Model | References |
|---|---|---|---|---|---|---|---|---|
| Euphorbiaceae | *Euphorbia heterophylla** | Wild poinsettia | X | | X | | Weibull | Cochavi et al. (2018) |
| Leguminosae | *Medicago lupulina* | Black medic | X | | X | | Weibull | Sharpe and Boyd (2019) |
| | *Vicia villosa* | Hairy vetch | X | X | X | X | Mechanistic | Renzi et al. (2018) |
| Lamiaceae | *Lamium amplexicaule* | Henbit | | X | | X | Weibull | Werle et al. (2014a) |
| | | | | X | X | | Gompertz | Hill et al. (2014) |
| Malvaceae | *Abutilon theophrasti* | Velvetleaf | | X | | X | Weibull | Werle et al. (2014b) |
| | | | X | X | X | X | Logistic/Gompertz | Masin et al. (2010) |
| | | | | X | X | | Weibull | Dorado et al. (2009) |
| | *Hibiscus trionum* | Venice mallow | | X | | X | Weibull | Werle et al. (2014b) |
| Orobanchaceae | *Orobanche cernua** | Broomrape | | X | X | | Weibull | Eizenberg et al. (2012) |
| Oxalidaceae | *Oxalis latifolia* | Broadleaf woodsorrel | | X | X | | Gaussian/linear | Royo-Esnal and López-Fernández (2010) |
| Papaveraceae | *Papaver rhoeas* | Corn poppy | | X | X | | Gompertz | Izquierdo et al. (2009) |
| Plantaginaceae | *Veronica peregrina* | Purslane speedwell | | X | | X | Weibull | Werle et al. (2014a) |
| Polygonaceae | *Persicaria maculosa** | Redshank | X | X | X | | Gompertz | Vleeshouwers and Kropff (2000) |
| | *Persicaria pensylvanica** | Pennsylvania smartweed | | X | | X | Weibull | Werle et al. (2014b) |
| | *Polygonum aviculare* | Prostrate knotweed | | X | X | | Gompertz | Yousefi et al. (2014) |
| | | | | X | | X | Weibull | Royo-Esnal et al. (2015b) |
| | | | X | X | X | | Tb, To, Tc estim | Kruk and Benech-Arnold (1998) |
| Portulacaceae | *Portulaca oleracea* | Common purslane | X | X | X | | Tb, To, Tc estim | Kruk and Benech-Arnold (1998) |

| | | | | | | | | |
|---|---|---|---|---|---|---|---|---|
| Ranunculaceae | *Ranunculus repens* | Creeping buttercup | | X | X | | Logistic | Harris et al. (1998) |
| Rubiaceae | *Galium aparine* | Bedstraw/cleaver | | X | X | X | Weibull | Royo-Esnal et al. (2010) |
| | *Galium spurium* | Bedstraw/false cleaver | | X | X | X | Weibull | Royo-Esnal et al. (2010) |
| | *Galium tricornutum* | Three-horn bedstraw | | X | X | X | Weibull | Royo-Esnal et al. (2010) |
| Solanaceae | *Datura ferox* | Fierce thornapple | | X | X | | Gompertz | Dorado et al. (2009) |
| | *Datura stramonium* | Jimsonweed | | X | | X | Weibull | Werle et al. (2014b) |
| | | | | X | X | | General-logistic | Dorado et al. (2009) |
| | *Solanum nigrum* | Black nightshade | | X | X | | General-logistic/ Gompertz | Dorado et al. (2009) |
| | *Solanum ptychanthum* | Eastern black nightshade | | X | | X | Weibull | Werle et al. (2014b) |
| Violaceae | *Viola bicolor* | Field pansy | | X | | X | Weibull | Werle et al. (2014a) |
| 22 | 58 | | 27 | 74 | 54 | 51 | 105 | 49 research articles |

The most popular models applied for each species are also provided, although in some of the provided reference other models could have been published for a certain species. The last row summarizes the total number for each column. All families and species are named as they appear in the official page 'theplantlist.org', consulted in April 2020; that means that some species' name have been changed with respect to the original publication. These species have been marked with an asterisk (*) and the synonyms are listed after the table

**Erigeron bonariensis* = *Conyza bonariensis; Euphorbia heterophylla* = *E. geniculata; Spergula arvensis* = *Spergularia arvensis; Orobanche cernua* = *O. cumana; Persicaria maculosa* = *Polygonum persicaria; Persicaria pensylvanica* = *Polygonum pensylvanicum*

**Table 5.2** Monocotyledonous families and species (with their common names) for which germination (Germ) and emergence (Emerg) models considering empirical and mechanistic approaches have been developed, based on either thermal time (TT) of hydrothermal time (HTT)

| Family | Species | Common name | Germ | Emerg | TT | HTT | Model | References |
|---|---|---|---|---|---|---|---|---|
| Cyperaceae | *Cyperus difformis* | Umbrella sedge | X | | X | | Probit | Pedroso et al. (2019) |
| | | | | X | X | | Sigmoidal | Lundy et al. (2014) |
| | *Cyperus rotundus* | Nut sedge | | X | X | | Weibull/logistic | Dorado et al. (2009) |
| Poaceae | *Alopecurus carolinianus* | Carolina foxtail | | X | | X | Weibull | Werle et al. (2014a) |
| | *Alopecurus myosuroides* | Black-grass | X | X | X | X | Weibull/Sigmoidal | Colbach et al. (2006) |
| | *Agropyron desertorum* | Wheatgrass | | X | X | | Weibull/Gompertz | Behtari and de Luis (2012) |
| | *Apera spica-venti* | | | X | X | | Log-logistic | Scherner et al. (2017) |
| | *Avena fatua* | Wild oat | | X | X | | Weibull/Gompertz | Aboutalebian et al. (2017) |
| | | | | X | X | | Gompertz | Yousefi et al. (2014) |
| | | | X | X | | X | Logistic/GA | Blanco et al. (2014) |
| | | | | X | | X | Weibull/ANN | Chantre et al. (2014) |
| | | | | X | | X | ANN | Chantre et al. (2012) |
| | | | | X | X | X | Weibull | Martinson et al. (2007) |
| | | | | X | X | | Weibull | Page et al. (2006) |
| | *Avena sterilis* | Wild oat | | X | X | | Logistic/Gompertz/GA | Haj Seyed Hadi and Gonzalez-Andujar (2009) |
| | | | | X | X | X | Weibull | Leguizamón et al. (2005) |
| | *Bromus diandrus* | Rigid brome | | X | | X | Chapman | Garcia et al. (2013) |
| | | | | X | | X | Non-parametric | Cao et al. (2013) |
| | *Bromus* sp. | Brome | | X | X | | Logistic/Gompertz/GA | Haj Seyed Hadi and Gonzalez-Andujar (2009) |
| | *Bromus tectorum* | Downy brome | | X | | X | Weibull | Werle et al. (2014a) |
| | | | X | | | X | Probit-regression | Taylor et al. (2007) |
| | *Cenchrus spinifex* | Field sandbur | | X | | X | Weibull | Werle et al. (2014b) |

| | | | | | | | | |
|---|---|---|---|---|---|---|---|---|
| | *Digitaria sanguinalis* | Large crabgrass | X | X | X | X | Logistic/Gompertz | Masin et al. (2010) |
| | | | | X | X | | Weibull/Gompertz | Dorado et al. (2009) |
| | | | | X | X | | Webull | Leguizamón et al. (2009) |
| | *Echinochloa colona* | Jungle rice | | X | X | | Weibull | Leguizamón et al. (2009) |
| | *Echinochloa crus-galli* | Barnyardgrass | | X | | X | Weibull | Werle et al. (2014b) |
| | | | | X | X | X | Weibull | Bagavathiannan et al. (2011) |
| | | | X | X | X | X | Logistic/Gompertz | Masin et al. (2010) |
| | | | | X | X | | General logistic | Dorado et al. (2009) |
| | *Echinochloa oryzoides** | | X | X | X | X | Log probit | Boddy et al. (2012) |
| | *Echinochloa* ssp. | Watergrass | | X | X | | Sigmoidal | Lundy et al. (2014) |
| | *Eleusine indica* | Goosegrass | | X | X | | Weibull | Leguizamón et al. (2009) |
| | *Eriochloa villosa* | Woolly cupgrass | | X | | X | Weibull | Werle et al. (2014b) |
| | *Festuca arundinacea* | Tall fescue | | X | X | | Weibull/Gompertz | Behtari and de Luis (2012) |
| | *Lolium rigidum* | Annual ryegrass | | X | X | X | Logistic/Gompertz/ Weibull | Sousa-Ortega et al. (2020a) |
| | | | | X | X | X | Gompertz | Izquierdo et al. (2013) |
| | | | | X | X | | Logistic/Gompertz/ GA | Haj Seyed Hadi and Gonzalez-Andujar (2009) |
| | *Oryza sativa* | Weedy red rice | | X | X | | Wang and Engel | Streck et al. (2011) |
| | *Panicum dichotomiflorum* | Fall panicum | | X | | X | Weibull | Werle et al. (2014b) |
| | *Phalaris minor* | Littleseed canarygrass | X | | X | | Weibull/beta-function | Derakhshan et al. (2014) |
| | *Phalaris paradoxa* | Hood canarygrass | | X | | X | Weibull/Gompertz/ Log | Gonzalez-Andujar et al. (2016b) |
| | *Poa annua* | | | X | X | | Log-logistic | Scherner et al. (2017) |
| | *Rottboellia cochinchinensis* | Itchgrass | | X | X | | Weibull/logistic | León et al. (2015) |

(continued)

**Table 5.2** (continued)

| Family | Species | Common name | Germ | Emerg | TT | HTT | Model | References |
|---|---|---|---|---|---|---|---|---|
| | *Setaria faberi* | Giant foxtail | | X | | X | Weibull | Werle et al. (2014b) |
| | *Setaria parviflora** | Marsh bristlegrass | | X | X | | Weibull | Leguizamón et al. (2009) |
| | *Setaria pumila* | Yellow foxtail | | X | | X | Weibull | Werle et al. (2014b) |
| | | | X | X | X | X | Logistic/Gompertz | Masin et al. (2010) |
| | *Setaria viridis* | Green foxtail | | X | | X | Weibull | Werle et al. (2014b) |
| | | | X | X | X | X | Logistic/Gompertz | Masin et al. (2010) |
| | *Sorghum bicolor* | Shattercane | | X | | X | Weibull | Werle et al. (2014b) |
| | *Sorghum halepense* | Aleppo grass | | X | X | | Logistic | Loddo et al. (2012) |
| | | | X | X | X | X | Logistic/Gompertz | Masin et al. (2010) |
| | | | | X | X | | Gompertz | Dorado et al. (2009) |
| | | | | X | X | | Weibull | Leguizamón et al. (2009) |
| | *Triticum aestivum* | Spring wheat | | X | | X | Gompertz | Bullied et al. (2012) |
| | *Vulpia myuros* | | | X | X | | Log-logistic | Scherner et al. (2017) |
| | *Urochloa panicoides* | Liver seed grass | X | X | X | | Gompertz | Ustarroz et al. (2016) |
| | *Urochloa* sp. | Signalgrass | | X | X | | Weibull | Leguizamón et al. (2009) |
| 2 | 37 | | 12 | 56 | 41 | 30 | 80 | 33 (7 combined with dicot models) |

The most popular models applied for each species are also provided, although in some of the provided reference other models could have been published for a certain species. The last row summarizes the total number for each column. All families and species are named as they appear in the official page 'theplantlist.org', consulted in April 2020; that means that some species' name have been changed with respect to the original publication. These species have been marked with an asterisk (∗) and the synonyms are listed after the table

**Echinochloa oryzoides* = *E. phyllopogon; Setaria parviflora* = *S. geniculata*

or hydrothermal-time concept proposed by Gummerson (1986) which assumes that seeds need to accumulate a certain amount of growing degree days (°Cd) before completing germination and emergence. Soil temperature and soil water potential data is usually obtained from in situ determinations using dataloggers, or alternatively, they can be estimated using free available software, such as the STM[2] (Soil Temperature and Moisture Model) developed by the USDA-ARS (Spokas and Forcella 2009).

NLR models can be classified as thermal, hydrothermal or photohydrothermal-time based. In any case, during the cropping season, daily mean soil temperature is accumulated (TT) above a specific soil water threshold (base water potential for germination, $\Psi_b$) (Roman et al. 2000) or a combination of water and photoperiod thresholds (Royo-Esnal et al. 2015a, c, 2019) until weed emergence is completed. In order to generate thermal, hydrothermal-time or photohydrothermal-time data, it is necessary to estimate 'species-specific' threshold parameters (base temperature, base water potential for seed germination/emergence). Threshold parameters for thermal-time sum are the base and ceiling temperatures ($T_b$ and $T_c$, respectively), while for HTT, usually, the base soil water potential is considered ($\psi_b$). Base temperature is considered that one above which thermal time is accumulated and below which not (Roberts 1988); in contrast, ceiling temperature is that one below which temperature is accumulated and above which not. With these considerations, for a given species, the cumulative TT ($\theta_T$) is estimated with the following formula (Eq. 5.1):

$$\theta_T = \sum_{i=1,n} (T_i - T_b) \quad \text{if} \quad T_i > T_b \quad \text{and} \quad < T_c$$
$$\theta_T = 0 \quad \text{Otherwise} \tag{5.1}$$

where $\theta_T$ is the cumulative thermal time at day $i$ and $T_i$ is the mean daily temperature at day $i$. This way of estimating TT was improved by introducing the hydrotime concept (Roman et al. 1999), where thermal time for emergence is accumulated only when a threshold moisture value ($\psi_b$) is available for a given species, and the thermal time (TT) derives in the hydrothermal-time (HTT) concept. Cumulative HTT ($\theta_{HTT}$) is estimated with the following formula (Eq. 5.2):

$$\theta_{HTT} = \sum_{i=1,n} (T_i - T_b)(\Psi_i - \Psi_b)$$
$$\Psi_i - \Psi_b = 1 \quad \text{if} \quad \Psi_i > \Psi_b \tag{5.2}$$
$$\Psi_i - \Psi_b = 0 \quad \text{if} \quad \Psi_i < \Psi_b$$

where $\Psi_i$ is the soil water potential at day $i$. This way of estimating either TT or HTT accumulation has been useful for emergence modelling (Tables 5.1 and 5.2).

Other approaches have also been considered, making corrections or calibrations to this basic TT and HTT estimations. For example, using the beta-function, Cochavi et al. (2018) considered that the germination/emergence rate increases from $T_b$ to $T_o$, while it decreases from $T_o$ to $T_c$. Therefore, to consider this variation rate, a correction factor $r$ (Eq. 5.3) is included to the thermal-time estimation, which is as follows:

$$r = \left( \left( \frac{T_i - T_b}{T_o - T_b} \right) \left( \frac{T_c - T_i}{T_c - T_o} \right)^{\frac{T_c - T_o}{T_o - T_b}} \right)^a \tag{5.3}$$

where $T_i$ is the mean daily temperature at day $i$, $T_b$ is the base temperature, $T_o$ is the optimal temperature, $T_c$ is the ceiling temperature and $a$ is the shape of the slope. Therefore, the newly estimated $\theta_T$ would be calculated as follows (Eq. 5.4):

$$\theta_T = \sum_{i=1,n} r(T_i - T_b) \tag{5.4}$$

This methodology was found useful in order to improve the accuracy of the germination model for *Euphorbia geniculata* (Cochavi et al. 2018).

As previously mentioned, photoperiod can also be a useful factor to improve the accuracy of the models. Attempts have been done to introduce this factor to achieve better predictions on the emergence of some weeds. Royo-Esnal et al. (2015a) have been able to describe the emergence of autumn and spring cohorts of *Thlaspi arvense* in a single accurate model by introducing photoperiod as a correcting factor for HTT turning it in photohydrothermal time (PhHTT) (Eq. 5.5) (Fig. 5.2).

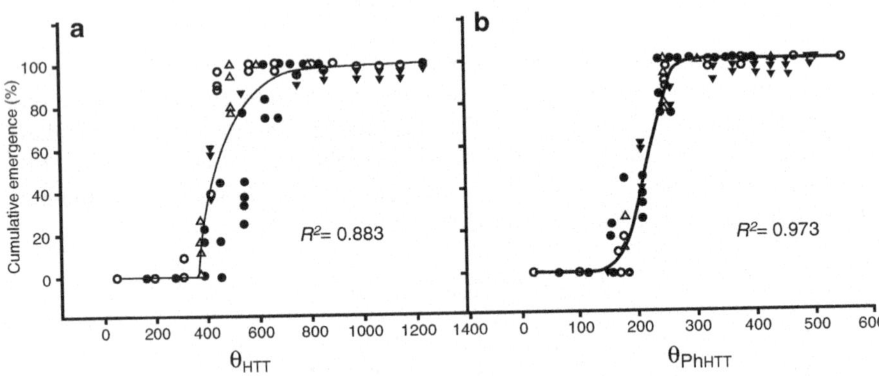


**Fig. 5.2** $\theta_{HTT}$ (**a**) and $\theta_{PhHTT}$ (**b**) based emergence models developed for *Thlaspi arvense*. Black dots (●) and triangles (▼) represent autumn-winter emergences in 2011–2012 and 2012–2013, and white dots (○) and triangles (Δ) represent spring emergences. The shorter photoperiod in autumn-winter corrects the HTT with a lower factor, while spring HTT are multiplied by a greater factor; hence, points from graph **a** get closer in graph **b**

$$\theta_{\mathrm{PhHTT}} = \sum_{i=1,n} \theta_{\mathrm{HTT}i} D_i \tag{5.5}$$

where $\theta_{\mathrm{PhHTT}}$ is the photohydrothermal-time sum at day $i$ and $D_i$ is the proportional day length in day $i$. This is the simplest way of estimating PhHTT, but other ways of integrating the light factor have also been suggested (Royo-Esnal et al. 2015c, 2019).

Soil management practices can also influence field emergence dynamics. For instance, soil disturbance is known to affect the amount of seedbank germination and the proportion of emerged seedlings in arable weed species (Torra et al. 2018), but usually the periodicity of these events seems to be less affected (Froud-Williams et al. 1984). As observed in Fig. 5.3a, under no-tillage system, the emergence of *Apera spica-venti* and *Vulpia myuros* is delayed with respect to mouldboard ploughing (Scherner et al. 2017). Conversely, the emergence patterns of *Poa annua* (Fig. 5.3b) or *Bromus diandrus* (Fig. 5.3c, d) were not influenced by the tillage system (Garcia et al. 2013; Scherner et al. 2017). Similarly, for *Ambrosia trifida* not-till vs. minimum tillage did not affect the emergence pattern (Barnes et al. 2017), which was clearly affected by cover crops and perennial crops (Goplen et al. 2018).

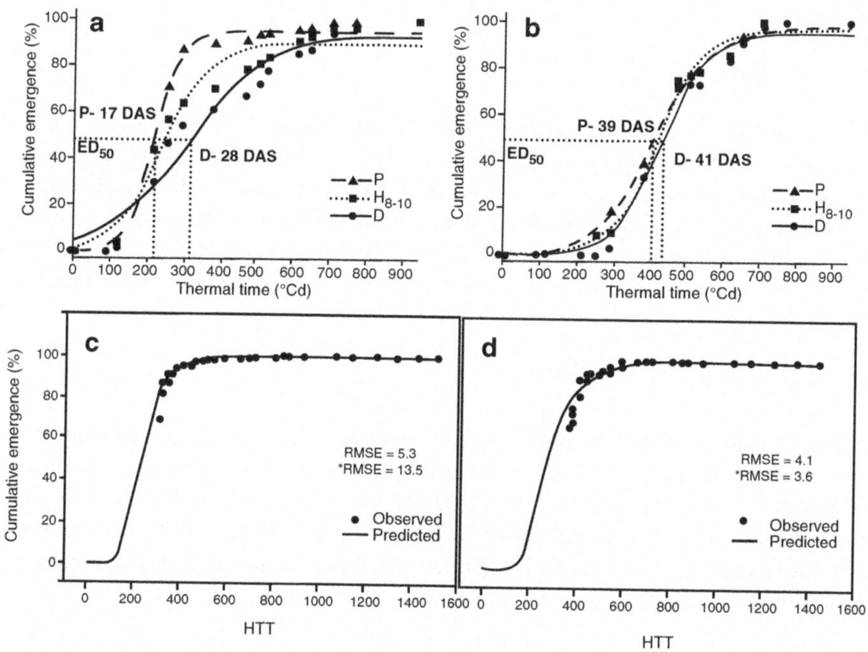


**Fig. 5.3** (**a**) Cumulative emergence patterns based on thermal time of *Apera spica-venti* and *Vulpia myuros* (full symbols, considered together) and (**b**) *Poa annua,* with D, direct drilling (straight black line); P, mouldboard ploughing (dashed line) and H8-10, cultivation down to 8–10 cm (grey line) (Scherner et al. 2017). Below, emergence patterns for *Bromus diandrus* based on hydrothermal time ($\theta_{\mathrm{HTT}}$) and adjusted to the developed model in (**c**) no till and (**d**) mouldboard ploughing (Garcia et al. 2013)

## 5.2.2 *NLR Drawbacks*

Despite their advantages, NLR models present various statistical and bioecological limitations (Gonzalez-Andujar et al. 2016a) that must be considered from both theoretical and practical perspectives (Onofri et al. 2010). For this reason, alternative approaches should be considered aiming to overcome some of the limitations described by Gonzalez-Andujar et al. (2016a).

### 5.2.2.1 Statistical Limitations

Initial Parameters

The initial parameter estimates determine the quality of the solutions (Gonzalez-Andujar et al. 2016a). Poor initial or biased estimates usually result on inadequate solutions (Holmström and Petersson 2002). For this reason, it is necessary to gather a good pool of data and proceed by an iterative method trying different sets of initial values until the optimum solution is final (see Chap. 2).

Algorithm Selection

The parameters of the model are estimated using optimization algorithms such as Marquardt–Levenberg or Gauss–Newton (Ratkowsky 1983). NLR models fitting routines are widely available in both free and commercial statistical software. The selection of the appropriate algorithm for a given problem is an important issue; however, 'default algorithm options' are commonly used mainly by unexperienced users. As a consequence of an inadequate algorithm selection, wrong solutions can be obtained.

Statistical Dependence of the Data

Field data sampling is performed sequentially over the same experimental units. Consequently, the percentage of emergences observed on one date will depend on the amount of emergences that occurred on the previous sampling dates. Thus, emergence observations are dependent values, when the NLR models would require statistically independent data to be applied. As a consequence, a positive autocorrelation of the residuals occurs, and erroneous predictions could be obtained.

Censored Data

Field emergence observation cannot be continuous, unless a video camera is placed. In practice, observations are performed periodically (usually on a weekly basis). Although emerged seedlings are counted in each date, the emergence process could have occurred at any time between the two sampling dates. For example, considering a weekly counting base, suppose that ten new seedlings emerged in these 7 days. As the sigmoidal emergence model is a continuous function, 'estimated emergence data' result from a linear interpolation among the two consecutive counts. However, we are aware that in a real system, it is impossible to know when the 'true emergences' have occurred (as they could have happened in the whole period in between or immediately after the previous sampling date or immediately before the present sampling date). And even they could have happened in different days and different amounts. These 'blind' periods obscure the real emergence timing, and these are known as 'censored data', which can lead to biased results.

In order to make the error associated to these predictions more realistic, some authors have suggested the application of other statistical methods to the emergence data. Recently, Onofri et al. (2019) proposed the use of *time-to-event* data, which considers an interval of time for a seedling to be emerged. In hydrothermal-time-to-event models, the standard errors obtained for the parameters describing emergence are larger than those obtained by the NLR models. The effect of the censored data is corrected by incorporating the uncertainty concept in accordance to the time lapse between sampling dates (Onofri et al. 2018, 2019).

#### 5.2.2.2 Bioecological Limitations

The use of cumulative TT, HTT or PhHTT as single explanatory variables in NLR models also involves certain bioecological shortcomings.

##### TT/HTT/PhHTT as Explanatory Variables

As highlighted by Chantre et al. (2018), such indexes are based on soil microclimatic variables (soil temperature and water potential) which are in turn dependent on many site-specific variables (e.g. soil texture, surface cover, seed burial depth). As a consequence, index calculation depends on (1) in situ soil temperature and moisture measurements, or, alternatively, site-specific soil microclimatic variable estimation using specific software (such as the STM[2]), and (2) the assumption that emergence rates are proportional to the amount by which soil temperature and soil water potential exceed a given threshold value (Bradford 2002). In addition, species-specific thresholds ($T_b$ and $\psi_b$ for seed germination/emergence) are generally obtained under laboratory-controlled conditions or alternatively by matching field data with TT or HTT following a nonlinear least-squares curve-fitting optimization procedure. Regardless of the method used for threshold estimation, certain statistical and biological assumptions underlie. As depicted by Bradford (2002), (1) population-based threshold parameters and soil microclimatic variables (soil temperature and water potential) are assumed independent; (2) thermal/hydrothermal time among individuals of a population is considered to follow a normal distribution. Last but not the least, such thresholds could not be simplified to a 'set of unique' parameters for a given weed population, even less for different populations of a given weed species, as many sources of variability exist.

### Intra- and Inter-population Variability

As mentioned in the previous topic, population's threshold values can be modified due to both internal (genetics) and external factors (environmental factors). For example, it is known that soil water potential and cumulative temperature suffered by the mother plants of *Alopecurus myosuroides* in the reproductive and maturation phases have an effect on the dormancy level of the progeny (Menegat et al. 2018). Similarly, maternal effects associated to the soil fertilization environment (Longas et al. 2016) proved to influence seed germinability on *Buglossoides arvensis*. Also variations on the physical characteristics of the fruit coat for the one-seeded fruit species *Neslia paniculata* were influenced by the maternal environment (Royo-Esnal et al. 2019). Seed dimorphism can also affect to dormancy and germination of seeds, as in *Polygonum aviculare* (Costea and Tardiff 2005).

Thus, at an intra-population level, these thresholds can vary over time due to genetic, environmental and gene-environment interactions. Variability among populations is also associated to genetic and environmental factors that translate into ecological adaptations (ecotypes) (Keller and Kollmann 1999). For example, a *Echinochloa crus-galli* population from Norway emerges differently from an Italian population, while populations from Latvia, Sweden and Poland behave similar to the Norwegian, and those from Spain, Turkey and Iran, more similar to the Italian one (Royo-Esnal et al. 2018a). But a *E. crus-galli* population from a maize field from Spain also behaves differently from a population from a Spanish rice field (Royo-Esnal et al. 2018a).

The reasons that can explain such differences are not few. For instance, different soil management (e.g. cultivation operation) vary the vertical distribution of the seeds within the soil profile (Grundy et al. 1996). As a result, seeds will have different temperature, soil moisture and light conditions, and the rate of dormancy release and germination will be different (Cao et al. 2011). In addition, deeply buried seeds need more time to emerge than seeds near the soil surface, thereby delaying and/or extending the emergence flushes.

Although TT/HTT/PhHTT models are of general applicability, validation results show that empirical models may not be accurate if environmental conditions vary significantly from the original conditions in which the experiment was conducted (Izquierdo et al. 2013). As a consequence, the accuracy of a given model is rarely extrapolable to other climatic areas or habitats (Dorado et al. 2009). As stated by Chantre et al. (2014), weed species ecological adaptations hinder the development of 'universal' weed emergence predictive models.

### Difficulties Establishing the Zero Moment for Thermal-Time Accumulation

A zero moment must be established prior to the onset of TT or HTT accumulation for weed emergence. Generally, this moment is not a big problem for summer weeds in temperate and Mediterranean climates, where soil moisture is not restrictive, thus establishing this starting point at some time during winter, or when soil disturbance for crop seeding is performed, which are reasonable options. For example, $T_b$ values for *Amaranthus retroflexus* (8.9 °C) (Guillemin et al. 2013), *Echinochloa crus-galli* (9.7 °C) (Bagavathiannan et al. 2011) and *Solanum nigrum* (11.6 °C) (Guillemin et al. 2013) usually emerge after crop sowing. However, other species such as *Chenopodium album* (3 °C) (Schutte et al. 2014) and *Bassia scoparia* (3.5 °C) (Werle et al. 2014b) show advanced emergence periods, usually starting before soil preparation for sowing and lasting till crop setting. Extended emergence periods are associated to the seedbank dormancy dynamics and its interaction with environmental germination/emergence cues (as described in Chap. 4).

In temperate and Mediterranean climates, soil moisture is seldom a limiting factor, as spring seasons are usually wet. Conversely, when the climatic conditions tend to be more oceanic with mild winters, or in tropical (rainy and drought periods) or arid/semiarid climates, soil moisture tends to be a limiting factor for TT/HTT accumulation. For example, for some winter annual weeds, which usually have low $T_b$ values, like *Avena sterilis* (0.8 °C) (Leguizamón et al. 2005), *Papaver rhoeas* (1 °C) (Izquierdo et al. 2009) and *Galium aparine* (0 °C) (Royo-Esnal et al. 2010), emergence occurs mainly during early autumn when soil temperature drops below $T_c$. In many cases, despite the appropriate temperature conditions, HTT accumulation for emergence does not start until a rainfall event occurs (e.g. first autumn rains) as it happens in *Bromus diandrus* (Garcia et al. 2013) where emergence can be delayed for more than 3 months (Royo-Esnal et al. 2018b).

### Lack of Knowledge on the Level of Infestation

As seen before, these models describe the cumulative percentage of emergence during a certain crop season (Figs. 5.2 and 5.3). Although they are valuable tools for weed emergence simulation/prediction and for the application of control methods from a short-time planning perspective, they also lack the capacity to predict the population dynamics in the long term, unless a previous seedbank estimation has been performed. Moreover, empirical models lack the mechanistic insight that reductionist approaches provide, and only simulations have been performed trying to combine emergence models with population dynamic bibliographic data, for example, for *A. fatua* (Gonzalez-Diaz et al. 2007). From a long-term management planning perspective, weed population bioecological studies aiming to characterize the demographic dynamics of weed species are of overwhelming importance.

### 5.2.3 New Modelling Approaches

New approaches have turn up as alternative methods to tackle some of the previously mentioned drawbacks (Gonzalez-Andujar et al. 2016a). Among them, algorithmic modelling (*sensu* Breiman 2001) is of special interest to weed modellers as it clearly departs from the concept of an 'ideal system' described by complete-and-precise information and heads towards a real, uncertain and complex system. Real-world problems are in fact typically ill-defined systems, difficult to model and with large-scale solution spaces (Bonissone 1997). From this perspective, soft computing-based models have proved their capacity to deal with such systems. As stated by Das et al. (2013), soft computing techniques (SCT) unlike conventional computing (also known as hard computing) are tolerant to uncertainty, imprecision and partial truth, not requiring strict mathematical definitions.

Among soft computing techniques, artificial neural networks (ANNs) as modelling framework (Chantre et al. 2012, 2014) and genetic algorithms (Haj Seyed Hadi and Gonzalez-Andujar 2009; Blanco et al. 2014) as optimization engines have been proposed for weed emergence modelling (Tables 5.1 and 5.2).

As reviewed by Gonzalez-Andujar et al. (2016a), ANN models are inspired by the operation of the biological networks of the animal brain. ANNs are generally represented as a system of interconnected processing units (neurons), which exchange signals (i.e. information) between each other. The connections have numeric values (i.e. weights) that are adjusted during the training (i.e. learning) process using a given algorithm. Therefore, an ANN model is characterized by (1) its architecture (i.e. the pattern of neuronal connections), (2) its learning process (i.e. the training function for weight estimation) and finally (3) its activation functions (i.e. mathematical functions that process input data).

In Chantre et al. (2018), a feed-forward ANN with three layers was implemented. The theoretical model, which is briefly shown in Fig. 5.4, has (1) two input variables ($x_1$, $x_2$) each connected to a given receptor neuron of an entrance layer, (2) an intermediate eight-neuron layer and (3) an output variable ($y$).

As observed in Fig. 5.4, each neuron of the entrance layer receives a given input variable ($x_1$, $x_2$) and broadcasts its value to each neuron of the hidden layer. Each neuron computes an activation function and generates an outcome ($z_1$, …, $z_8$) which is further transmitted to the output layer neuron which finally yields the network output ($y$). The output signal of each neuron in the hidden layer ($z_j$) is calculated as (Eq. 5.6)

$$z_j = f\left(\sum_{i=1,2} v_{ij} x_i + v_{0j}\right) \quad j = 1,\ldots,8 \tag{5.6}$$

while the output of the network is given by (Eq. 5.7)

$$y = f\left(\sum_{j=1,8} w_j z_j + w_0\right) \tag{5.7}$$

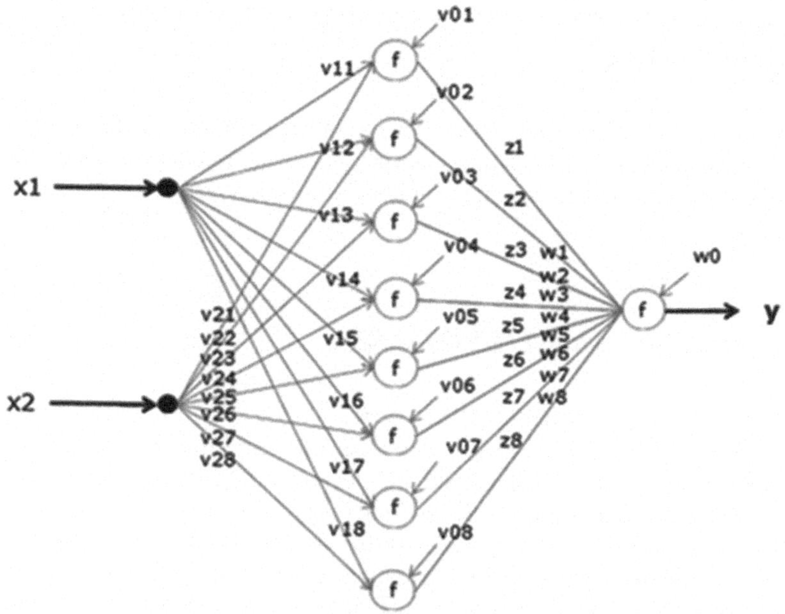


**Fig. 5.4** ANN architecture with three layers (entrance, intermediate and exit layer). $x_1$ and $x_2$ represent the input variables (each neuron of the first layer receives a given input variable); a hidden layer with eight nodes (i.e. processing units) and a unique exit neuron that produces the final response (output variable). *f*(.) represents the activation functions of the neurons; $v_{ij}$ are the connection weights between input-hidden layer neurons; $w_j$ are the connection weights between hidden and output layer neurons; $v_{0j}$ is the bias of each hidden neuron *j*, while $w_0$ is the output neuron bias; $z_j$ is the output signal of each hidden neuron. *y* stands for the output variable. Extracted from Chantre et al. (2018)

where *f*(.) represents the transfer (activation) function, $v_{ij}$ are the weights of the connections between the input and the intermediate neurons, $v_{0j}$ is the bias on neuron, $w_j$ represent the weights of the connections between the intermediate and output neurons and $w_0$ is the bias on the output neuron.

The modelling framework proposed by Chantre et al. (2018) showed reliable predictions for *A. fatua, L. multiflorum* and *V. villosa* ssp. *villosa*. The main strength of the proposed ANN approach is the absence of specific underlying modelling assumptions (e.g. normal/log-normal distribution of the cumulative emergence function) and the direct use of commonly available field meteorological data. Although SCT are well known for their flexibility and uncertainty tolerance, they also have some limitations, such as the following: they have (1) very low extrapolation capacity; thus, a wide range of observed scenarios are needed to capture data variability; and (2) limited interpretable biological meaning of input-output relationships.

## 5.3 Decision Support System Developments: Websites and Software

In this regard, few attempts have been done to integrate empirical emergence models in software or computer programs aiming to provide practical decision-making for agronomists, producers and stakeholders. However, some developments can be mentioned.

### *5.3.1 WeedCast Model (USDA-ARS)*

Forcella (1998) developed a tool called WeedCast to assess weed species emergence in real time. This program, which is downloadable from the USDA-ARS website (https://www.ars.usda.gov/research/software/download/?softwareid=112), predicts the emergence potential, timing and seedling height of up to 20 weed species.

To use this program, some previous information must be added, such as soil structure, initial soil water content, previous season's crop and type of tillage. Weather data must also be provided to the software, as well as the user's required time lapse of simulation.

In Fig. 5.5, the interface of the software is shown. A hypothetical scenario consists of a field from Morris (Minnesota, USA) with chisel plough management, corn crop during the previous season and a sandy loam soil under wet conditions (Fig. 5.5a). *Echinochloa crus-galli* (barnyardgrass), *Solanum nigrum* (black nightshade), *Xanthium strumarium* (common cocklebur) and *Chenopodium album* (common lambsquarters) were the selected species (Fig. 5.5b). With this information, the program predicts cumulative emergence patterns as well as the expected height of the seedlings. Additional data, such as soil temperature (thermal time) and moisture (soil water potential), are provided as model's output (Fig. 5.5c).

### *5.3.2 AlertInf (Masin et al. 2012, 2014)*

Masin et al. (2012, 2014) in the Veneto region of Italy developed a useful tool that predicts the emergence of *A. theophrasti*, *A. retroflexus*, *C. album*, *P. persicaria*, *S. nigrum* and *Sorghum halepense* in maize and soybean (Fig. 5.6). The emergence of these species is modelled based on the soil temperature from 0 to 10 cm of several sites of Veneto's region.

AlertInf allows the user to select the meteorological station available nearby, the date when the seedbed was prepared and the date for which the emergence percentage of a given species is required (Fig. 5.6). Despite this valuable information, the user must be aware that the program only provides the percentage of emergence, but information about the phenological stage is lacking; thus, in order to apply herbi-

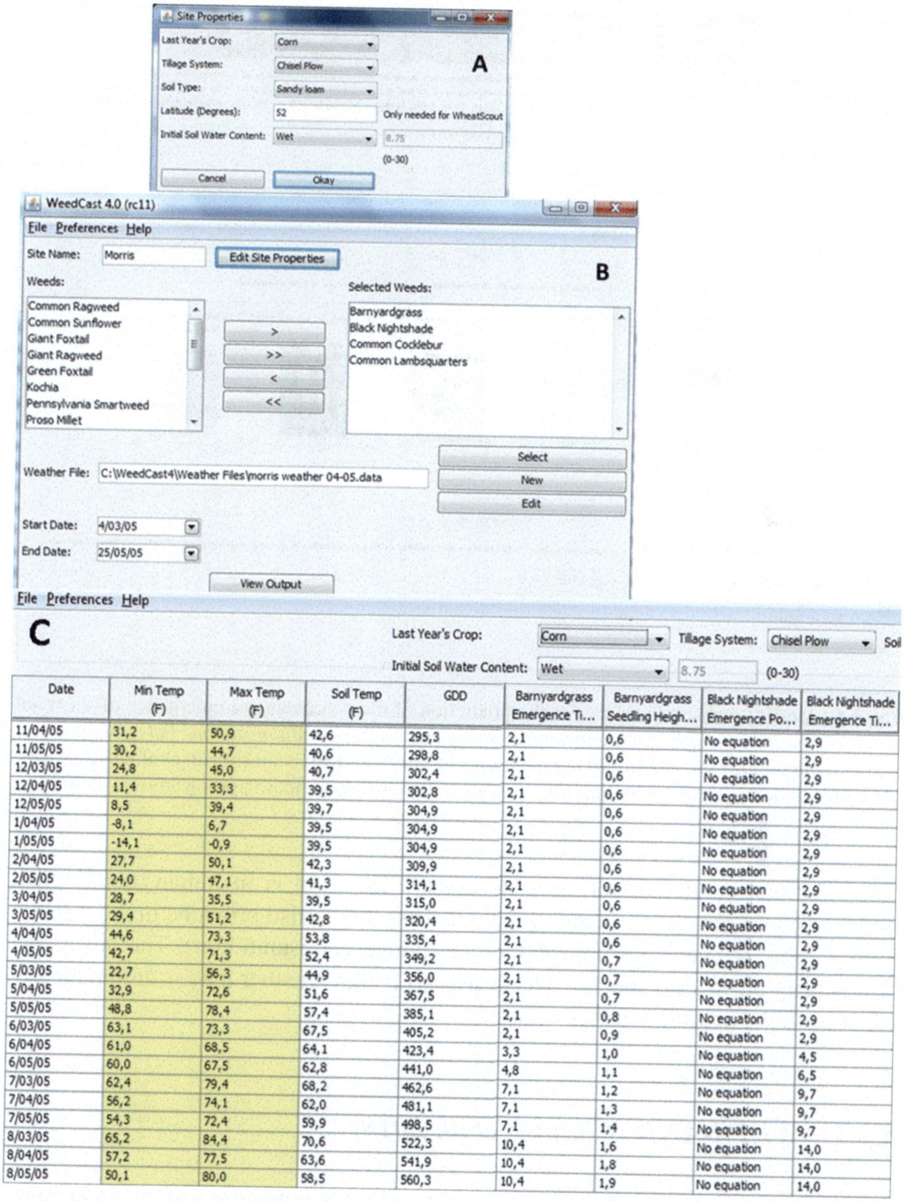


| Date | Min Temp (F) | Max Temp (F) | Soil Temp (F) | GDD | Barnyardgrass Emergence Ti... | Barnyardgrass Seedling Heigh... | Black Nightshade Emergence Po... | Black Nightshade Emergence Ti... |
|---|---|---|---|---|---|---|---|---|
| 11/04/05 | 31,2 | 50,9 | 42,6 | 295,3 | 2,1 | 0,6 | No equation | 2,9 |
| 11/05/05 | 30,2 | 44,7 | 40,6 | 298,8 | 2,1 | 0,6 | No equation | 2,9 |
| 12/03/05 | 24,8 | 45,0 | 40,7 | 302,4 | 2,1 | 0,6 | No equation | 2,9 |
| 12/04/05 | 11,4 | 33,3 | 39,5 | 302,8 | 2,1 | 0,6 | No equation | 2,9 |
| 12/05/05 | 8,5 | 39,4 | 39,7 | 304,9 | 2,1 | 0,6 | No equation | 2,9 |
| 1/04/05 | -8,1 | 6,7 | 39,5 | 304,9 | 2,1 | 0,6 | No equation | 2,9 |
| 1/05/05 | -14,1 | -0,9 | 39,5 | 304,9 | 2,1 | 0,6 | No equation | 2,9 |
| 2/04/05 | 27,7 | 50,1 | 42,3 | 309,9 | 2,1 | 0,6 | No equation | 2,9 |
| 2/05/05 | 24,0 | 47,1 | 41,3 | 314,1 | 2,1 | 0,6 | No equation | 2,9 |
| 3/04/05 | 28,7 | 35,5 | 39,5 | 315,0 | 2,1 | 0,6 | No equation | 2,9 |
| 3/05/05 | 29,4 | 51,2 | 42,8 | 320,4 | 2,1 | 0,6 | No equation | 2,9 |
| 4/04/05 | 44,6 | 73,3 | 53,8 | 335,4 | 2,1 | 0,6 | No equation | 2,9 |
| 4/05/05 | 42,7 | 71,3 | 52,4 | 349,2 | 2,1 | 0,7 | No equation | 2,9 |
| 5/03/05 | 22,7 | 56,3 | 44,9 | 356,0 | 2,1 | 0,7 | No equation | 2,9 |
| 5/04/05 | 32,9 | 72,6 | 51,6 | 367,5 | 2,1 | 0,7 | No equation | 2,9 |
| 5/05/05 | 48,8 | 78,4 | 57,4 | 385,1 | 2,1 | 0,8 | No equation | 2,9 |
| 6/03/05 | 63,1 | 73,3 | 67,5 | 405,2 | 2,1 | 0,9 | No equation | 2,9 |
| 6/04/05 | 61,0 | 68,5 | 64,1 | 423,4 | 3,3 | 1,0 | No equation | 4,5 |
| 6/05/05 | 60,0 | 67,5 | 62,8 | 441,0 | 4,8 | 1,1 | No equation | 6,5 |
| 7/03/05 | 62,4 | 79,4 | 68,2 | 462,6 | 7,1 | 1,2 | No equation | 9,7 |
| 7/04/05 | 56,2 | 74,1 | 62,0 | 481,1 | 7,1 | 1,3 | No equation | 9,7 |
| 7/05/05 | 54,3 | 72,4 | 59,9 | 498,5 | 7,1 | 1,4 | No equation | 9,7 |
| 8/03/05 | 65,2 | 84,4 | 70,6 | 522,3 | 10,4 | 1,6 | No equation | 14,0 |
| 8/04/05 | 57,2 | 77,5 | 63,6 | 541,9 | 10,4 | 1,8 | No equation | 14,0 |
| 8/05/05 | 50,1 | 80,0 | 58,5 | 560,3 | 10,4 | 1,9 | No equation | 14,0 |

**Fig. 5.5** WeedCast 4.0 Windows interface showing soil site property description (**a**), weed species selection and weather file (**b**) and the final output (**c**) including the cumulative emergence and height of the weeds, soil temperature, growing degree days (GDD), soil moisture and soil water potential are also provided. This software can be downloaded at https://www.ars.usda.gov/research/software/download/?softwareid=112. For clarity of Figure **c**, the order of the columns in the output has been modified

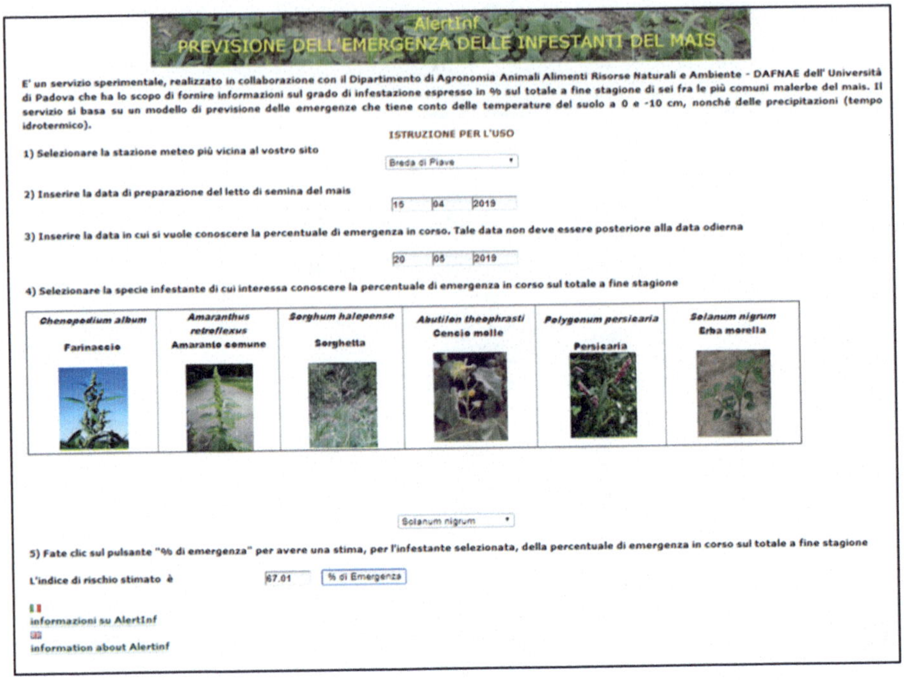


**Fig. 5.6** AlertInf website interface for the prediction of the percentage of emergence of six weed species (*Abutilon theophrasti*, *Amaranthus retroflexus*, *Chenopodium album*, *Polygonum persicaria*, *Solanum nigrum* and *Sorghum halepense*) in maize and soybean (Masin et al. 2012, 2014). This information is available at http://www.arpa.veneto.it/upload_teolo/agrometeo/infestanti.htm

cide solutions, field monitoring is necessary. This tool is in Italian, but English information of how to use it and of technical aspects is also provided in the website (http://www.arpa.veneto.it/upload_teolo/agrometeo/infestanti.htm). The advantage of this program with respect to WeedCast 4.0 is that weather data is automatically taken from the nearest weather station.

### 5.3.3 *WEPS-ANN (Chantre et al. 2018)*

An artificial neural network approach was proposed by Chantre et al. (2018) to predict the emergence patterns of weed species in the semiarid Pampean region of Argentina (Austral Pampas). Unlike traditional empirical weed emergence models, the ANN approach allows a direct input–output relationship between daily generated meteorological information and field emergence data without the necessity of soil microclimatic derived indexes (i.e. thermal/hydrothermal time) or 'species-specific' population thresholds. (see Sect. 5.2.3 for further details).

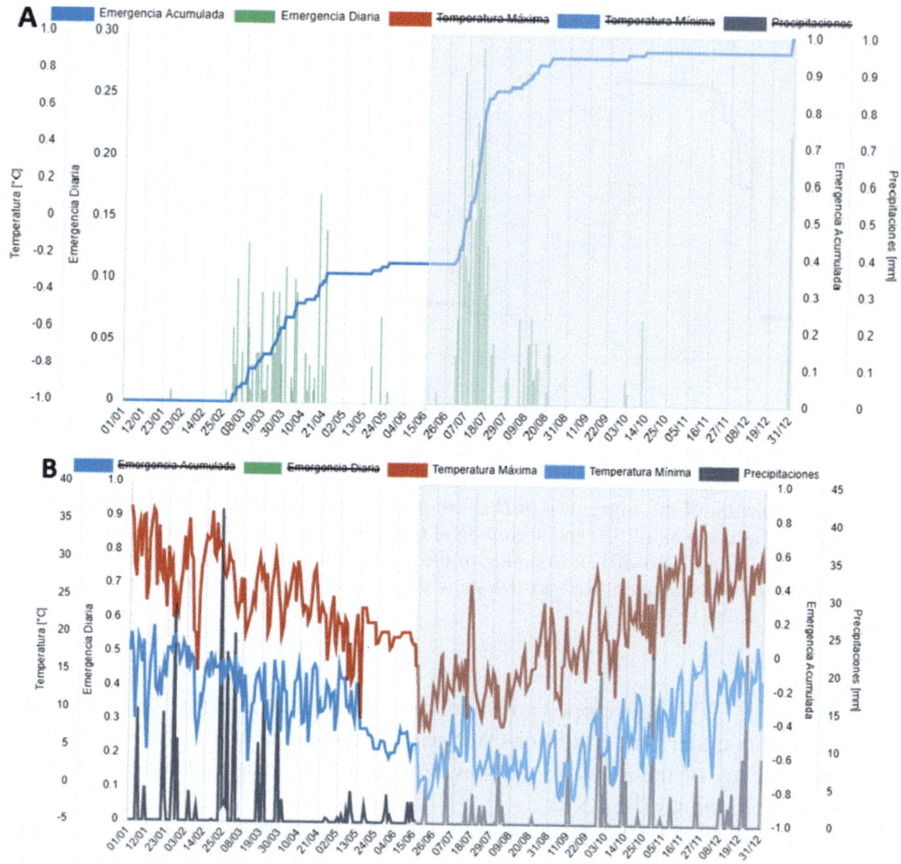


**Fig. 5.7** WEPS-ANN website interface. Daily and cumulative emergence data of *Avena fatua* (**a**) and meteorological data (**b**). Available at http://pronostico-malezas.frbb.utn.edu.ar/

As observed in Fig. 5.7a, both the daily and cumulative emergence pattern of *A. fatua* can be easily monitored. The website also provides a chart with site-specific meteorological data (Fig. 5.7b). A website version including *A. fatua*, *L. multiflorum* and *Vicia villosa* Roth. is currently available at http://pronostico-malezas.frbb.utn.edu.ar/.

## 5.4 Practical Implementation of Emergence Models

Apart from the available platforms (websites and software), practical examples of the use of these models in commercial fields are very scarce. An example of real field implementation of these models is provided by Royo-Esnal et al. (2018b).

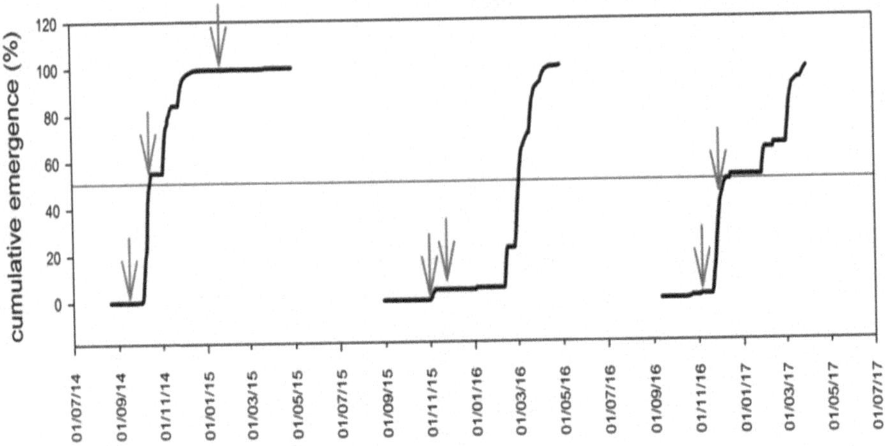


**Fig. 5.8** Simulation of the emergence of *Bromus diandrus* based on the HTT emergence model developed by Garcia et al. (2013) and according to the climatic conditions in each season (2014–2015, 2015–2016 and 2016–2017). Counting of the HTT started after the first important rains in late summer and early autumn. Red arrows show the sowing dates each season (modified from Royo-Esnal et al. 2018b)

Garcia et al. (2013) developed an emergence model that describes the emergence of *B. diandrus* under arid conditions of Spain. Royo-Esnal et al. (2018a, b) evaluated the efficacy of the delay in the sowing date of the crop and the effect in the control of *B. diandrus* (Fig. 5.8).

Royo-Esnal et al. (2018b) observed that the first sowing of rapeseed (canola) in September 2014 was not useful for mechanical control, but delay seeding (on 31 October 2014) allowed the elimination of more than 50% of the *B. diandrus* population within the season (Fig. 5.8). When sowing was delayed to January, 100% of the emergences had already occurred and were killed.

In Fig. 5.8, it can be observed that the sowing delay was a valuable tool in order to reduce the initial *B. diandrus* infestation in autumn-winter 2014 and 2016, but not in 2015 (due to a severe drought condition), as the peak of emergences was delayed to February 2016. In the latter case, an earlier sowing would have even been more appropriate in order to favour the crop with more time to grow before the emergence of *B. diandrus*.

## 5.5 Conclusions

Although the number of examples proving the usefulness of these tools is few, the interest to optimize the prediction of emergence timing of weed species is still an actual issue, and ongoing research is being performed in different countries.

Additional efforts should be directed to the implementation of these models into technological platforms (e.g. software, web platforms, smartphone apps) which can help to unleash the potential of DSS tools.

Future projects should try to develop these platforms, integrating emergence models, and in close collaboration with companies, which can lead farmers to a better management of their fields, improving the weed control and avoiding unnecessary economic losses due to inadequate application of control methods.

## References

Aboutalebian MA, Nazari S, Gonzalez-Andujar JL (2017) Evaluation of a model for predicting *Avena fatua* and *Descurainia sophia* seed emergence in winter rapeseed. Span J Agric Res 15:e03SC01

Ali HH, Kebaso L, Manalil S, Chauhan BS (2020) Emergence and germination response of and to different temperatures and moisture stress regimes . Plant Species Biology 35:16–23

Allen BL, Vigil MF, Jabro JD (2014) Camelina growing degree hour and base temperature requirements. Agron J 106:940–944

Bagavathiannan MV, Norsworthy JK, Smith KL, Burgos N (2011) Seedbank Size and Emergence Pattern of Barnyardgrass (*Echinochloa crus-galli*) in Arkansas. Weed Sci 59:359–365

Barnes ER, Werle R, Sandell LD, Lindquist JL, Knezevic SZ, Sikkema PH, Jhala AJ (2017) Influence of tillage on common ragweed (*Ambrosia artemisiifolia*) emergence pattern in Nebraska. Weed Technol 31:623–631

Behtari B, de Luis M (2012) Seedling emergence of tall fescue and wheatgrass under different climate conditions in Iran. Span J Agric Res 10:183–190

Beltran JC, Pannell DJ, Doole GJ, White B (2012) A bioeconomic model for analysis of integrated weed management strategies for annual barnyardgrass (*Echinochloa crus-galli* complex) in Philippine rice farming systems. Agr Syst 112:1–10

Blanco AM, Chantre GR, Lodovichi MV, Bandoni A, López RL, Vigna MR, Gigón R, Sabbatini MR (2014) Modeling seed dormancy release and germination for predicting Avena fatua L. field emergence: a genetic algorithm approach. Ecol Model 272:293–300

Boddy LG, Bradford KJ, Fischer AJ (2012) Population-based threshold models describe weed germination and emergence patterns across varying temperature, moisture and oxygen conditions. J Appl Ecol 49:1225–1236

Bonissone PP (1997) Soft computing: the convergence of emerging reasoning technologies. Soft Comput 1:6–18

Bradford KJ (2002) Applications of hydrothermal time to quantifying and modeling seed germination and dormancy. Weed Science 50:248–260

Breiman L (2001) Statistical modelling: the two cultures (with comments and a rejoinder by the author). Stat Sci 16(3):199–231

Bullied WJ, Van Acker RC, Bullock PR (2012) Hydrothermal modeling of seedling emergence timing across topography and soil depth. Agron J 104:423–436

Cao R, Francisco-Fernandez M, Anand A, Bastida F, Gonzalez-Andujar JL (2011) Computing statistical indices for hydrothermal times using weed emergence data. J Agric Sci 149:701–771

Cao R, Francisco-Fernández M, Anand A, Bastida F, Gonzalez-Andujar JL (2013) Modeling *Bromus diandrus* seedling emergence using nonparametric estimation. J Agric Biol Environ Stat 18:64–86

Cardina J, Herms CP, Herms DA, Forcella F (2007) Evaluating phenological indicators for predicting giant foxtail (*Setaria faberi*) emergence. Weed Sci 55:455–464

Cardoso VJM, Pereira FJM (2008) Germination of *Drymaria cordata* (L.) Willd. ex Roem and Schult. seeds: effect of water potential. Rev Bras Bot 31:253–261
Chantre GR, Batlla D, Sabbatini MR, Orioli G (2009) Germination parameterization and development of an after-ripening thermal-time model for primary dormancy release of *Lithospermum arvense* seeds. Ann Bot 103:1291–1301
Chantre GR, Sabbatini MR, Oiroli GA (2010) An after-ripening thermal-time model for *Lithospermum arvense* seeds based on changes in population hydrotime parameters. Weed Res 50:218–227
Chantre GR, Blanco AM, Lodovichi MV, Bandoni AJ, Sabbatini MR, López RL, Vigna MR, Gigón R (2012) Modeling *Avena fatua* seedling emergence dynamics: an artificial neural network approach. Comput Electr Agric 88:95–102
Chantre GR, Blanco AM, Forcella F, Van Acker RC, Sabbatini MR, Gonzalez-Andujar JL (2014) A comparative study between non-linear regression and artificial neural network approaches for modelling wild oat (*Avena fatua*) field emergence. J Agric Sci 152:254–262
Chantre GR, Vigna MR, Renzi JP, Blanco AM (2018) A flexible and practical approach for real-time weed emergence prediction based on artificial neural networks. Biosyst Eng 170:51–60
Clay SA, Davis A, Dille A, Lindquist J, Ramírez AHM, Sprague C, Reicks G, Forcella F (2014) Common sunflower seedling emergence across the U.S. Midwest. Weed Sci 62:63–70
Cochavi A, Goldwasser Y, Horesh A, Igbariya K, Lati RN (2018) Impact of environmental factors on seed germination and emergence of wild poinsettia (*Euphorbia geniculata* Ortega). Crop Prot 114:68–75
Colbach N, Dürr C, Roger-Estrade J, Chauvel B, Caneill J (2006) AlomySys: Modelling black-grass (Alopecurus myosuroides Huds.) germination and emergence, in interaction with seed characteristics, tillage and soil climate. European Journal of Agronomy 24:95–112
Costea M, Tardiff FJ (2005) The biology of Canadian weeds 131. Polygonum aviculare L. Can J Plant Sci 85:481–506
D'Amico MB, Calandrini GL, González-Andujar JL, Chantre GR (2018) Analysis of different management strategies for annual ryegrass (Lolium rigidum) based on a population dynamic model. Int J Bifurcation Chaos *28*(12):1830041
Das SK, Kumar A, Das B, Burnwal A (2013) On soft computing techniques in various areas. Comput Sci Inform Technol 3:59–68
Davis AS, Clay S, Cardina J, Dille A, Forcella F, Lindquist J, Sprague C (2013) Seed burial physical environment explains departures from regional hydrothermal a model of giant ragweed (*Ambrosia trifida*) seedling emergence in U.S. Midwest Weed Sci 61:415–421
Derakhshan A, Gherekhloo J, Vidal RA, De Prado R (2014) Quantitative description of the germination of littleseed canarygrass (*Phalaris minor*) in response to temperature. Weed Sci 62:250–257
Dorado J, Sousa E, Calha IM, Gonzalez-Andujar JL, Fernández-Quintanilla C (2009) Predicting weed emergence in maize crops under two contrasting climatic conditions. Weed Res 49:251–260
Eizenberg H, Hershenhorn J, Achdari G, Ephrath JE (2012) A thermal time model for predicting parasitism of *Orobanche cumana* in irrigated sunflower-Field validation. Field Crop Res 137:49–55
Ekeleme F, Forcella F, Archer DW, Akobundu IO, Chikoye D (2005) Seedling emergence model for tropic ageratum (*Ageratum conyzoides*). Weed Res 53:55–61
Forcella F (1998) Real time assessment of seed dormancy and seedling growth for weed management. Seed Sci Res 8:201–209
Forcella F, Benech Arnold RL, Sanchez R, Ghersa CM (2000) Modeling seedling emergence. Field Crops Research 67:123–139
Froud-Williams RJ, Chancellor RJ, Drennan DSH (1984) The effects of seed burial and soil disturbance on emergence and survival of arable weeds in relation to minimal cultivation. J Appl Ecol 21:629–641
Garcia AL, Torra J, Recasens J, Forcella F, Royo-Esnal A (2013) Hydrothermal emergence model for *Bromus diandrus*. Weed Sci 61:146–153. https://doi.org/10.1614/WS-D-12-00023.1

Gardarin A, Dürr C, Colbach N (2012) Modeling the dynamics and emergence of multispecies weed seed bank with species traits. Ecol Model 240:123–138

Goldwasser Y, Miryamchik H, Rubin B, Eizenberg H (2016) Field dodder (*Cuscuta campestris*)—a new model describing temperature-dependent seed germination. Weed Sci 64:53–60

Gonzalez-Andujar JL, Chantre GR, Morvillo C, Blanco AM, Forcella F (2016a) Predicting field weed emergence with empirical models and soft computing techniques. Weed Res 56:415–423

Gonzalez-Andujar JL, Francisco-Fernández M, Cao R, Reyes M, Urbano JM, Forcella F, Bastida F (2016b) A comparative study between nonlinear regression and nonparametric approaches for modelling *Phalaris paradoxa* seedling emergence. Weed Res 56:367–376

Gonzalez-Diaz L, Leguizamon E, Forcella F, Gonzalez-Andujar JL (2007) Integration of emergence and population dynamic models for long term weed management using wild oat (*Avena fatua* L.) as an example. Span J Agric Res 5:199–203

Goplen JJ, Sheaffer CC, Becker RL, Moon RD, Coulter JA, Breitenbach FR, Behnken LM, Gunsolus JL (2018) Giant Ragweed (*Ambrosia trifida*) emergence model performance evaluated in diverse cropping systems. Weed Sci 66:36–46

Grundy A (2003) Predicting weed emergence: a review of approaches and future challenges. Weed Res 43:1–11

Grundy A, Mead A, Bond W (1996) Modelling the effect of weed-seed distribution in the soil profile on seedling emergence. Weed Res 36:375–384

Grundy A, Mead A, Burston S (2003a) Modelling the emergence response of weed seeds to burial depth: interactions with seed density, weight and shape. J Appl Ecol 40:757–770

Grundy AC, Peters NCB, Rasmussen IA, Hartmann KM, Sattin M, Andersson L, Mead A, Murdoch AJ, Forcella F (2003b) Emergence of *Chenopodium album* and *Stellaria media* of different origins under different climatic conditions. Weed Res 43:163–176

Guillemin JP, Gardarin A, Granger S, Reibel C, Munier-Jolain N, Colbach N (2013) Assessing potential germination period of weeds with base temperatures and base water potentials. Weed Res 53:76–87

Gummerson R (1986) The effect of constant temperatures and osmotic potentials on the germination of sugar beet. J Exp Bot 37:729–741

Haj Seyed Hadi MR, Gonzalez-Andujar JL (2009) Comparison of fitting weed seedling emergence models with nonlinear regression and genetic algorithm. Comput Electr Agric 65:19–25

Harris SM, Doohan DJ, Gordon RJ, Jensen KIN (1998) The effect of thermal time and soil water on emergence of *Ranunculus repens*. Weed Res 38:405–412

Hill EC, Renner KA, Sprague CL (2014) Henbit (*Lamium amplexicaule*), common chickweed (*Stellaria media*), shepherd's-purse (*Capsella bursa-pastoris*), and field pennycress (*Thlaspi arvense*): fecundity, seed dispersal, dormancy, and emergence. Weed Sci 62:97–106

Holmström K, Petersson J (2002) A review of the parameter estimation problem of fitting positive exponential sums to empirical data. Applied Mathematics and Computation 126:31–61

Izquierdo J, Gonzalez-Andujar JL, Bastida F, Lezaún JA, Arco MJS (2009) A thermal time model to predict corn poppy (*Papaver rhoeas*) emergence in cereal fields. Weed Sci 57:660–664

Izquierdo J, Bastida F, Lezaún JM, Sánchez del Arco MJ, Gonzalez-Andujar JL (2013) Development and evaluation of a model for predicting *Lolium rigidum* emergence in winter cereal crops in the Mediterranean area. Weed Res 53:269–278

Juroszek P, Neuhoff D, Köpke U (2017) Night-time tillage revisited: the delayed soil desiccation process in night-time tilled plots may promote unexpected weed germination. Weed Res 57:213–217

Keller M, Kollmann J (1999) Effects of seed provenance on germination of herbs for agricultural compensation sites. Agric Ecosyst Environ 72:87–99

Kruk BC, Benech-Arnold RL (1998) Functional and quantitative analysis of seed thermal responses in prostrate knotweed (*Polygonum aviculare*) and common purslane (*Portulaca oleracea*). Weed Sci 46:83–90

Lammoglia SK, Moeys J, Barriuso E, Larsbo M, Marín-Benito JM, Justes E, Alleto L, Ubertosi M, Nicolardot B, Munier-Joalin N, Mamy L (2017) Sequential use of the STICS crop model and of the MACRO pesticide fate model to simulate pesticides leaching in cropping systems. Environ Sci Pollut Res 24:6895–6909

Leblanc ML, Cloutier DC, Stewart KA, Hamel C (2003) The use of thermal time to model common lambsquarters (*Chenopodium album*) seedling emergence in corn. Weed Sci 51:718–724

Leblanc ML, Cloutier DC, Stewart KA, Hamel C (2004) Calibration and validation of a common lambsquarters (*Chenopodium album*) seedling emergence model. Weed Sci 52:61–66

Leguizamón ES, Fernández-Quintanilla C, Barroso J, Gonzalez-Andujar JL (2005) Using thermal and hydrothermal time to model seedling emergence of *Avena sterilis* ssp. *ludoviciana* in Spain. Weed Res 45:149–156

Leguizamón ES, Rodríguez N, Rainero H, Pérez M, Pérez L, Zorza E, Fernández-Quintanilla C (2009) Modelling the emergence pattern of six summer annual weed grasses under no tillage systems in Argentina. Weed Res 49:98–106

León RG, Izquierdo J, Gonzalez-Andujar JL (2015) Characterization and modeling of itchgrass (*Rottboellia cochinchinensis*) biphasic seedling emergence patterns in the tropics. Weed Sci 63:623–630

Loddo D, Masin R, Otto S, Zanin G (2012) Estimation of base temperature for *Sorghum halepense* rhizome sprouting. Weed Res 52:42–49

Lodovichi MV, Blanco AM, Chantre GR, Bandoni JA, Sabbatini MR, López R, Vigna M, Gigón R (2013) Operational planning of herbicide-based weed management. Agr Syst 121:117–129

Longas MM, Chantre GR, Sabbatini MR (2016) Soil nitrogen fertilisation as a maternal effect on *Buglossoides arvensis* seed germinability. Weed Res 56:462–469

Lundy ME, Hill JE, van Kessel C, Owen DA, Pedroso RM, Boddy LG, Fischer AJ, Linquist BA (2014) Site-specific, real-time temperatures improve the accuracy of weed emergence predictions in direct-seeded rice systems. Agr Syst 123:12–21

Martinson K, Durgan B, Forcella F, Wiersma J, Spokas K, Archer D (2007) An emergence model for wild oat (*Avena fatua*). Weed Sci 55:584–591

Masin R, Loddo D, Benvenuti S, Zuin MC, Macchia M, Zanin G (2010) Temperature and water potential as parameters for modeling weed emergence in Central-Northern Italy. Weed Sci 58:216–222

Masin R, Loddo D, Benvenuti S, Otto S, Zanin G (2012) Modelling weed emergence in Italian maize fields. Weed Sci 60:254–259

Masin R, Loddo D, Gasparini V, Otto S, Zanin G (2014) Evaluation of weed emergence model AlertInf for maize in soybean. Weed Sci 62:360–369

Masin R, Onofri A, Gasparini V, Zanin G (2017) Can alternating temperatures be used to estimate base temperature for seed germination? Weed Res 57:390–398

McGiffen M, Spokas K, Forcella F, Archer D, Poppe S, Figueroa R (2008) Emergence prediction of common groundsel (*Senecio vulgaris*). Weed Sci 56:58–65

Menalled F, Schonbeck M (2011) Manage the weed seed bank: minimize "Deposits" and maximize "Withdrawals". http://www.extension.org/pages/18527

Menegat A, Milberg P, Nilsson ATS, Andersson L, Giulia V (2018) Soil water potential and temperature sum during reproductive growth control of seed dormancy in *Alopecurus myosuroides* Huds. Ecol Evol 8:7186–7194

Norsworthy JK, Oliveira MJ (2007) A model for predicting common cocklebur (*Xanthium strumarium*) emergence in soybean. Weed Sci 55:341–345

Oerke EC (2006) Crop losses to pests. J Agric Sci 144:31–43

Onofri A, Benincasa P, Mesgaran M, Ritz C (2018) Hydrothermal-time-to-event models for seed germination. Eur J Agron 101:129–139

Onofri A, Piepho HP, Kozak M (2019) Analyzing censored data in agricultural research: a review with examples and software tips. Ann Appl Biol 174:3–13

Onofri A, Gresta F, Tei F (2010) A new method for the analysis of germination and emergence data of weed species. Weed Research 50:187–198

Oryokot JOE, Murphy SD, Thomas AG, Swanton CJ (1997) Temperature- and moisture-dependent models of seed germination and shoot elongation in green and redroot pigweed (*Amaranthus powellii, A. retroflexus*). Weed Sci 45:488–496

Page ER, Gallagher RS, Kemanian AR, Zhang H, Fuerst EP (2006) Modeling site-specific wild oat (*Avena fatua*) emergence across a variable landscape. Weed Sci 54:838–846
Pedroso RM, Neto DD, Filho VR, Fischer AJ, Al-Khatib K (2019) Modeling germination of smallflower umbrella sedge (*Cyperus difformis* L.) seeds from rice fields in California across suboptimal temperatures. Weed Technology 33:733–738
Pannell DJ, Stewart V, Bennett A, Monjardino M, Schmidt C, Powles SB (2004) RIM: a bioeconomic model for integrated weed management. Agr Syst 79:305–325
Ratkowsky D (1983) Nonlinear regression analysis.(ed. D Ratkowsky). Marcel 484 Decker Inc., New York, USA
Renzi JP, Chantre GR, Cantamutto MA (2018) Vicia villosa ssp. villosa Roth field emergence model in a semiarid agroecosystem. Grass Forage Sci 73:146–158
Renzi JP, Chantre GR, González-Andújar JL, Cantamutto MA (2019) Development and validation of a simulation model for hairy vetch (Vicia villosa Roth) self-regeneration under different crop rotations. Field Crop Res *235*:79–86
Roberts EH (1988) Temperature and seed germination. Symp Soc Exp Biol 42:109–132
Roman ES, Thomas AG, Murphy SD, Swanton CJ (1999) Modelling germination and seedling elongation of common lambsquarters (*Chenopodium album*). Weed Sci 47:149–155
Roman ES, Murphy SD, Swanton CJ (2000) Simulation of seedling emergence. Weed Science 48:217–224
Royo-Esnal A, López-Fernández ML (2010) Modelling leaf development in *Oxalis latifolia*. Span J Agric Res 8:419–424
Royo-Esnal A, Torra J, Conesa JA, Forcella F, Recasens J (2010) Modeling the emergence of three bedstraw (*Galium*) species. Weed Sci 58:10–15
Royo-Esnal A, Necajeva J, Torra J, Recasens J, Gesch RW (2015a) Emergence of field pennycress (*Thlaspi arvense* L.): comparison of two accessions and modelling. Ind Crop Prod 66:161–169
Royo-Esnal A, Garcia AL, Torra J, Forcella F, Recasens J (2015b) Describing *Polygonum aviculare* emergence in different tillage systems. Weed Res 55:387–395
Royo-Esnal A, Gesch RW, Forcella F, Torra J, Recasens J, Necajeva J (2015c) The role of light in the emergence of weeds: using *Camelina microcarpa* as an example. PLoS One 10:e0146079
Royo-Esnal A, Loddo D, Necajeva J, Jensen PK, de Mol F, Economou G, Taab A, Bochenek A, Synowiec A, Calha I, Andersson L, Uludag A, Uremis I, Onofri A, Torresen K (2018a) Emergence of *Echinochloa crus-galli* (L.) P. Beauv. populations along Europe and the Middle East. In: 18th Symposium of the European Weed Research Society, Ljubljana, Slovenia, 18–22 June 2018
Royo-Esnal A, Recasens J, Garrido J, Torra J (2018b) Rigput brome (*Bromus diandrus* Roth.) management in a no tilled field in Spain. Agronomy 8:251. https://doi.org/10.3390/agronomy8110251
Royo-Esnal A, Gesch RW, Necajeva J, Forcella F, Edo-Tena E, Recasens J, Torra J (2019) Germination and emergence of *Neslia paniculata* (L.) Desv. Ind Crop Prod 129:455–462
Scherner A, Melander B, Jensen PK, Kudsk P, Avila LA (2017) Reducing tillage intensity affects the cumulative emergence dynamics of annual grass weeds in winter cereals. Weed Res 57:314–322
Schutte BJ, Regnier EE, Harrison SK, Schmoll JT, Spokas K, Forcella F (2008) A hydrothermal seedling emergence model for Giant Ragweed (*Ambrosia trifida*). Weed Sci 56:555–560
Schutte BJ, Tomasek BJ, Davis AS, Andersson L, Benoit DL, Cirujeda A, Dekker J, Forcella F, Gonzalez-Andujar JL, Graziani F, Murdoch AJ, Neve P, Rasmussen IA, Sera B, Salonen J, Tei F, Tørresen KS, Urbano JM, Lutman P (2014) An investigation to enhance understanding of the stimulation of weed seedling emergence by soil disturbance. Weed Research 54:1– 12
Sester M, Durr C, Darmency H, Colbach N (2007) Modelling the effects of cropping systems on the seed bank dynamics and the emergence of weed beet. Ecol Model 204:47–58
Sharpe SM, Boyd NS (2019) Black medic (*Medicago lupulina*) germination response to temperature and osmotic potential, and a novel growing degree-day accounting restriction for heat-limited germination. Weed Sci 67:246–252

Shrestha A, Roman ES, Thomas SG, Swanton CJ (1999) Modeling germination and shoot-radicle elongation of *Ambrosia artemisiifolia*. Weed Sci 47:557–562

Soltani E, Gonzalez-Andújar JL, Oveisi M, Salehi N (2018) Development and validation of a predictive model for seedling emergence of volunteer canola (*Brassica napus*) under semi-arid climate. Int J Plant Prod 12:53–60

Sousa-Ortega C, Chamber E, Urbano JM, Izquierdo J, Loureiro I, Marí AI, Cordero F, Vargas M, Saavedra M, Lezaun JA, Paramio JA, Fernández JL, Torra J, Royo-Esnal A (2020a) Should emergence models for Lolium rigidum be changed throughout climatic conditions? The case of Spain. Crop Protection 128:105012

Sousa-Ortega C, Royo-Esnal A, DiTommaso A, Izquierdo J, Loureiro I, Marí AI, Cordero F, Vargas M, Saavedra M, Paramio JA, Fernández JL, Torra J, Urbano JM (2020b) Modeling the Emergence of North African Knapweed (*Centaurea diluta*), an Increasingly Troublesome Weed in Spain. Weed Science:1–30

Spokas K, Forcella F (2009) Software tools for weed seed germination modeling. Weed Sci 57:216–227

Streck NA, Lago I, Oliveira FB, Heldwein AB, de Ávila LA, Bosco LC (2011) Modelling the development of cultivated rice and weedy red rice. Trans ASABE 54:371–384

Taylor JR, Roundy BA, Allen PS, Meyer SE, Egget DL (2007) Soil water sensor accuracy for predicting seedling emergence using a hydrothermal time model. Arid Land Res Manag 21:229–243

Torra J, Recasens J, Royo-Esnal A (2018) Seedling emergence response of rare arable plants to soil tillage varies by species. PLoS One 13:e0199425

Ustarroz D, Kruk BC, Satorre EH, Ghersa CM (2016) Dormancy, germination and emergence of *Urochloa panicoides* regulated by temperature. Weed Res 56:59–68

Valencia-Gredilla F, Supiciche ML, Chantre GR, Recasens J, Royo-Esnal A (2020) Germination behaviour of to constant and alternating temperatures across different populations. Annals of Applied Biology 176:36–46

Vleeshouwers LM, Kropff MJ (2000) Modelling field emergence patterns in arable weeds. New Phytologist 148:445–457

Werle R, Bernards ML, Arkebauer TJ, Lindquist JL (2014a) Environmental triggers of winter annual weed emergence in the Midwestern United States. Weed Sci 62:83–96

Werle R, Sandell LD, Buhler DD, Hartzler RG, Lindquist JL (2014b) Predicting emergence of 23 summer annual weed species. Weed Sci 62:267–279

Wu L, Boyd N, Cutler GC, Olson AR (2013) Spreading dogbane (*Apocynum androsaemifolium*) development in wild blueberry fields. Weed Sci 61:422–427

Yousefi AR, Oveisi M, Gonzalez-Andujar JL (2014) Prediction of annual weed seed emergence in garlic (*Allium sativum* L.) using soil thermal time. Sci Hortic 168:189–192

Zambrano-Navea C, Bastida F, Gonzalez-Andujar JL (2013) A hydrothermal seedling emergence model for *Conyza bonariensis*. Weed Res 53:213–220

# Chapter 6
# Weed Interference Models

**Manpreet Singh, Simerjeet Kaur, and Bhagirath S. Chauhan**

**Abstract** Weeds are a major biotic constraint in agricultural systems. Farmers need to quantify the damage that weeds cause to crops, and many models have been developed to predict yield loss. Empirical functions are the most commonly used models, which additionally provide information for weed threshold values. The limitations of such models are that they are based on statistical functions and usually do not consider biological insights for crop-weed interference. Conversely, mechanistic models take into account various underlying processes but are rather complex in nature; thus, their major utility lies in generating information for weed studies under different locations/conditions. Mechanistic models are based on simulation models that mingle both explanatory and descriptive features, with well-known plant processes studied in a mechanistic fashion, and poorly understood processes considered as a descriptive approach. Weed interference models are an important part of the decision support systems to establish recommendations based on the economic quantification of different weed management strategies. Thus, these models are very useful tools for the development of integrated weed management. In this chapter, we present empirical and mechanistic models that are currently in use for studying crop-weed interference.

**Keywords** Crop-weed competition · Weed interaction · Empirical models · Mechanistic models · Weed growth models · Yield loss models · Critical period · Weed biomass models · Allelopathic interactions

---

M. Singh (✉) · S. Kaur
Punjab Agricultural University, Ludhiana, Punjab, India
e-mail: agrimanpreet@pau.edu; simer@pau.edu

B. S. Chauhan
The Centre for Crop Science, Queensland Alliance for Agriculture and Food Innovation (QAAFI) and School of Agriculture and Food Sciences (SAFS), The University of Queensland, Gatton, QLD, Australia
e-mail: b.chauhan@uq.edu.au


G. R. Chantre, J. L. González-Andújar (eds.), *Decision Support Systems for Weed Management*, https://doi.org/10.1007/978-3-030-44402-0_6

## 6.1 Introduction

Weeds are a major biotic constraint of agricultural systems worldwide. These are undesirable plants interfering with crop production and use of natural resources. Interference denotes direct and indirect effects, which two neighbouring plants impose upon each other. Weed science generally includes the study of competition and allelopathy as main crop-weed interactions. Crop productivity reduction depends upon weed species present, severity and period of infestation, crop vigour and climatic conditions. Annually, weeds cause an economic loss of more than US$100 billion (Appleby et al. 2000), and herbicides worth US$25 billion are being used for weed management on a global scale (Agrow 2003). A 10% yield loss has been estimated despite the adoption of various weed control measures by farmers (Appleby et al. 2000). Further, weed-related crop yield losses are estimated at 5 and 25% for developed and underdeveloped countries, respectively. In India, for example, annual yield losses in ten major crops are about US$11 billion (Gharde et al. 2018), while a sum of US$43 billion has been reported in corn and soybean in the USA and Canada (Kansas State University 2016).

Competition and allelopathy are the two main components of weed interference. Resource competition, a form of physical interference, occurs when a number of plant species adversely affect each other while utilizing common resources, which are in short supply. In allelopathy, a chemical interference, a crop or weed species produces phytotoxic or growth-inhibiting substances in spite of abundant nutrient supply. Although weed allelopathy has been demonstrated in many studies, the effect of this chemical interference has been minor relative to competition. Moreover, it may be very difficult to separate allelopathic from competitive effects in field studies (Bertholdsson 2010). This difficulty in assessing allelopathic interactions has led to fewer real field investigations compared to weed competition studies, which alone has been referred to as 'weed interference' in most cases.

Regarding crop-weed competition, two main principles should be mentioned: (1) more aggressive individuals usually dominate in the intermixed community of weeds and crops, and (2) the first individual to occupy a given area has an advantage over latecomers. The latter principle is of great importance in practical weed management, where agronomic practices are always focused on a rapid crop establishment before weed settlement. In general, weeds with a similar growth pattern to crops are more serious competitors than those species with a dissimilar growth pattern.

Field weed-crop competition is always more severe in the early stages of a crop than at later stages. Early weed infestations lead to a decrease in the crop photosynthetic rate, thus overruling a crop's asymmetric growth advantage. This leads to crop smothering due to competition and ultimately results in yield losses. However, there is a critical competition period (CCP) during which impact of weeds on crop yield is highest (i.e. the shortest time span during crop growth when weeding results in the highest economic returns). Therefore, the crop has to be maintained in weed-free conditions during the CCP. An attempt for creating weed-free conditions throughout the crop ontogeny may involve unnecessary additional expenditure without an equivalent increase in yield. Invariably, a crop of 100 days must be kept weed-free for the first 35 days of sowing. The crop can compete successfully with late-emerging weeds primarily through shading; however, these late-emerging weeds would produce propagules increasing the soil seedbank. In general, the intensity of weed-crop competition is most severe in slow-growing rate crops (i.e. 6–8 weeks after sowing) such as sunflower, cotton and sugarcane.

Yield reduction is a widely accepted variable for the quantification of crop-weed competition. The relationship between crop yield and weed density is clearly nonlinear; thus, increasing weed density results in a yield decline following, in general, a hyperbolic model or other nonlinear functions (i.e., exponential). In addition, weeds interfere with crop handling and act as a reservoir of insect pests and diseases.

## 6.2 Factors Affecting Weed-Crop Competition

Competition can be interspecific (among weed and crop) or intraspecific (within the same species), while the characteristic crop-weed association is the final result of competition (Tominaga and Yamasue 2004). The imbalanced nature of crop-weed competition can be manipulated in order to favour crops by modifying soil and cropping conditions. Association of weed(s) with a particular crop may be due to a congenial environment [e.g. *Cichorium intybus* L. in Egyptian clover/*berseem* (*Trifolium alexandrinum* L.)], morphological similarities [e.g. *Phalaris minor* L. in wheat (*Triticum aestivum* L.)], seed shedding behaviour (e.g. *P. minor* in wheat), seed escape removal through sieving (e.g. *Malva neglecta* L. seeds in wheat), germination/emergence flushes (e.g. *P. minor* in wheat) or continuous use of herbicides with the same mechanism of action [e.g. dominance of *Caesulia axillaris* in rice (*Oryza sativa* L.) due to repeated anilofos herbicide applications] (Walia 2010).

As a general rule, competition remains up to the time when the crop covers ≥80% of the ground (Rasmussen 1992). The competitive ability of crops depends upon many factors, such as (1) crop type and cultivar (variety) selection, sowing date, row spacing and tillage systems; (2) weed density (abundance) and composition; (3) edapho-climatic factors (soil type, environmental conditions); and (4) crop rotation scenarios.

### 6.2.1 *Crop Density and Spatial Uniformity*

Both factors influence the ecological niche of weeds. As the crop density increases, weed growth decreases accordingly, although very high crop densities may lead to interspecific competition (Bleasdale and Nelder 1960). A crop planted at wide row spacing may result in dense weed growth and both inter- and intraspecific competition (Chauhan and Johnson 2010a), whereas a crop grown in square planting is ideal for reducing intraspecific competition (Hashem et al. 1998).

### 6.2.2 Weed Density and Composition

Weed density affects the length of the CCP, while weed emergence timing relative to the crop determines density thresholds (Dunan et al. 1995). Weed species architecture and growth habitat (i.e. grass, broadleaves or sedges) also affect weed-crop competitive relationships. Weed diversity has a variable effect on crop yield, and this difference may be due to various weed ecological adaptations acquired over time. For example, annual weeds which present a high leaf area development rate (e.g. *Echinochloa crus-galli* (L.) P. Beauv. in rice and *Setaria viridis* in corn) greatly influence weed-crop competition reducing crop yield significantly. *Xanthium strumarium* L. and wild mustards (*Brassica* and *Sinapis* spp.) are better competitors than grass weeds at early growth stages due to their capability for rapid soil cover. Also, *Brassica* spp. develop a more extensive root system than grass species after 3 weeks of emergence (Snapp et al. 2005). Many perennial weeds have an extensive deep-root system which feeds from deep soil layers (e.g. *Sorghum halepense* L., *Acroptilon repens* L., *Diplotaxis tenuifolia*). In dry areas, perennials are more competitive than annual weeds. For similar weed densities, a composite stand of weed species is always more competitive than a solid single weed stand (Liebman et al. 2004).

### 6.2.3 Crop Types and Varieties

It has been estimated that the use of vigorous crop species (or cultivars) can reduce costs involved in weed management by 30% (Snapp et al. 2005). Some crops such as sorghum (*Sorghum bicolor* L. Moench), corn (*Zea mays*), pearl millet (*Pennisetum glaucum* (L.) R.Br.) and cowpea (*Vigna unguiculata* (L.) Walp.) have the capacity for rapid growth and fast soil cover showing tolerance to weed competition. On the other hand, crops such as sugarcane (*Saccharum officinarum* L.), cotton (*Gossypium arboreum* L.) and sunflower (*Helianthus annuus* L.) have slow initial growth requiring a longer time for soil surface cover.

Choice of a given crop variety depends on its adaptation to local edapho-climatic and management conditions. Barley (*Hordeum vulgare)* has a higher competitive ability against weeds compared to other cereal crops like rye, wheat and oat. Such high competitive ability is attributed to the development of a deep extensive root system within 3 weeks of sowing. Crops differ in their competitive ability due to differences in crop vigour (Mohler 2004). Tall and rapid canopy-forming crops (or varieties) usually have lower 'weed pressure' than short-statured, slow-growing crops. However, such tall varieties usually lodge and are low yielders. Crops (or varieties) with a short stature are usually more susceptible to weed competition than taller ones; therefore, dwarf and high yielding varieties perform better only when chemical weed management is used (Saito et al. 2010).

### 6.2.4 Edapho-climatic Conditions

Soil texture, pH, fertility and moisture influence crop-weed competition by affecting the vigour of both crops and weeds. Fertile soils usually favour weed growth, thus reducing crop yield. For example, *Buglossoides arvensis* vegetation coverage and seed production per plant increase significantly in nitrogen-enriched soils (Bischoff and Mahn 2000). Thus, weed control measures must be highly efficient to minimize weed pressure. The soil-type reaction is also an important factor that influences weed-crop competition. In general, a pH $\approx 7$ is adequate for crop growth, while acid or alkaline soils negatively influence crop stands, thus increasing weed pressure. Some species like *Rumex acetosella* L. are typical of acidic soils, while others such as *Taraxacum officinale* F.H. Wigg. are more abundant on alkali-reactive soils (Tominaga and Yamasue 2004). Soil-water relations, particularly the quantity of rainfall and its distribution, influence weed-crop competition. Weeds are more adapted to moisture stress (both drought and inundation effects) than crops. The time of irrigation might also influence the weed-crop balance as weeds get more benefit when a 'weedy crop' is irrigated. As the inherent capacity of crops to compete against weeds is weakened by climatic and soil stresses (Mohler 2004), various farm operations can be adjusted in such a way to suppress weed growth.

### 6.2.5 Agronomic Practices

Crop sowing time, seed rate, row spacing, fertilization and water management influence weed-crop competition (Swanton et al. 2015). Regarding sowing time, weed-crop competition will be higher if the crop is sown before or after the recommended time for optimal growth (Walia 2010). Early wheat sowing suffers less competition from *P. minor* infestation due to early crop establishment (Mehra and Gill 1988). At the right sowing time, a crop attains adequate vigour, thus being more competitive than weeds. Similarly, if a good initial crop stand is obtained, the competition ability of the crop will be enhanced.

The time of weed emergence relative to the crop is the most important variable for deciding the impact of weed-crop competition (Dew 1972; O'Donovan et al. 1985; Kropff and Spitters 1991). If the first weed emergence flush takes place along with crop emergence, then intense weed-crop competition is expected. The threshold levels of *Amaranthus retroflexus* and *E. crus-galli* in corn, soybean and dry bean crops are two to ten times lower for weeds that emerge with the crops than those weeds emerging 3–4 weeks after crop emergence (Knezevic et al. 1994; Cardina et al. 1995; Dieleman et al. 1996). Sowing methods such as 'dust mulching' that dries the topsoil can be used for weed control. By using this practice, the weed seedbank present in the topsoil cannot absorb moisture, and thus, weed emergence will be delayed. In such situations, early seedling emergence is avoided, and the crop individuals will be more competitive. The tillage method, frequency and depth of cultivation can influence weed establishment and growth (Pekrun and Claupein 2004; Chauhan and Johnson 2009). For example, mouldboard ploughing may bring buried weed seeds to the soil surface, thus affecting seedbank distribution and weed composition. Deep ploughing in summer months may also be helpful for controlling deep-rooted perennials like *Saccharum spontaneum* and *Cyperus rotundus* (Bàrberi 2002). In non-tillage systems, cover crops are being used for weed control (Price and Norsworthy 2013).

In addition, crop rotations, as well as alternating row and broadcast crops, can be helpful in reducing weed-crop competition. Competitive ability of crops can be increased, and weeds can be reduced by following a crop rotation, for example, *P. minor* infestation in wheat can be reduced by rotating rice-wheat with rice-berseem (*Trifolium alexandrinum* L.) (Malik and Singh 1995). In addition, this practice will help to reduce other weeds in the succeeding year following the wheat crop.

## 6.3 Harnessing Weed-Crop Competition: A Weed Management Option

The objective of weed management is to evolve methods for different situations to ensure a sustainable ecosystem and minimum weed interference. Weed management programs require constant reviewing and improvement for modern agricultural systems due to weed diversity and herbicide resistance issues.

Integrated weed management including sanitation, mechanical methods, biological means and herbicide use reduction is gaining importance due to the well-known negative environmental and social effects of more narrow, herbicide-dependent strategies. Weed preventative methods such as the use of weed-free crop seed, clean farm machinery, well-rotted farmyard manure, sanitation and legal measures can be adopted to prevent the new entry of weeds in an area. These weed management efforts emphasize reducing crop-weed competition in the early crop growth stage. Good 'crop husbandry' methods include selective stimulation of crops, stale seedbed, smother cropping, crop rotation, summer fallowing, zero tillage and soil solarization. Weed control practice cannot substitute good crop husbandry methods. As stated by Chauhan et al. (2012), if good crop husbandry methods are applied, half of the required weed control is achieved.

An example of this is the selective stimulation of crop growth leading to vigorous crop plants that better compete with weeds through rapid ground cover. Competitive crops/varieties sown at proper planting time, the implementation of bidirectional, narrow spacing or the ridge/bed sowing of crops may also result in a low weed density. Early crop seedling vigour can also be maximized by maintaining a proper crop stand, or by split applications of inorganic fertilizers, especially in sandy soils. Adequate application of N, P and K fertilizers in a band (or as side dressing) improves crop growth. Weeds growing 20 cm or more away from the fertilizer band usually fail to make use of even a mobile nutrient like N. Foliar fertilization in wide row crops such as maize, sugarcane and cotton might also help in selective stimulation. Addition of N fixers, phosphorus solubilizing cultures and soil amendments (like gypsum or lime) are important steps for favouring crop growth.

Stale seedbed methods allow for weed management before crop planting. In general, one or two weed flushes can be managed using chemical or mechanical interventions. Otherwise, weeds may overtake the crop at the early crop growth period. Soil tillage affects vertical distribution of weed seeds in arable soils, with more than 56% of the seedbank in the upper soil layer in zero tillage, while only 5% in the top layer under conventional tillage (Chauhan and Johnson 2009). In countries like the USA and the UK, crops are sown using zero tillage to avoid seed burial and reduce the persistence of annual weeds. However, crops sown under zero tillage have more problems with perennial weeds. Zero tillage with residue retention practiced over a 5- to 6-year period of wheat monoculture results in a decrease in the density of *P. minor* with an incremental increase in perennial and broadleaf weeds (Simerjeet-Kaur personal observation).

Smother crops (such as *Vigna unguiculata, Medicago sativa, Trifolium alexandrinum, Pennisetum* sp., *Sorghum bicolor, Sorghum* × *drummondii, Brassica juncea* and *Hordeum vulgare*) can be used as intercrops for weed suppression. In addition, crop rotation is effective in controlling weeds which are associated with a particular crop. With each crop, certain typical weeds appear which are less serious in some other crops. Some weeds increase their numbers quickly if a favourable crop is raised continuously (i.e. monoculture). For example, parasitic weeds like *Cuscuta* in *Medicago sativa* or *Trifolium alexandrinum* can be reduced by rotating the crop with cereal crops. *Phalaris minor* density in wheat may be reduced by rotations with potato or mustard/berseem. *Avena fatua* L. (wild oat) can be controlled in rotation with pea and chickpea for 2–3 years. Many perennial weeds, including *S. halepense, C. rotundus, Cynodon dactylon* and *Cirsium arvense*, were managed with the introduction of paddy rice in the Indian Punjab (Walia 2010).

Fallow in summer has been a common weed management practice for decades, especially for perennial weeds in India, as well as many other countries with tropical climates. In this practice, the soil is heated during the hot summer months (April–June) through a process called solarization. Soil solarization with the help of a polyethylene sheet increases the surface temperature of the soil reaching 40–45 °C. Thus, seeds, rhizomes and tubers of *S. halepense* and *C. rotundus* are easily desiccated which is sufficient to kill propagules in the top 5 cm soil layer. Some weeds are more susceptible to flooding than others, for example, the latter technique can be used in combination with glyphosate applications for the control of the invasive perennial *A. repens* in Argentinian pastures (Gajardo 2019). Conversely, water drainage may be employed for controlling aquatic and semi-aquatic weeds.

## 6.4 Need for Crop-Weed Competition Models

Weeds represent a continuous problem in agricultural production due to their dynamic and resilient nature. Mathematical models offer a significant tool for understanding and predicting the crop yield losses incurred due to weed-crop interference. Weed-crop competition models help to inform weed management decisions, both on a short-term basis to tackle the present weed population and in the long term to plan sustainable weed management strategies (Renton et al. 2015). For example, herbicide-based weed management has helped the agricultural community in a big way; however, after prolonged use, problems of evolution of herbicide-resistant weeds, shifts in weed flora and environmental pollution have surfaced (Johnson et al. 2009). Weed scientists are interested in reducing reliance on herbicides, creating the challenge of developing sustainable weed management practices, which could then be incorporated into present practices. Weed management practices require the integration of two objectives: first to prevent crop yield loss due to weed competition in the short term and second in the long term to avoid the addition of weed seeds or asexual propagules to the soil seedbank (Battle et al. 1996). Competition models can be integrated within the framework of a decision support system (DSS) (Renton and Lawes 2009; Lawes and Renton 2010). Modelling of crop-weed competition can also help to generate the in-depth scientific basis of various processes and to understand interactions.

### 6.4.1 *Empirical Versus Mechanistic Models*

Various types of models have been used to predict crop-weed competition. The approach used in some models is empirical, while other models are based on mechanistic processes. Empirical models may help to predict the crop yield loss (or crop yield) in response to variable weed density (or biomass) in certain environmental conditions. Conversely, mechanistic models may be useful for understanding bio-ecophysiological processes and also for predictive purposes.

Most competition studies are based on empirical models. Linear and nonlinear regression models have been developed (Patterson 1995; Zimdahl 2004), such as the negative power function (Shinozaki and Kira 1956; Bleasdale and Nelder 1960; Watkinson 1980) or the hyperbolic function (Cousens 1985). More complex empirical models have been developed by taking into account variables such as weed emergence times (Cousens et al. 1987; Neve et al. 2003), multiple weed species in simultaneous competition (Firbank and Watkinson 1985; Pantone and Baker 1991; Park et al. 2002; Diggle et al. 2003), and weeds with multiple emergence periods and variable degrees of overlapping with the crop (Peltzer et al. 2012).

Crop-weed competition is a complex phenomenon, and to understand this, a detailed mechanistic model offers better insights than an empirical model. Mechanistic or explanatory models take into account all underlying processes or mechanisms and their dependence on each other with respect to time and external drivers. These modelling methods are process based and dynamic and also referred to as ontological, mechanistic or bottom-up approaches.

Many of the existing competition models have been proposed for use in both research and application, and there is no distinction between them. Simulation models provide avenues for conducting crop-weed experiments under variable climatic conditions and different hypotheses for testing. Simulation models use historical weather data and need validation for each location. A mechanistic simulation approach studies the reasons for a particular response. Sensitivity analysis may also be performed to identify the most important factor for that response. With modelling approaches, various hypotheses may be tested about the type of complex relationships among variables and uncovering knowledge gaps. Plant processes such as light interception and photosynthesis, which are well known, are studied in a mechanistic way, while plant processes such as resource allocation, which are poorly understood, are considered in the descriptive approach.

### 6.4.2 Basics of Empirical Competition Modelling

Empirical crop-weed competition models are derived from agronomic studies comprising crops and weed species in diverse experimental designs. Most important designs used to study competition are mixtures of additive and replacement series (Gibson et al. 1999; Freckleton and Watkinson 2000; Swanton et al. 2015). Replacement series comprise two species grown in different proportions while maintaining overall constant stand density (de Wit 1960). It is not a favoured approach because of its dependence on total stand density (Inouye and Schaffer 1981; Connolly 1986) and failure to differentiate the effects of intra- and interspecific competition (Snaydon 1991; Watkinson and Freckleton 1997). In additive series, both the population and proportion of crop and weed species are varied in mixtures. In field experiments, the density of crop species is kept constant, while weed density is varied. A complete additive design may help in the estimation of both intra- and interspecific competition when analysed using a two-species regression model (Pantone and Baker 1991; Park et al. 2002).

Commonly used single species models (e.g. the equation given below) can be extended for studying two or more species by using the relationship:

$$w_i w_{m,i} \left(1 + \Sigma \alpha_{ij} N_j\right)^{-b} \tag{6.1}$$

where $w$ is a measure of plant performance; $w_m$ is the performance of an isolated plant; $\alpha$ represents the per capita effects of intra- ($\alpha_{ii}$) and interspecific ($\alpha_{ij}$) competition (Watkinson 1985); $b$ is the parameter, which determines yield-density relationship ($b > 1$ represents over-turning; $b = 1$ represents asymptotic; $b < 1$ represents uniform increase); and $i$ and $j$ are the species.

One such preliminary study described the crop yield loss due to weed competition with the hyperbolic model (Cousens 1985, see Sect. 6.4.4.2).

An alternative approach for studying competition is a neighbourhood approach (Mack and Harper 1977). It comprises evaluating the performance of a target species in relation to the density of neighbouring species in proximity. This approach assumes that the production of target plants is related to number, biomass and proximity of the neighbouring plants. Thus, such models incorporate a spatial arrangement in addition to the density of weed species individuals; however, such designs may be resource intensive due to extensive data requirements. This approach has been little used for the competition quantification context in agricultural studies.

### *6.4.3 Basics of Mechanistic Competition Modelling*

Existing crop growth models (e.g. CERES) were modified to include weed species. Models are used to simulate crop and weed growth in two distinct productive scenarios. The first one is the potential production scenario in which a crop is grown under a stress-free environment and crop growth is entirely determined by soil, climatic and crop factors, with inputs supplied in ample quantities. The instantaneous $CO_2$ assimilation rate of the canopy is measured, and the daily growth rate is obtained after subtracting respiration costs. In the second one, the crop suffers from water stress. This effect is simulated using a soil water balance in the model. The potential $CO_2$ assimilation rate will reduce in the case of water shortage. Leaf area development as affected by temperature is simulated by using relative growth rate and specific leaf area. The relative distribution of radiation over the species is also simulated, and potential transpiration is measured. Plant growth reduction is calculated on the basis of potential transpiration and soil moisture content for both species separately.

### *6.4.4 Examples of Competition Models*

Spitters and Aerts (1983) introduced mechanistic dynamic simulation models for crop-weed competition on the basis of the distribution of resources like light, water and nutrients within the species. The growth of crop and weed is determined by their dry matter accumulation which is calculated from the amount of resources (light, water, nutrients, etc.) assimilated by the competing species. These models describe the background of competition and may provide more insight into the crop-weed system. Such models are effective for a wide range of environments and can be used to simulate competition effects for new environments after thorough validation. These simulation models have been developed for diverse scenarios such as stress-free potential production situations and also for production scenarios in which plants face water and/or $N$ stress (Spitters and Aerts 1983; Spitters 1984).

Most of the weed-crop simulation models (Ryel et al. 1990; Weaver et al. 1994; Weaver 1996; Chikoye et al. 1996; Park et al. 2003; Deen et al. 2003) have been developed as products of crop simulation models in which single species models are extrapolated to two species.

#### 6.4.4.1 Weed Growth Models

One of the most comprehensive weed growth models of crop-weed competition is INTERCOM (Kropff and Spitters 1992; Kropff and van Laar 1993), which is based on the earlier work done on simulation modelling of crop growth (de Wit et al. 1978; Spitters et al. 1989) and competition (Spitters and Aerts 1983; Spitters 1989). The growth of each species is measured from day 1 and expressed at the population level, that is, in kg biomass $ha^{-1}$, and measured till maturity. Light interception and resource distribution among competing species is the main approach used in this model. INTERCOM has been calibrated and validated successfully over locations, different crops, weed species, crop-weed competition and contrasting climatic conditions. However, Kropff et al. (1993) observed that in an extremely dry year, yield losses in maize due to *E. crus-galli* were underestimated. The model did not take into account the effects of water stress on crop morphological development, especially stem elongation.

Another mechanistic model is SOYWEED developed by Wilkerson et al. (1990), which is derived from the crop growth model SOYGRO (Jones et al. 1987), and it simulates soybean [*Glycine max* (L.) Merr.] and *Xanthium strumarium* L. competition for light and water. The heterogeneity in leaf area distribution was considered in the first version of the SOYWEED model, and variables crop and weed height were not. SOYWEED also does not simulate weed seed production. Later, another sub-model (LTCOMP) was incorporated into SOYWEED by Wiles and Wilkerson (1991), which takes into account the competition for light. Simulation of light interception was done as a function of plant height, leaf area and the extinction coefficient of crop and weed species. This LTCOMP-SOYWEED model simulated the combined growth of crops and weeds with improved efficacy.

Another model, ALMANAC, simulates competition for light, water and nutrients (N and P) between two plant species (Kiniry et al. 1992). It was developed, parameterized and evaluated for two competing species, *S. halepense* and *Setaria faberi* in soybean and wheat. Practically, this model can be used for simulating weed-crop interference along with its various other uses such as modelling hydrology, climate change, erosion, plant community dynamics, soil carbon, pesticide fate, nutrient cycling and phenology studies.

NTRM-MSC (Nitrogen, Tillage, Residue Management—Multiple Species Competition) model developed by Bail and Shaffer (1993) simulates competition between at least ten plant species. It simulates light interception, soil water and *N* dynamics and competition for these resources by the competing species. This model was parameterized for *Zea mays-A. retroflexus* competition and predicted leaf area development and biomass of each species in a mixture. However, the model did not consider the morphological plasticity of weeds grown in monoculture and mixture, also underestimating light interception by weeds in monoculture.

A crop growth model for rice was constructed by Graf et al. (1990a) and expanded to simulate crop-weed competition for light and *N* in the presence of multiple weed species (Graf et al. 1990b). Weed species were grouped on the basis of height, leaf shape, growth form and phenology. The competition model developed by Spitters and Aerts (1983) was used for the simulation of competition for light. The proportion of the soil profile exploited by the crop and each of the weed groups was used for simulation of competition for *N*. This model was later parameterized and validated for competition between *E. crus-galli* and rice by Graf and Hill (1992) and simulated the effect of densities of both crop and weed on rice yield. Lotz et al. (1990) used a similar model for describing competition between *Triticum aestivum* and weeds for light and water. Ryel et al. (1990) developed a model to simulate competition for light in a mixed canopy and predicted instantaneous capture of incident radiation and net photosynthesis of each species. This simulation is based on incident radiation, canopy structure and photosynthetic characteristics. Barnes et al. (1990) and Beyschlag et al. (1990) used this model to simulate crop-weed competition between irrigated wheat and *A. fatua*. Dunan et al. (1994) developed a model to simulate crop-weed competition for light, similar to that of Spitters and Aerts (1983), but plant height was not considered a parameter. It included an economical sub-model to analyse different weed management strategies and was applied for the study of *A. fatua*-barley (*H. vulgare*) competition.

#### 6.4.4.2 Yield Loss Models

For the decision on weed management to be economical, knowledge on how uncontrolled weeds impact the crop yield is required. Therefore, competition models are an integral part of the weed management profitability assessment. Most of the developed models are empirical in nature. Crop-weed competition models have been developed on biologically sound principles, which consider the following points (Kropff 1988):

1. When weeds are added one by one to the crop stand, each plant may have a competitive effect, which can be measured by crop yield loss. At low density, the effect of competition could be additive, as there would be little or no intra-specific competition.
2. Yield loss can never be beyond 100%, and it can be expected that yield loss may approach a certain upper limit as weed density increases.
3. At high densities, the distance between weed plants would be less, and they start to interfere with each other due to intra-specific competition as a result. This may lead to a decrease in impact that each weed plant would make on the crop yield.

In crop-weed competition modelling, the rectangular hyperbolic model (Fig. 6.1) best defines the relationship between crop yield loss and weed density (or biomass)

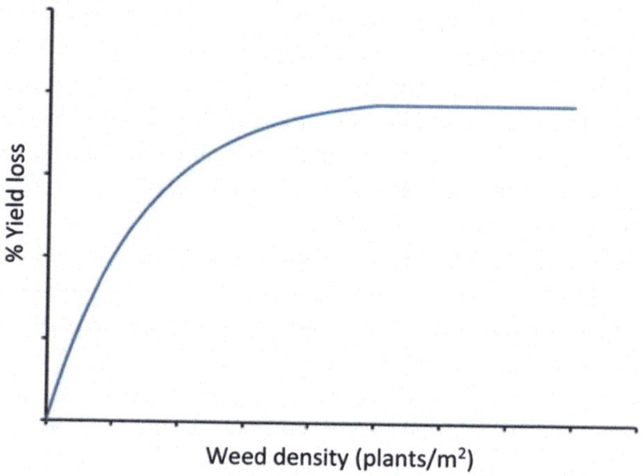


**Fig. 6.1** The rectangular hyperbolic model for the relation of crop yield loss with weed density

irrespective of the weed species, crop and location (Cousens 1985). Such models provide a good explanation of data for a variety of weeds in various crops and can be easily interpreted in agronomic and biological terms. As we can only measure yield and yield loss cannot be observed directly, such hyperbolic models should be designed in order to be able to express in terms of yield. Even though experimental designs contain weed-free plots to observe yield losses, the observations in these plots are also open to error, like any other observations. Cousens' (1985) model describes crop yield per unit area ($Y$) as a negative function of weed density per unit area ($N_{weed}$):

$$Y = Y_{wf}\left[1-(I.N_{weed})/(1+I.N_{weed}/A)\right] \tag{6.2}$$

where $Y_{wf}$ represents the crop yield under weed-free conditions, and the parameter $A$ represents the upper limit of proportional crop yield loss as weed density approached infinity, whereas the parameter $I$ can be taken as the initial slope of the curve, i.e. the amount of proportional yield loss attributable to a single weed per unit area as weed density approaches zero.

There are other similar equations for a rectangular hyperbola in literature, which may define the relation of increasing crop yield loss with increasing weed density. Crop yield loss as a function of weed population can also be shown by using the following simple formula that follows an increasing hyperbola:

$$\mathrm{YL} = \mathrm{UYL}.N_{weed}/(N_{50}\mathrm{YL}+N_{weed}) \tag{6.3}$$

where YL is the absolute yield loss at a weed density $N_{weed}$, and the parameters UYL is the upper limit of yield loss as $N_{weed}$ approaches infinity, and $N_{50}$YL is the weed density at which 50% of the upper limit of yield loss occurs. $Y$ becomes equal to YL when $Y = Y_{wf} - \mathrm{YL}$, UYL $= Y_{wf}.A$ and $N_{50}$ YL $= A/I$.

The relative competitiveness of a weed species and environmental conditions are important factors determining the value of $N_{50}$ for a weed species. Where crop yield loss quickly approaches the maximum value with increase in weed density, low values of $N_{50}$ are observed, while higher values of $N_{50}$ signify that each additional weed plant per unit area has a relatively smaller effect on crop yield loss. This advocates that weed species with low values of $N_{50}$ would be relatively more competitive than weed species having more values of $N_{50}$.

Weed emergence timings may affect this hyperbolic yield density equation. Cousens et al. (1987) revised this simple empirical model to account for the timing of weed emergence relative to the crop along with weed density. A similar two-parameter model was suggested by Kropff et al. (1995) which describes crop yield loss as dependent on the relative weed leaf area. This model takes into account the timing of weed emergence relative to crop emergence; thus, it was more acceptable than the density-based model. Lotz et al. (1990) reported that predicted yield loss seemed to be mainly affected by weed emergence timing. Although these dynamic prediction models help to explain variation in yield loss due to weed density and relative weed emergence time, additional research experiments might increase the practicality of such models for the study of crop-weed interactions and for use in advisory systems. In some simulation models, all weed plants of a particular species are assumed to emerge on one date; therefore, those approaches might overestimate weed competitiveness.

Besides regular hyperbolic functions, other nonlinear functions have been developed to model the weed-crop competition, such as the negative exponential model (Fig. 6.2). Torner et al. (1991) used an exponential model to demonstrate the rela-

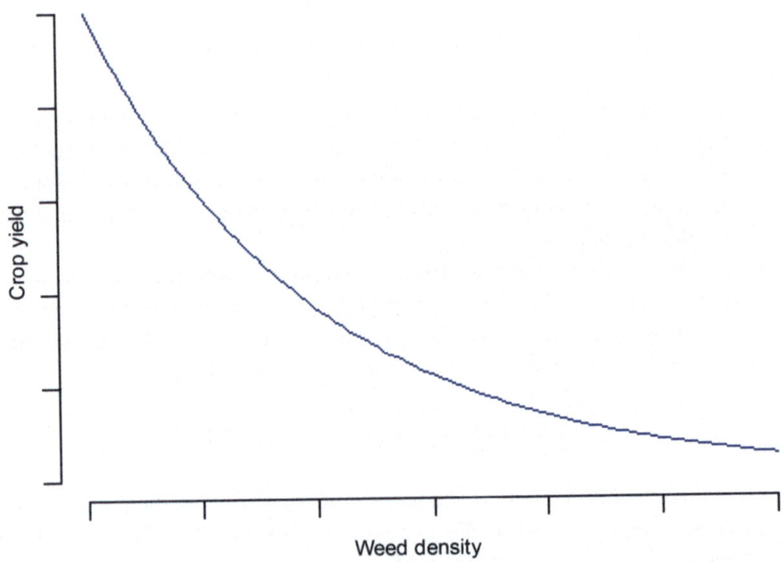


**Fig. 6.2** Generalized negative exponential relation between crop yield and weed density

tion of barley yield with *Avena sterilis*. The relationship was described using the equation

$$Y = a.\exp^{-bx} \tag{6.4}$$

where $Y$ is barley yield, $x$ is wild oat density, $a$ is an estimate of barley yield in the absence of wild oats and $b$ is the estimate of the rate of reduction in barley yield with increase in wild oat density. A similar function was proposed for the competitive relationship between the growth of coffee plantations in competition with different weed species (Ronchi and Silva 2006). Variations in predictions of yield loss models have been attributed to different environmental conditions, limiting their use over a wide range of conditions.

#### 6.4.4.3 Crop Yield and Weed Biomass Functions

The qualitative measure of weed competitiveness is weed biomass, and crop yield loss is believed to have a linear relationship with increasing weed biomass (over a wide range of weed density). Weeds accumulate dry matter by utilizing the same resources that would otherwise be used by crop plants (Spitters and Aerts 1983). Over a range of weed densities, weed biomass replaces a constant proportion of crop yield (which can be represented by weed biomass produced divided by crop biomass lost due to weeds), and this replacement is generally linear (Fig. 6.3). Most of the studies relating crop yield and weed biomass showed a negative linear function. For instance, linear regression models explained about 95% of the variation in grain yield of rice due to the biomass of grass weeds and sedges associated in the crop, when grown in a dry direct-seeded system (Singh et al. 2014). In the same study, a linear relation of grain yield with broadleaf weed dry matter described 91–93% variation in grain yield.

#### 6.4.4.4 Critical Period of Crop-Weed Competition

To study the critical period of crop-weed competition, two sets of treatments are commonly used (Nieto et al. 1968; Weaver and Tan 1987; Hall et al. 1992). In the first treatment set, the crop is kept weed-free after sowing for increasing lengths of time to determine the period when the crop must be kept free of weeds to avoid yield loss. No weed control measures are required beyond this point to achieve maximum crop production. In the second treatment set, the crop is kept weedy for increasing lengths of time to determine the maximum period for which a crop can tolerate weeds without any significant yield loss. The critical period of weed control is the combination of these two periods (Fig. 6.4), and the presence of weeds before and after this period would not reduce crop yield.

The Gompertz equation is generally used to model the effect of the weed-free period (treatment set I) on grain yield, whereas the logistic equation is used to model

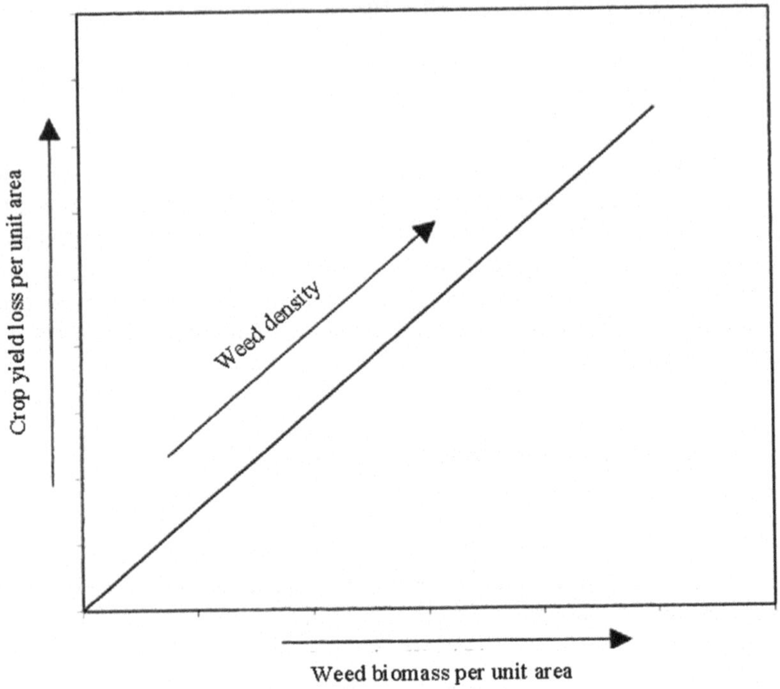


**Fig. 6.3** Linear replacement of crop yield/biomass loss with weed biomass over increasing weed density range

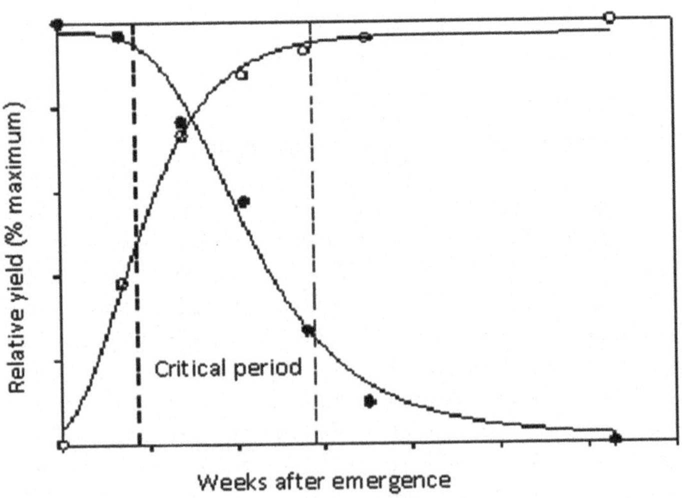


**Fig. 6.4** The influence of time of weed emergence or weed removal on percent maximum crop yield

the influence of weed duration (treatment set II) on yield. In most crops, this period can be clearly marked from the studies, where the early and late association of weeds does not affect the crop. For example, Singh et al. (2014) reported that the critical period for crop-weed competition starts at 17–20 days and ends at 48–57 days for short duration rice cultivars in dry-seeded rice systems. In other crops or situations, there may not be a clear period, requiring further investigation into the period up to which the weeds must be controlled before being allowed to emerge/grow in the crop. For example, in a study on soybean, Almarie (2017) suggested weed removal for the first 5 weeks to get relatively unaffected yields. The critical periods and weed threshold density levels provide good information to producers, but emphasis must be given on controlling weed seed production which involves the use of weed control measures beyond this critical period. Weaver et al. (1992) demonstrated the use of eco-physiological (mechanistic) models for prediction of the critical period of crop-weed competition in sugar beet and transplanted tomato. The simulation model suggested that weed density determines the start of the critical period, such as early-season competition, and also the total length of the critical period to avoid crop yield losses.

#### 6.4.4.5 Crop-Weed Allelopathy Interaction Models

Although weed-crop allelopathy has been recognized as a part of crop-weed interference, most of the studies developing crop-weed interference models pertain to crop-weed competition only. The reason for this may be difficulty in separating this type of chemical interference from the physical competition. Liu et al. (2005) proposed a simple mathematical equation for separating the effect of allelopathy from competition, which was derived from the equation for calculating the performance of one species in a two-species mixture. This equation was given as

$$y_{\mathrm{m}} = y_i f\left(\delta_j\right) \tag{6.5}$$

where $y_{\mathrm{m}}$ is the yield of main species in mixed-species scenario; $y_i$ is the yield of main species in isolation, i.e. in the absence of other species; and $f(\delta_j)$ is the function (effect) of other species on main species. $f(\delta_j)$ may be equal to 1 if there is no effect, greater than 1 if there is stimulation and less than 1 if there is interference. As the interference is dependent on the densities of two species, $f(\delta_j)$ was included in the equation as

$$y_{\mathrm{m}} = y_i \left(y_{\mathrm{m}} / y_i\right)^{m/n} \tag{6.6}$$

where $m$ and $n$ are the densities per unit area of main species and other species, respectively.

This equation of $f(\delta_j)$ was modified to two separate equations for allelopathy and competition:

$$\text{Equation1}\left(\text{for allelopathy}\right): R_{\text{a}} = \left(y_{\text{m/a}} - y_i\right) / y_i$$

$$\text{Equation2}\left(\text{for competition}\right): R_{\text{c}} = \left(y_{\text{m/c}} - y_i\right) / y_i$$

where $R_a$ and $R_c$ refer to the magnitude of allopathic effect and competition effect in two-species mixture.

Other approaches were also suggested to separate the effects of allelopathy from competition in literature. Bertholdsson (2011) suggested the use of partial least square regression (PLSR). Using PLSR models, it was predicted that with improvement in both crop biomass and allelopathy in wheat (to the levels of triticale), up to 60% reduction in weed biomass can be observed, while allelopathy improvement alone can suppress weeds by only 18–28%.

Many studies suggest that allelopathy and competition could act synergistically (Reigosa et al. 1999), as having even a little allelopathic effect can change the balance of competition between the species. Similarly, allelochemicals have been believed to release under stress conditions, which may occur due to competition for important resources. This compounded effect of allelopathy and competition becomes difficult to measure in models which makes estimations on the basis of a simple additive effect.

## 6.5 Challenges and Future Work Direction

Models of crop-weed interference can contribute to improved weed management strategies and evaluation of weed control programs (Orwick et al. 1978; Lotz et al. 1995; Debaeke et al. 1997); however, most of the models have been developed using a limited set of experimental data usually from a single site and rarely being validated over a number of locations. Crop-weed competition models should be linked to weed density dynamics models and simulation of weed seed production over time. Linkage of such models can result in useful integrated decision support system (DSS) for the management of weeds in crops. Morphological and physiological plasticity in weed species is another challenge for models that are developed on the basis of weed growth. Recently, research on weed biology and ecology has been undertaken, but it needs to be strengthened in a more systematic way to elucidate suitable weed management decisions and simulation models (Van Acker 2009; Chauhan and Johnson 2010b). Cousens (1999) reported that weed threshold levels need to be exploited practically for estimating crop yield losses. Many underlying processes in weed science are still not understood; extensive studies should be conducted to shift our focus from 'what occurs' to 'why things happen in this way'. It is important to enhance the practical utility of these models and strengthen the research efforts for more scientific insights of processes involved in crop-weed interference.

Numerous eco-physiological models of crop-weed competition have been developed to improve understanding of underlying processes. Crop-weed competition models may be used at different locations to deduce crop yield loss due to weeds (Lotz et al. 1995; Vitta and Satorre 1999), to understand the crop-weed-environment-management interactions (Lindquist and Kropff 1996) and, in situations where there is no experimental data, to extrapolate the model (Kropff 1988). Even with these models, sensitivity analysis can help us identify the plant traits that confer competitiveness, thus giving guidelines to breeders for developing competitive crops (Lindquist and Kropff 1996).

The process-based competition models can be used to predict yield losses (Lotz et al. 1995); the attempts in this direction need to be undertaken by extrapolation outside the experimental data. The interdisciplinary experimentation should be undertaken for constant feedback for testing of hypotheses about mechanisms included in the model, rather than just validating the model or making predictions from it.

In the dominion of descriptive modelling, this area has been well studied, and various empirical functions are available. Models may be refined with time by using improved available data. However, an intensive data collection is highly expensive and unjustifiable. New ways by which statistical models are being parameterized or validated can be improved, for example, Bayesian methods allow users to generate a semi-automated process where parameters in crop-weed competition models are updated as, and when, new data for a given location or weed/crop species is available (Albert 2009). The meta-modelling methods may also permit statistical models to be validated using the outputs of mechanistic models, thus partly averting the need for collection of more empirical data (Conti and O'Hagan 2010; Renton 2011a; Renton and Savage 2015).

The development of mechanistic simulation approaches will decide the future of weed management scenarios and lead to a new scientific vision to tackle emerging problems like superweeds and weed shifts. Such models can be integrated into DSS for better understanding of crop-weed interactions and management of weeds under threshold levels. Research on integrated weed management approaches will get strengthened with the development of more mechanistic simulation modelling approaches. Improved theoretical understanding with models will likely result in practical outcomes. Mechanistic models will represent spatial and temporal details by using advanced computational skills. Adaptive mechanisms and plant processes in response to a change in environment will be understood with more studies based on process-based models (Evers et al. 2010, 2011; Bongers et al. 2014; Zhu et al. 2015).

Model development, validation, sensitivity analyses and documentation are time-consuming processes, and useful models need frequent updating with new information. Mechanistic models are sometimes too complex to be comprehended by new users (Renton 2011a). Efforts must be taken to develop and link complementary models rather than attempting to put all the details into one 'supermodel'

(Renton 2011b; Holzworth et al. 2014; Lawes and Renton 2015). For modelling biological systems such as crop-weed competition, the quality of data and insights will undergo frequent testing and refinement (Haefner 2005). There are many available programming languages for writing the models, which would then help in sharing the model due to accessible structure and available documentation of such languages.

## 6.6 Conclusions

Modelling studies provide the basic framework which can be utilized in decision support systems for effective management of weeds. Many models suggest the threshold levels of specific weed species, where the yield of the crop could be affected significantly. Moreover, the models for the critical period of crop-weed competition, which provide information on the exact timing of weed control, are required to reduce the yield loss in crops. Such models can be incorporated in DSS to provide help in adjusting the weed control methods and timing as per the crop needs. There are several models which provide information on crop-weed competition based on morphological traits of weeds and crops. Breeders may benefit from such models for developing competitive crop cultivars. Bio-economic models (which link crop and weed density models with the economics of weed control) may play a role in the promotion and adoption of new weed management technologies.

Crop-weed competition models achieved notable success in demonstrating the effect of competition on crop yield and profitability of using weed management strategies. However, our understanding of prediction under diverse environments, that is, spatial and temporal variability, in model parameters has to be enhanced for a wide range of weeds and crops. Integrated weed management strategies can be devised using these models, and weed biodiversity in current and future crops may be studied. Against the backdrop of climate change, models will continue to help us simulate and understand crop-weed competition scenarios and their effects on crop growth and yield in general and on agriculture at regional/national/international levels.

## References

Agrow (2003) Agrochemical sales flat in 2002. Agrow: World Crop Protection News. http://ipm.osu.edu/trans/043_141.htm

Albert J (2009) Bayesian computation with R. Springer, New York

Almarie AA (2017) The critical period for weed competition in soybean (*Glycine max* L. Merr) under Iraqi irrigated areas. ARPN J Agril Biol Sci 4:128–132

Appleby AP, Muller F, Carpy S (2000) Weed control. In: Muller F (ed) Agrochemicals. Wiley, New York, pp 687–707

Bail DA, Shaffer MJ (1993) Simulating resource competition in multispecies agricultural plant communities. Weed Res 33:299–310

Bàrberi P (2002) Weed management in organic agriculture: are we addressing the right issues? Weed Res 42:177–193

Barnes PW, Beyschlag W, Ryel R, Flint SD, Caldwell MM (1990) Plant competition for light analysed with a multispecies canopy model. III. Influence of canopy structure in mixtures and monocultures of wheat and wild oat. Oecologia 82:560–566

Battle M, Bender M, Sowers T, Tans PP, Butler JH, Elkins JW et al (1996) Atmospheric gas concentrations over the past century measured in air from fin at the South Pole. Nature 383:231–235. https://doi.org/10.1038/383231a0

Bertholdsson NO (2010) Breeding spring wheat for improved allelopathic potential. Weed Res 50:49–57

Bertholdsson NO (2011) Use of multivariate statistics to separate allelopathic and competitive factors influencing weed suppression ability in winter wheat. Weed Res 51:273–283

Beyschlag W, Barnes PW, Ryel R, Caldwell MM, Flint SD (1990) Plant competition for light analyzed with a multispecies canopy model. II. Influence of photosynthetic characteristics on mixtures of wheat and wild oat. Oecologia 82:374–380

Bischoff A, Mahn E-G (2000) The effects of nitrogen and diaspore availability on the regeneration of weed communities following extensification. Agric Ecosyst Environ 77:237–246

Bleasdale JKA, Nelder JA (1960) Plant population and crop yield. Nature 188:342

Bongers FJ, Evers JB, Anten NPR, Pierik R (2014) From shade avoidance responses to plant performance at vegetation level: using virtual plant modelling as a tool. New Phytol 204:268–272. https://doi.org/10.1111/nph.13041

Cardina J, Regnier EE, Sparrow DH (1995) Velvetleaf (*Abutilon theophrasti*) competition and economic thresholds in conventional- and no-tillage corn (*Zea mays*). Weed Sci 43:81–87

Chauhan BS, Johnson DE (2009) Influence of tillage systems on weed seedling emergence pattern in rainfed rice. Soil Tillage Res 106(1):15–21

Chauhan BS, Johnson DE (2010a) Implications of narrow crop row spacing and delayed *Echinochloa colona* and *Echinochloa crus-galli* emergence for weed growth and crop yield loss in aerobic rice. Field Crop Res 117:177–182

Chauhan BS, Johnson DE (2010b) The role of seed ecology in improving weed management strategies in the tropics. Adv Agron 105:221–262. https://doi.org/10.1016/S0065-2113(10)05006-6

Chauhan BS, Singh RG, Mahajan G (2012) Ecology and management of weeds under conservation agriculture: a review. Crop Prot 38:57–65. https://doi.org/10.1007/s11356-016-7794-7

Chikoye D, Hunt LA, Swanton CJ (1996) Simulation of competition for photosynthetically active radiation between common ragweed (*Ambrosia artemisiifolia*) and dry bean (*Phaseolus vulgaris*). Weed Sci 44:545–554

Connolly J (1986) On difficulties with replacement-series methodology in mixture experiments. J Appl Ecol 23:125–137

Conti S, O'Hagan A (2010) Bayesian emulation of complex multi-output and dynamic computer models. J Stat Plan Inference 140:640–651. https://doi.org/10.1016/j.jspi.2009.08.006

Cousens R (1985) A simple model relating yield loss to weed density. Ann Appl Biol 107:239–252

Cousens R (1999) Weed science doesn't have to be a contradiction in terms. In: Bishop A, Boersma M, Barnes C (eds) Proceedings of the 12th Australian Weeds Conference. Hobart, TAS, Tasmanian Weed Society, pp 364–371

Cousens R, Brain P, O'Donovan JT, O'Sullivan PA (1987) The use of biologically realistic equations to describe the effects of weed density and relative time of emergence on crop yield. Weed Sci 35:720–725

de Wit CT (1960) On competition. Verslagen van Landboukundigonderzoekingen (Netherlands) 66:1–82

de Wit CT, Goudriaan J, van Laar HH, Penning de Vries FWT, Rabbinge R, van Keulen H, Sîbma L, de Jonge C (1978) Simulation of assimilation, respiration, and transpiration of crops. Simulation monographs. Pudoc, Wageningen, p 14

Debaeke P, Caussanel JP, Kiniry JR et al (1997) Modelling crop: weed interactions in wheat with ALMANAC. Weed Res 37:325–341

Deen W, Cousens R, Warringa J, Bastiaans L, Carberrys P, Rebel K, Riha S, Murphy C, Benjamin LR, Cloughley C, Cussans J, Forcella F, Hunt T, Jamieson P, John L, Wangs E (2003) An evaluation of four crop-weed competition models using a common data set. Weed Res 43:116–129

Dew DA (1972) An index of competition for estimating crop loss due to weeds. Can J Plant Sci 52:921–927

Dieleman A, Hamill AS, Fox GC, Swanton CJ (1996) Decision rules for postemergence control of pigweed (*Amaranthus* spp.) in soybean (*Glycine max*). Weed Sci 44:126–132

Diggle AJ, Neve PB, Smith FP (2003) Herbicides used in combination can reduce the probability of herbicide resistance in finite weed populations. Weed Res 43:371–382. https://doi.org/10.1046/j.1365-3180.2003.00355.x

Dunan CM, Moore FD III, Westra P (1994) A plant process-economic model for wild oats management decisions in irrigated barley. Agr Syst 45:355–368

Dunan CM, Westra P, Schweizer EE, Lybecker DW, Moore FD (1995) The concept and application of early economic period threshold: the case of DCPA in onions (*Allium cepa*). Weed Sci 43:634–639

Evers JB, Vos J, Yin X, Romero P, van der Putten PEL, Struik PC (2010) Simulation of wheat growth and development based on organ-level photo- synthesis and assimilate allocation. J Exp Bot 61:2203–2216. https://doi.org/10.1093/jxb/erq025

Evers JB, van der Krol AR, Vos J, Struik PC (2011) Understanding shoot branching by modelling form and function. Trends Plant Sci 16:464–467. https://doi.org/10.1016/j.tplants.2011.05.004

Firbank LG, Watkinson AR (1985) A model of interference within plant monocultures. J Theor Biol 116:291–311

Freckleton RP, Watkinson AR (2000) Designs for greenhouse studies of interactions between plants: an analytical perspective. J Ecol 88:386–391

Gajardo O (2019) Effects of soil water saturation and chemical control on the growth and multiplication parameters of *Acroptilon repens* L. in plots under irrigation. Doctoral Thesis, Dpto. de Agronomía, Universidad Nacional del Sur. 138 pp

Gharde Y, Singh PK, Dubey RP, Gupta PK (2018) Assessment of yield and economic losses in agriculture due to weeds in India. Crop Prot 107:12–18. https://doi.org/10.1016/j.cropro.2018.01.007

Gibson JD, Connoll J, Hartnett DC, Weidenhamer JD (1999) Designs for greenhouse studies of interactions between plants. J Ecol 87:1–16

Graf B, Hill JE (1992) Modelling the competition for light and nitrogen between rice and *Echinochloa crus-galli*. Agr Syst 40:345–359

Graf B, Rakotobe O, Zahner P, Delucchi V, Gutierrez AP (1990a) A simulation model for the dynamics of rice growth and development: Part I—The carbon balance. Agr Syst 32:341–365

Graf B, Gutierrez AP, Rakotobe O, Zahner P, Delucchi V (1990b) A simulation model for the dynamics of rice growth and development: Part II—The competition with weeds for nitrogen and light. Agr Syst 32:367–392

Haefner JW (2005) Modeling biological systems: principles and applications, 2nd edn. Springer, Berlin

Hall MR, Swanton CJ, Anderson GW (1992) The critical period of weed control in grain corn (*Zea mays*). Weed Sci 40:441–447

Hashem A, Radosevich SR, Roush ML (1998) Effect of proximity factors on competition between winter wheat (*Triticum aestivum*) and Italian ryegrass (*Lolium multiflorum*). *Weed Sci* 46:181–190

Holzworth DP, Huth NI, de Voil PG et al (2014) APSIM evolution towards a new generation of agricultural systems simulation. Environ Model Softw 62:327–350. https://doi.org/10.1016/j.envsoft.2014.07.009

Inouye RS, Schaffer WM (1981) On the ecological meaning of ratio (de Wit) diagrams in plant ecology. Ecology 62:1679–1681

Johnson WG, Davis VM, Kruger GR, Weller SC (2009) Influence of glyphosate-resistant cropping systems on weed species shifts and glyphosate-resistant weed populations. Eur J Agron 31:162–172. https://doi.org/10.1016/j.eja.2009.03.008

Jones JW, Boote KJ, Jagtap SS, Hoogenboom G, Wilkerson GG (1987) SOYGRO V5.4: Soybean crop growth simulation model user's guide. Agric Eng Dep and Agron Dep, University of Florida, Gainesville, Florida Agric. Exp. Stn. J. No. 8304, Gainesville, FL

Kansas State University (2016) Left uncontrolled, weeds would cost billions in economic losses every year. *Science Daily*. www.sciencedaily.com/releases/2016/05/160516130720.htm. Accessed 21 Sept 2019

Kiniry JR, Williams JR, Gassman PW, Debaeke P (1992) A general, process-oriented model for two competing plant species. Trans ASAE 35:801–810

Knezevic SV, Weise SF, Swanton CJ (1994) Interference of redroot pigweed (*Amaranthus retroflexus*) in corn (*Zea mays*). Weed Sci 42:568–573

Kropff MJ (1988) Modelling the effects of weeds on crop production. Weed Res 28:465–471

Kropff MJ, Spitters CJT (1991) A simple model of crop loss by weed competition from early observations on relative leaf area of weeds. Weed Res 31:97–105

Kropff MJ, Spitters CJT (1992) An eco-physiological model for interspecific competition, applied to the influence of *Chenopodium album* L. on sugar beet. I. Model description and parameterization. Weed Res 32:437–450

Kropff MJ, van Laar HH (1993) Modelling crop-weed interactions. CAB International in Association with the International Rice Research Institute, Wallingford, p 274

Kropff MJ, Weaver SE, Lotz LAP, Lindquist JL, Joenje W, Schnieders BJ, van Keulen NC, Migo TR, Fajardo FF (1993) Understanding crop-weed interactions in field situations. In: Kropff MJ, van Laar HH (eds) Modelling crop-weed interactions. CAB International, Wallingford, pp 105–136

Kropff MJ, Lotz LAP, Weaver SE, Bos HJ, Wallinga J, Migo T (1995) A two parameter model for prediction of crop loss by weed competition from early observations of relative leaf area of the weeds. Ann Appl Biol 126:329–346

Lawes R, Renton M (2010) The Land Use Sequence Optimiser (LUSO): a theoretical framework for analysing crop sequences in response to nitrogen, disease and weed populations. Crop Pasture Sci 61:835–843

Lawes R, Renton M (2015) Gaining insight into the risks, returns and value of perfect knowledge for crop sequences by comparing optimal sequences with those proposed by agronomists. Crop Pasture Sci 66:622–633

Liebman M, Mohler CL, Staver CP (2004) Ecological management of agricultural weeds. Cambridge University Press, Cambridge, p 525

Lindquist JL, Kropff MJ (1996) Applications of an ecophysiological model for irrigated rice (*Oryza sativa*)—*Echinochloa* competition. Weed Sci 44:52–56

Liu DL, An M, Johnson IR, Lovett JV (2005) Mathematical modelling of allelopathy: IV. Assessment of contributions of competition and allelopathy to interference by barley. Nonlinearity in Biology. Toxicol Med 3:213–224

Lotz LAP, Kropff MJ, Groeneveld RMW (1990) Modelling weed competition and yield losses to study the effect of omission of herbicides in winter wheat. Neth J Agric Sci 38:711–718

Lotz LAP, Wallinga J, Kropff MJ (1995) Crop–weed interactions: quantification and prediction. In: Glen DM, Greaves MP, Anderson HM (eds) Ecology and integrated farming systems. Wiley, Chichester, pp 31–47

Mack R, Harper JL (1977) Interference in dune annuals: spatial patterns and neighbourhood effects. J Ecol 65:345–363

Malik RK, Singh S (1995) Littleseed canarygrass (*Phalaris minor* Retz.) resistance to isoproturon in India. Weed Technol 9:419–425

Mehra SP, Gill HS (1988) Effect of temperature on germination of *Phalaris minor* Retz. and its competition in wheat. J Res Punjab Agric Univ 25:529–533

Mohler CL (2004) Enhancing the competitive ability of crops. In: Liebman M, Mohler CL, Staver CP (eds) Ecological management of agricultural weeds. Cambridge University Press, Cambridge, pp 269–321

Neve P, Diggle AJ, Smith FP, Powles SB (2003) Simulating evolution of glyphosate resistance in *Lolium rigidum* I: population biology of a rare resistance trait. Weed Res 43:404–417

Nieto JN, Brondo MA, Gonzalez JT (1968) Critical periods of the crop growth cycle for competition from weeds. PANS (C) 14:159–166

O'Donovan JT, de St. Remy EA, O'Sullivan PA, Dew DA, Sharma AK (1985) Influence of the relative time of emergence of wild oat (*Avena fatua*) on yield loss of barley (*Hordeum vulgare*) and wheat (*Triticum aestivum*). Weed Sci 33:498–503

Orwick PL, Schreiber MM, Holt DA (1978) Simulation of foxtail (*Setaria viridis* var. *robusta-alba*, *Setaria viridis* var. *robusta-purpurea*) growth: the development of SETSIM. Weed Sci 26:691–699

Pantone DJ, Baker JB (1991) Reciprocal yield analysis of red rice (*Oryza sativa*) competition in cultivated rice. Weed Sci 39:42–47

Park SE, Benjamin LR, Watkinson AR (2002) Comparing biological productivity in cropping systems: a competition approach. J Appl Ecol 39:416–426

Park SE, Benjamin LR, Watkinson AR (2003) The theory and application of plant competition models: an agronomic perspective. Ann Bot 92:741–748. https://doi.org/10.1093/aob/mcg204

Patterson DT (1995) Effects of environmental stress on weed-crop interactions. Weed Sci 43:483–490

Pekrun C, Claupein W (2004) The effect of stubble tillage and primary tillage on population dynamics of Canada thistle (*Cirsium arvense*) in organic farming. J Plant Dis Prot 19:483–490

Peltzer S, Douglas A, Diggle AJ, George W, Renton M (2012) The Weed Seed Wizard: have we got a solution for you! In: Eldershaw V (ed) Proceedings of the 18th Australasian Weeds Conference. Presented at the 18th Australasian Weeds Conference. Weed Society of Victoria, Melbourne, pp 99–102

Price AJ, Norsworthy JK (2013) Cover crops for weed management in southern reduced-tillage vegetable cropping systems. Weed Technol 27:212–217

Rasmussen J (1992) Experimental approaches to mechanical weed control in field peas. In: Proceedings of the 9th International Symposium on the Biology of Weeds, 16–18 September 1992, Dijon, France. European Weed Research Society, Paris, pp 129–138

Reigosa MJ, Sanchez-Moreiras A, Gonzalez L (1999) Ecophysiological approach in allelopathy. Crit Rev Plant Sci 18:577–608

Renton M (2011a) How much detail and accuracy is required in plant growth sub-models to address questions about optimal management strategies in agricultural systems? AoB Plants 2011:plr006. https://doi.org/10.1093/aobpla/plr006

Renton M (2011b) RIM, LUSO, PERTH and the Wizard: a complementary family of models for supporting weed management decisions in cropping systems. In: Weed Management in a Changing World, 23rd Asian-Pacific Weed Science Society Conference, Cairns, Queensland, Australia, 26–29 September 2011, pp 454–460

Renton M, Lawes RA (2009) Land-Use Sequence Optimiser (LUSO): a simulation model for analysing strategic and tactical decisions regarding "break crops" in agricultural rotations. In: Anderssen RS, Braddock RD, Newham LTH (eds) 18th World IMACS Congress and MODSIM09 International Congress on Modelling and Simulation. Presented at the MSSANZ and IMACS, MSSANZ and IMACS, Cairns, Australia, pp 581–587

Renton M, Savage D (2015) Statistical emulators of simulation models to inform response to new biological invasions. In: Jarrad F, Low Choy S, Mengersen K (eds) Biosecurity surveillance: quantitative approaches. CABI, Wallingford

Renton M, Lawes R, Metcalf T, Robertson M (2015) Considering long-term ecological effects on future land-use options when making tactical break-crop decisions in cropping systems. Crop Pasture Sci 66:610–621

Ronchi CP, Silva AA (2006) Effects of weed species competition on the growth of young coffee plants. Planta Daninha 24:415–423
Ryel RJ, Barnes PW, Beyschlag W, Caldwell MM, Flint SD (1990) Plant competition for light analysed with a multispecies canopy model. Oecologia 82:304–331
Saito K, Azoma K, Rodenburg J (2010) Plant characteristics associated with weed competitiveness of rice under upland and lowland conditions in West Africa. Field Crop Res 16:308–317
Shinozaki K, Kira T (1956) Intraspecific competition among higher plants. VII. Logistic theory of the C-D effect. J Inst Polytech Osaka City Univ D7:35–72
Singh M, Bhullar MS, Chauhan BS (2014) The critical period for weed control in dry-seeded rice. Crop Prot 66:80–85
Snapp SS, Swinton SM, Labarta R, Mutch D, Black JR, Leep R, Nyiraneza J, O'Neil K (2005) Evaluating cover crops for benefits, costs and performance within cropping system niches. Agron J 97:322–332
Snaydon RW (1991) Replacement or additive designs for competition studies? J Appl Ecol 28:930–946
Spitters CJT (1984) A simple simulation model for crop-weed competition. In: Proceedings 7th International Symposium on Weed Biology. Ecology and Systematics, Paris, France, pp 355–366
Spitters CJT (1989) Weeds: population dynamics, germination and competition. In: Rabbinge R, Ward SA, van Laar HH (eds) Simulation and systems management in crop protection. Simulation monographs. Pudoc, Wageningen, pp 182–216
Spitters CJT, Aerts R (1983) Simulation of competition for light and water in crop-weed associations. Asp Appl Biol 4:467–484
Spitters CJT, van Keulen H, van Kraalingen DWG (1989) A simple and universal crop growth simulator: SUCROS87. In: Rabbinge R, Ward SA, van Laar HH (eds) Simulation and systems management in crop protection. Simulation monographs. Pudoc, Wageningen, pp 147–181
Swanton CJ, Nkoa R, Blackshaw RE (2015) Experimental methods for crop–weed competition studies. Weed Sci 63:2–11. https://doi.org/10.1614/WS-D-13-00062.1
Tominaga T, Yamasue Y (2004) Crop-associated weeds. In: Inderjit (ed) Weed biology and management. Springer, Dordrecht, pp 47–63. https://doi.org/10.1007/978-94-017-0552-3_3
Torner C, González-Andújar JL, Fernández-Quintanilla C (1991) Wild oat (*Avena sterilis* L) competition with winter barley: plant density effects. Weed Res 31:301–307
Van Acker RC (2009) Weed biology serves practical weed management. Weed Res 49:1–5. https://doi.org/10.1111/j.1365-3180.2008.00656.x
Vitta JI, Satorre EH (1999) Validation of a weed: crop competition model. Weed Res 39:259–269
Walia US (2010) Weed Management, 3rd edn. Kalyani Publishers, Ludhiana. 373 pp
Watkinson AR (1980) Density dependence in single species populations of plants. J Theor Biol 83:345–357
Watkinson AR (1985) Plant responses to crowding. In: White J (ed) Studies on plant demography. Academic Press, London, pp 275–289
Watkinson AR, Freckleton RP (1997) Quantifying the impact of arbuscular mycorrhiza on plant competition. J Ecol 85:541–545
Weaver SE (1996) Simulation of crop-weed competition: models and their applications. Phytoprotection 77:3–11. https://doi.org/10.7202/706096ar
Weaver SE, Tan CS (1987) Critical period of weed interference in field-seeded tomatoes and its relation to water stress and shading. Can J Plant Sci 67:573–581
Weaver SE, Kropff MJ, Groeneveld RMW (1992) Use of ecophysiological models for crop-weed interference: the critical period of weed interference. Weed Sci 40:302–307
Weaver SE, Kropff MJ, Cousens R (1994) A simulation model of competition between winter wheat and *Avena fatua* for light. Ann Appl Biol 124:315–331
Wiles LJ, Wilkerson GG (1991) Modeling competition for light between soybean and broadleaf weeds. Agr Syst 35:37–51

Wilkerson GG, Jones JW, Coble HD, Gunsolus JL (1990) SOYWEED: a simulation model of soybean and cocklebur growth and competition. Agron J 82:1003–1010

Zhu J, van der Werf W, Anten NPR, Vos J, Evers JB (2015) The contribution of phenotypic plasticity to complementary light capture in plant mixtures. New Phytol 207:1213–1222. https://doi.org/10.1111/nph.13416

Zimdahl RL (2004) Weed-crop competition: a review. Wiley-Blackwell, Hoboken, NJ, 211 pp. https://doi.org/10.1002/9780470290224.ch9

# Chapter 7
# Site-Specific Based Models

**Cesar Fernández-Quintanilla, José Dorado, Dionisio Andújar, and J. M. Peña**

**Abstract** This chapter reviews the major conceptual approaches and specifications for the design of site-specific weed management decision support systems (SSWM-DSS), recent advances in the use of remote and ground platforms and sensors for information gathering and processing, and initial experiences translating this information into chemical and physical weed control actuations through decision algorithms and models.

**Keywords** Site-specific · Weed management · DSS design · Prescription maps · Online decisions · Sensors · Aerial images

## 7.1 Introduction

The advent of geospatial technologies (global positioning system (GPS), geographic information system (GIS)), information and communications technologies (ICT), new soil and plant sensors, and advanced agricultural machinery has opened the possibility of careful tailoring of soil and crop management to fit the different conditions found in each field. This concept has received different names: precision agriculture, precision farming, and smart agriculture. Site-specific weed management (SSWM) is the application of this concept to one particular aspect of agricultural production: weed control. Site-specific weed management is based on the fact that weed populations are commonly irregularly distributed within crop fields and it implies applying chemical and/or physical weed control measures only where and when they are really needed (Christensen et al. 2009).

Decision models for weed management can be divided into either efficacy based or population based. The efficacy-based systems assist decision-makers in choosing herbicide products and doses. Population-based models incorporate weed biology

C. Fernández-Quintanilla (✉) · J. Dorado · D. Andújar · J. M. Peña
Consejo Superior de Investigaciones Cientificas (CSIC), Instituto de Ciencias Agrarias, Madrid, Spain
e-mail: cesar@ica.csic.es; jose.dorado@csic.es; d.andujar@csic.es; jmpena@ica.csic.es


G. R. Chantre, J. L. González-Andújar (eds.), *Decision Support Systems for Weed Management*, https://doi.org/10.1007/978-3-030-44402-0_7

and ecology through simple, deterministic models (e.g. threshold models where plant densities above a specified economic threshold are controlled). The efficacy-based systems comprise large databases with herbicide performances in different crops, weed species, growth stages, etc., enabling ranking and recommendations of the most efficient product and dose against a weed mixture. In the population-based systems, the estimated yield loss or changes in the soil seed bank without weed control define the need for weed control and determine whether weed control will be cost-effective.

Although a large number of weed management decision support systems (DSS) have been developed in the past in various European countries (Rydahl et al. 2008; Parsons et al. 2009; Sønderskov et al. 2016) and in the USA (Neeser et al. 2004), all of them have ignored the spatial variation of weed populations within a field. This is a serious limitation. The use of field-scale mean density estimates in spatially heterogeneous weed populations results in underprediction of yield loss at locations where weed density is high and overprediction in parts of the field where weed densities are low or weeds are absent. Consequently, uniform herbicide application based on fixed economic weed thresholds is likely to result in under-application in some parts of the field and over-application in others.

In order to avoid this problem, it would be required to integrate site-specific information about weed species composition and density, knowledge about crop-weed competition, the effect on crop yield and quality, and the species-specific efficacies of possible control methods. The effect of soil conditions, crop husbandry, and machinery are also variables that have significant influence on weed emergence, competition, and propagation of different species and are also important for decision-making. The infinite combination of biological and agronomic variables with the range of efficacies of all possible control methods generates a need for SSWM-DSS that optimizes economic goals and meets environmental constraints.

Building SSWM-DSS needs, as a logical starting point, defining specific decisions (e.g., patch spraying) and the minimum spatial and temporal datasets needed to make those decisions, at both temporal and spatial scales—in other words, providing "the right data at the right time" (Yost et al. 2019). Different types of data may be needed for real-time decisions (e.g., for online spraying) than for generation of weed maps (e.g., for prescription maps based on historical data). Furthermore, decisions and definitions of minimum data will be different for each scale (e.g., individual plant, management zone, field, and farm).

## 7.2 Conceptual Approaches for the Design of SSWM-DSS

Several commercial and public DSS have been developed to be used in precision agriculture (Yost et al. 2019). Although they are mainly focused on N fertilization and irrigation practices, some of the basic concepts used in these systems could be applied to weed management. According to Yost et al. (2019), usable models and DSS for on-farm subfield management decisions should include:

1. Improved understanding of underlying mechanisms. Fundamental research on how incremental herbicide rates input changes affect outcomes is needed as differential responses may be accurate while absolute outcomes may not.
2. More synergy and synchrony between data generators (e.g., synergistic use of diagnostic environmental sensors and models for accurate and ongoing parameterization).
3. Better incorporation of end-user input and data. This may include translation of collected data to management parameters and decisions from grower's perspective in order to connect them to their property and feedback loop from farmer to consultant for better utilization of models to ensure farmer observations are incorporated.

The first decision algorithm specifically designed for patch spraying (DAPS) was developed by Christensen et al. (2003). The main components of DAPS were:

1. A crop-weed competition model that estimates yield loss as a function of weed species and their densities.
2. A model that estimates yield gain and net return as a function of herbicide dose and weed species composition and responses.
3. An algorithm that finds the economically optimal herbicide dose of a given weed mixture.

This model, in combination with an appropriate sensor system for weed recognition and classification as well as an improved application technology, allows variable rates and herbicide mixtures in real time. The results of a 5-year field experiment designed to assess this model under field conditions showed that optimization of the dosage to the local weed species composition and densities every year reduced herbicide usage 45–67% without reducing the crop yield or increasing the density of the weed population.

Lamastus-Stanford and Shaw (2004) adapted HADSS, a computerized yield loss and post-emergence herbicide selection decision aid, to the variable weed populations present in various soybean fields. To determine the effect of an SSWM program on the net returns and amount of herbicide applied to the fields, weed populations for each sample location within each field were subjected to HADSS, using various sampling scales (10 × 10 m, 40 × 40 m, 60 × 60 m, and 80 × 80 m). These researchers demonstrated the potential value of SSWM from an economic standpoint. The differences between projected net returns from SSWM and the broadcast applications ranged from \$10.42 to \$14.1 ha$^{-1}$. Results from larger, less-intense sampling scales were not significantly different.

Gutjar and Gerhards (2010) developed HPS-ONLINE, another decision model developed for site-specific herbicide application. This model can be divided into two parts:

1. Knowledge before application. Although HPS-ONLINE offers the possibility for the implementation of population dynamics aspects for weed control, in this program, the model's user is able to choose the importance of this aspect himself. If the user is planning to use long crop rotation intervals or narrow row

spacings, the population dynamics aspects may be disregarded. Regarding weather conditions, if the user considers that conditions are optimal for herbicide application, HPS-ONLINE offers the possibility for a general reduction of herbicide dose. This implementation of users' expert knowledge and experience is a valuable component of this system.

2. Knowledge during application. The system uses the weed coverage at the time of weed control for the estimation of weed competition. This data is obtained automatically using a bispectral sensor system. Plant species are classified in four categories (crop and three weed types) by their shape features using an automatic classifier. Based on this information, a separate application decision is made for each weed class.

Fernandez-Quintanilla et al. (2011) proposed the basic specifications for the design of a DSS based on spatiotemporal information on weed infestations. This proposal, designed to be used by a fleet of patch-spraying robots, was structured in five modules (Fig. 7.1):

1. Field inspection. Monitoring weed populations at various times of the growing season by using unmanned aerial vehicles (UAV).
2. Long-term decisions. The objective of this module is to optimize the choice of crop and herbicide rotations as well as the tillage system throughout a rotation defined by the user, trying to find the best long-term strategy. Historic (legacy)

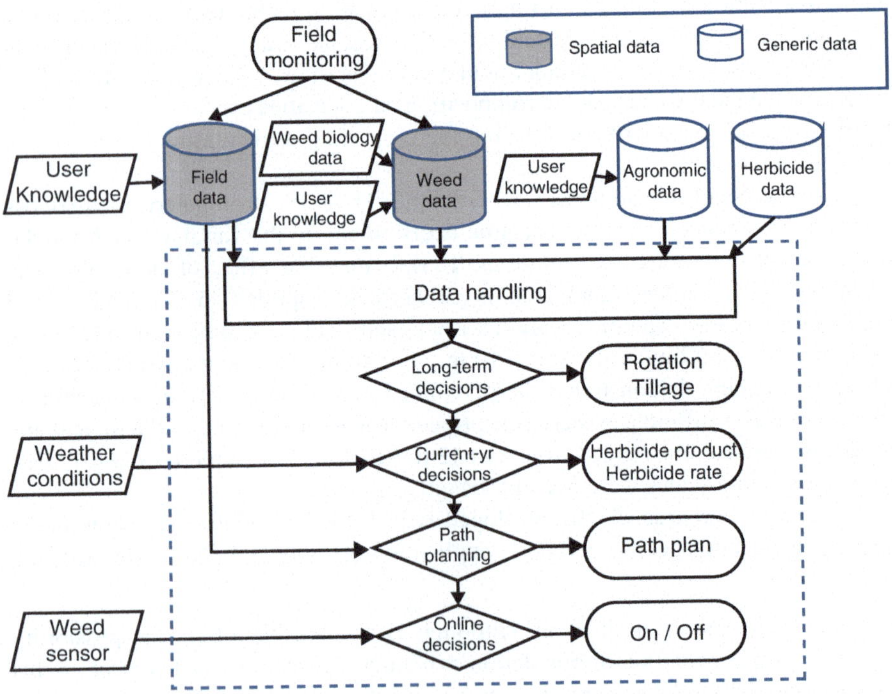


**Fig. 7.1** Proposed system architecture for a SSWM-DSS for a fleet of robots

maps, constructed using the weed distribution data obtained in the previous step, may provide one major element for this process. In addition, empirical knowledge of the farmer, gathered through many years of working the field, may also be a valuable data source that should be exploited.
3. Current year decisions. This is a complex decision requiring the integration of information on weed biology, expected crop yields, and potential yield losses caused by different weeds, herbicide options, timing and efficacy of each herbicide, influence of climatic conditions, economic profitability of the treatment, and herbicide resistance risks. Nowadays, agricultural growers and consultants can manage the integration of these complex factors by using available DSS (Parsons et al. 2009; Sønderskov et al. 2016).
4. Unit distribution and path planning. The objective of this module is to plan the routes to be followed by each individual robotic unit, taking into consideration the geometry of the field, the strategy of the operation, and the spatial distribution of weeds.
5. Online decisions. Although prescription maps may provide the basic information of the field areas that should be sprayed, this information needs to be contrasted with that obtained at spraying time with cameras or sensors that detect weed presence and discriminate different weed types. Once the detected weed patch has been considered as a suitable target for spraying, a fast-response controller could regulate discharge of the different herbicides in each individual nozzle.

Decisions derived from DSS should not be considered as compulsory operations to be conducted by the farmer but as recommendations to be considered in his final, personal decision. In any case, it is desirable to assess the results obtained from the application of those final decisions. This assessment should take into consideration agronomic, economic, and environmental criteria. In order to do that, it would be required to collect data at the field scale (e.g., total herbicide use, total yield, etc.) and at the subfield scale (e.g., yield mapping, weed mapping at harvest time, etc.).

## 7.3 Using Remote-Sensed Images to Construct Prescription Maps

Remote sensing technology can provide spatial and temporal information on the presence and absence of weeds at the field scale as key input data to feed a DSS. The primary phase consists of classifying the remote-sensed images into a weed map that shows the weed plants or the weed patches. Prior to the widespread use of drones, remote images taken with piloted aircrafts or satellite platforms allowed to detect large weed patches (at least 2 m by 10 m in size) with a low resolution. This fact limited the potential implementation of this technology to late-season inspection of crop fields with large weed patches (Brown and Noble 2005). Early studies used high-altitude images for late-season weed discrimination in cereal and legume crops, soybean and sunflower (Gómez-Candón et al. 2012a; de Castro et al. 2012,

2013; Gray et al. 2008; Peña-Barragán et al. 2007). However, although weed control treatments at a late crop stage are generally inappropriate or ineffective, these weed maps may be useful for weed control in the following year if weed patches are spatially stable over time, such as those of *Avena sterilis* (Barroso et al. 2004), *Ridolfia segetum* (Peña-Barragán et al. 2007), and *Alopecurus myosuroides* (Lambert et al. 2017). These late-season maps may also have other uses, e.g., relating weed infestations and crop yields (Gutiérrez et al. 2008; Peña-Barragán et al. 2010), assessing the impact or efficiency of specific weed control treatments (Franco et al. 2017;. Huang et al. 2018; Rasmussen et al. 2013), and studying weed population dynamics (Castillejo-González et al. 2019). Spatial resolution of the remote images (i.e., pixel size) is a key parameter to the success of weed detection and mapping. Hengl (2006) considered that pixel size should be at least a quarter of the target element (in our case, the weed plant or patch). Currently, drones generally provide remote images of a few centimeters and, in some cases, of less than 1 mm. This feature allows weed detection at very early stages of the crop and weeds, the critical period for weed treatments (Pflanz et al. 2018; Torres-Sánchez et al. 2013). Several investigations have demonstrated the capability of early-season drone-based images to detect weeds in maize (Castaldi et al. 2017; Gao et al. 2018; Peña et al. 2013; Pérez-Ortiz et al. 2016), sunflower (López-Granados et al. 2016; Pérez-Ortiz et al. 2015, 2016), barley (Franco et al. 2017; Rasmussen et al. 2013), wheat (Jurado-Expósito et al. 2019; Pflanz et al. 2018), vineyards (Jiménez-Brenes et al. 2019), and rice (Huang et al. 2018). High-resolution weed maps can be generated by combining advanced object-based image analysis (OBIA) techniques and machine learning algorithms (de Castro et al. 2018; Gao et al. 2018; Pérez-Ortiz et al. 2016; Pflanz et al. 2018) (Fig. 7.2). These procedures usually incorporate spatial and spectral information of each plant previously segmented as objects within the images. Classification algorithms are usually more effective in row crops, where the position of the weeds relative to the crop row is generally a decisive factor (Louargant et al. 2018; Peña et al. 2013).

In order to practice SSWM, a subsequent phase consists of converting weed maps to prescription maps. These maps are a set of grids with the corresponding weed infestation values (weed coverage or weed density). The prescription maps also provide additional information on the crop field and crop development. This information is important to design and apply DSS. Peña et al. (2013) developed an OBIA procedure based on a drone-based weed map in maize. This procedure consists of three levels of information according to the spatial scale of field observation. The upper level provides global information of the crop field, including field dimensions, number of crop rows, crop row orientation, average crop row separation, and total weed-free and weed-infested areas, including total area of three different categories of weed coverage (low, moderate, and high). The intermediate level provides detailed information on each crop row, including identification number, length, width, coordinates of the extremes, and number and category of the weed-infested grids of each row. Finally, the lower level provides detailed information on each grid unit, including identification number, coordinates, dimensions, relative position within the crop row, distance to the start and the end of the crop row, weed

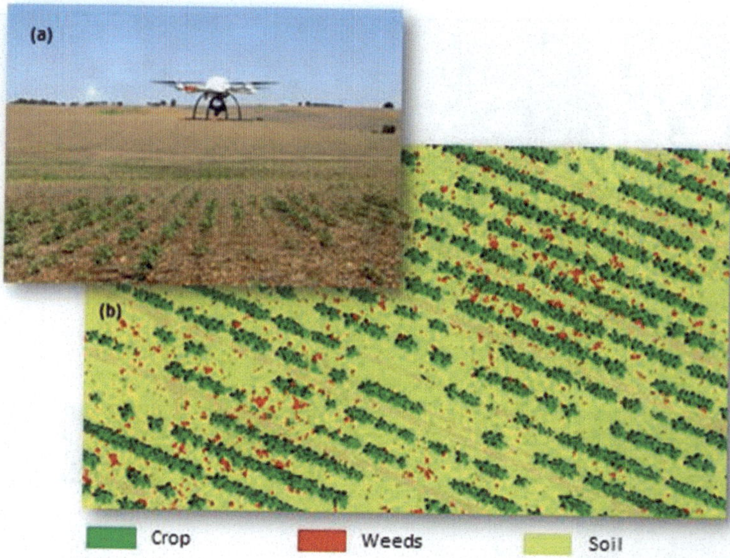


**Fig. 7.2** (**a**) Using an Unmanned Aerial Vehicle (UAV) flying at 30 m altitude over a sunflower field to detect weed infestations. (**b**) Classified image by applying an auto-trained Random Forest classifier Object-Based Image Analysis (RF-OBIA) algorithm (de Castro et al. 2018)

coverage percentage, and weed coverage category of each grid. For the development of a DSS, this dataset at the three levels enables to estimate the total area that needs weed treatment and, therefore, calculate the herbicide volume needed prior to its application, as well as the location of the weed-infested grids, the orientation of the crop rows, and the general pattern of the crop field, which is fundamental to plan treatment routes and optimize the tractor trajectory (Gonzalez-de-Santos et al. 2016).

Generally, the prescription maps are exported to the sprayer to conduct the treatment according to the position of each grid and the weed-treatment decision (e.g., spray or do not spray) following an SSWM strategy. The ultimate objective is to decrease the amount of herbicide in comparison to a uniform weed treatment. In their study, Peña et al. (2013) determined that the area free of weeds and with low weed coverage (<5% weeds) was 23% and 47%, respectively, which demonstrated the high potential for reducing herbicide applications in this case study. Castaldi et al. (2017), working also in maize crops, designed prescription maps of 2 m × 2 m in size from drone images and reported herbicide savings between 14 and 39.2% for patch spraying as compared to a blanket application.

However, overall herbicide savings would vary depending on the level of weed infestation, the criteria established for herbicide application, and the size of the spray grid considered. Gómez-Candón et al. (2012a, b) evaluated the impact of these variables on herbicide savings by using aerial images for mapping *Avena sterilis* in wheat fields and reported that the herbicide savings increased from 20% to 90% for weed treatment thresholds of 0% and 30% of weed coverage, respectively

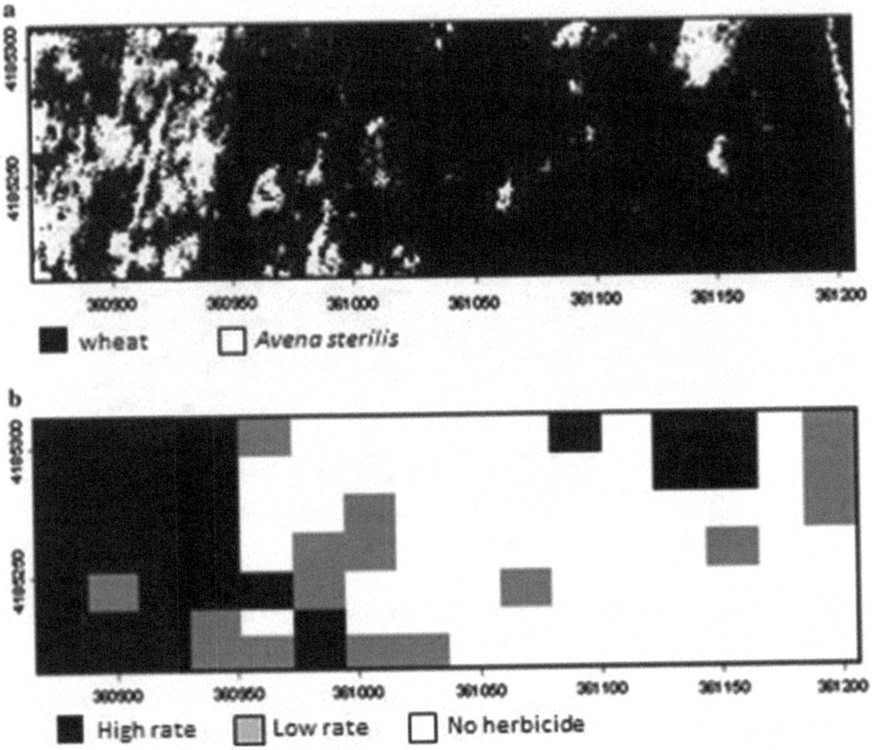


**Fig. 7.3** Using aerial images to design herbicide application: (**a**) NVDI image view of a winter wheat field infested with *Avena sterilis*. Original RGB images obtained from a plane flying at 1500 m altitude; (**b**) herbicide prescription map with three classes: high rates when >26% infested pixels, low rates when 11 to 26% infested pixels, and no herbicide when <11% infested pixels (Gómez-Candón et al. 2012a, b)

(Fig. 7.3). These authors also indicated that treatment efficiency increases three times as much if the grid size is reduced from 20 m × 6 m (i.e., for the entire sprayer platform) to 1.2 m × 1.5 m (i.e., for individual nozzles). López-Granados et al. (2016), using images from two sunflower fields, studied the variability of herbicide treatment maps generated from UAV images using weed thresholds ranging from 0% to 15%. The results obtained showed that the total area of weed treatment decreased from a maximum area of 46% at a weed threshold of 0% to a minimum area of 3% at a weed threshold of 15%. This study also evaluated the impact of flight altitude (i.e., image spatial resolution) and type of camera (i.e., RGB vs. multispectral) in the accuracy of the prescription maps as affected by the studied weed thresholds. The results showed that the multispectral camera was better than the RGB camera in all cases and that reasonable accuracy (i.e., overall accuracy >85%) was obtained from weed thresholds above 2.5–5% on average with both cameras. Huang et al. (2018), using UAV-based images on rice fields for the generation of prescription maps, also quantified high herbicide savings depending on the treatment threshold used, reporting savings between 58 and 71% for weed coverage thresholds of 0 to 25%, respectively.

## 7.4 Using Ground Sensing Imagery to Make Online Decisions

Farmers are more likely to adopt embodied technologies that do not require acquiring additional skills (e.g., smart online sprayers) than information intensive technologies that require special skills (e.g., weed mapping from aerial images) (Lutman and Miller 2007; Griffin 2016). The data required to take online decisions can be obtained from different types of sensors or cameras.

The simplest devices are spectral reflectance sensors. This technology, developed originally by Felton and McCloy (1992), is based on the fact that the spectral curve of plants differs significantly from the reflectance of soil. Consequently, if these sensors are located in crop-free areas (e.g., inter-rows, tramlines, fallow land), the decision to be made is simple and can be taken in milliseconds: all green detected objects are supposed to be weeds and should be sprayed. This principle has been widely used in the past for real-time patch spraying of herbicides (Felton and Mccloy 1992; Dammer and Wartenberg 2007; Dammer 2016). However, since crops and weeds cannot be discriminated with these sensors, this approach has important limitations.

Weed plants can be discriminated from other elements in the image (soil, crop) by using three sequential processes (Fernandez-Quintanilla et al. 2018):

1. Segmentation of the original image, obtaining an image with white pixels representing plant cover and black pixels depicting soil.
2. Identification of zones corresponding to crop rows, quantifying crop cover, and eliminating these pixels.
3. Estimation of weed cover after the improvement of image by filtering noise and errors from previous steps.

Up to now, the most widely used technique for crop-weed discrimination is based on imaging with sensitive sensors within the range of the visible light (Peteinatos et al. 2014). Relatively low-cost and easily operated RGB cameras can acquire images with proper spatial resolution to allow the identification of plant species based on their location, shape, color, and texture features. Using expensive hyperspectral images may result in substantial improvements in this process. The integration of these images with a machine-learning procedure can achieve high recognition levels for crop vs. weed discrimination (Zhang and Slaughter 2011; Zhang et al. 2012a).

In order to discriminate individual plant species, it is possible to create image databases for all the species of interest, using their spectral signatures and/or their characteristic shape features (Gerhards and Oebel 2006; Weis and Gerhards 2007; López-Granados et al. 2008). Berge et al. (2008) used an object-oriented algorithm ("WeedFinder") for the automatic detection of broad-leaved weeds in cereals. The results obtained in two experimental fields show that the estimates by the program and the corresponding true values of total broad-leaved weed density and weed cover were positively correlated, but there were serious dispersion and discrepancy from the 1:1 relationship (Fig. 7.4).

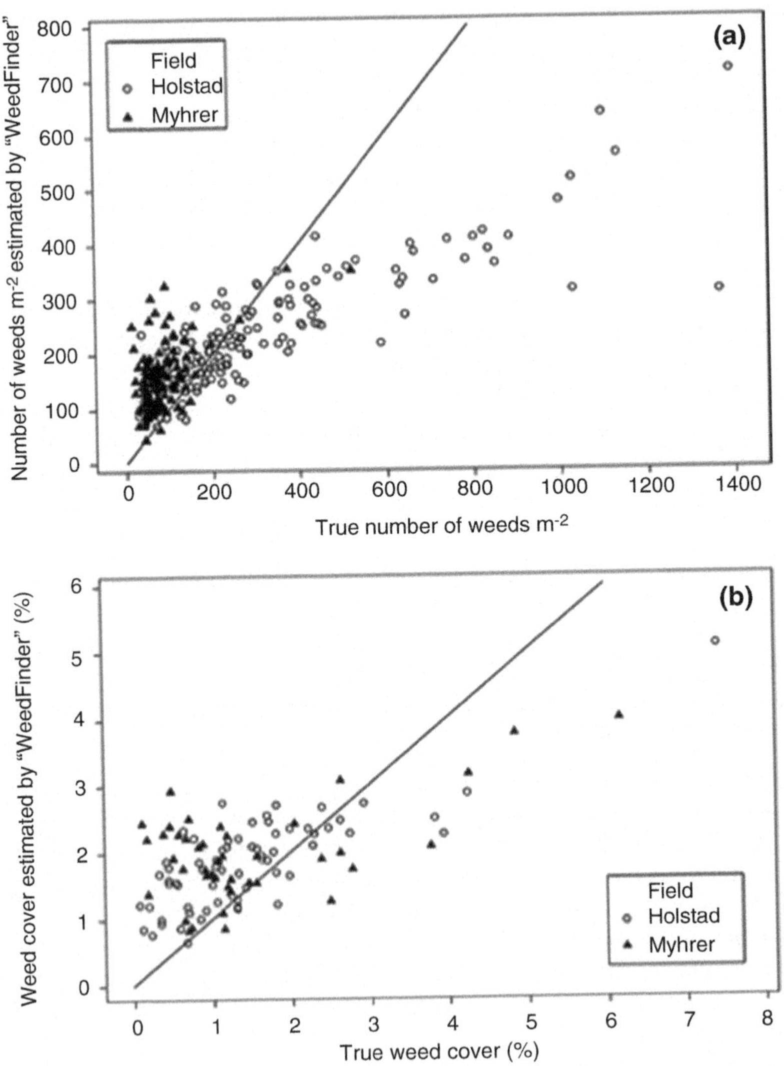


**Fig. 7.4** True values versus "WeedFinder" estimated values of (**a**) total broad-leaved weed density and (**b**) total broad-leaved weed cover in two wheat fields (Myhrer and Holstad) (Berge et al. 2008)

3D modeling has been recently proposed for the morphological characterization of weed plants (Andújar et al. 2018). These techniques can be based on visible images (Arvidsson et al. 2011), LiDAR (Guo et al. 2017), structured light (Nguyen et al. 2015), spectroscopy (Gutierrez et al. 2016), thermal images (Ludovisi et al. 2017), ultrasound (Andújar et al. 2011), etc.

In order to practice SSWM, in addition to the weed sensing system, it is necessary to implement a weed management DSS and have a precision weed control device, such as a boom sprayer having independent boom sections or nozzles or a precise physical weed removal actuator (Christensen et al. 2009; Fennimore et al. 2016).

According to a relatively simple approach, weed control actuator, either chemical or physical, will be started automatically when estimated total weed cover in a given area is higher than experimentally determined thresholds. However, and due to the relatively long processing time of all these processes, this approach has some limitations to be used for online actuation. In addition, some aspects such as leaf overlapping or plant biomass quantification still need further research.

The most powerful method capable of robust, automated in-field discrimination of individual plant species is based upon hyperspectral imaging. This principle has been used in various horticultural crops for automated weed control (Zhang et al. 2012b; Fennimore et al. 2016). To translate weed maps into spray control maps, the predominant object classification (weed, crop, soil) in each region of the hyperspectral image determined the spray decision for that zone.

Currently, the use of artificial intelligence models is replacing some of the last decade developments. The concept is wide and many machine-learning processes can provide novel tools for online weed identification through image processing (Liakos et al. 2018; Yu et al. 2019). These fast analysis algorithms can be used in conjunction with smart sprayers to accurately apply herbicides site-specifically. Various commercial initiatives using leading-edge hardware, software, and artificial intelligence have already yielded equipment capable to detect weeds, decide the action to be taken, and act immediately. Blue River Technology, a Silicon Valley startup, used computer vision and machine learning to identify plant species—both crops and weeds—with a high accuracy and then artificial intelligence algorithms to make spraying decisions on the spot.

Custom nozzle designs enabled spraying individual plants. EcoRobotix, a small Swiss company, has designed, constructed, and commercialized a small robotic unit equipped with a camera for weed recognition, a powerful computer with data processing using artificial intelligence algorithms, and a set of movable nozzles that deposit herbicide microdoses in a targeted way. Deepfield Robotics, a Bosch startup, has developed an autonomous weeding machine equipped with a weed detection camera, a machine learning system, and a mechanical weeding mechanism that destroy individual plants. Bosch has recently teamed up with Bayer, the giant German chemical firm, for a "smart spraying" research project using the detection and decision-making technologies developed for BoniRob.

## References

Andújar D, Escola A, Dorado J et al (2011) Weed discrimination using ultrasonic sensors. Weed Res 51:543–547

Andújar D, Calle M, Fernández-Quintanilla C et al (2018) Three-dimensional modeling of weed plants using low-cost photogrammetry. Sensors 18:1077. https://doi.org/10.3390/s18041077

Arvidsson S, Perez-Rodriguez P, Mueller-Roeber B (2011) A growth phenotyping pipeline for Arabidopsis thaliana integrating image analysis and rosette area modeling for robust quantification of genotype effects. New Phytol 191:895–907

Barroso J, Fernàndez-Quintanilla C, Ruiz D et al (2004) Spatial stability of Avena sterilis ssp. ludoviciana populations under annual applications of low rates of imazamethabenz. Weed Res 44:178–186

Berge T, Aastveit A, Fykse H (2008) Evaluation of an algorithm for automatic detection of broad-leaved weeds in spring cereals. Precis Agric 9:391–405

Brown RB, Noble SD (2005) Site-specific weed management: sensing requirements—what do we need to see? Weed Sci 53:252–258

Castaldi PF, Pascucci S et al (2017) Assessing the potential of images from unmanned aerial vehicles (UAV) to support herbicide patch spraying in maize. Precis Agric 18:76–94

Castillejo-González IL, de Castro AI, Jurado-Expósito M et al (2019) Assessment of the persistence of Avena sterilis l. patches in wheat fields for site-specific sustainable management. Agronomy 9:30

Christensen S, Heisel T, Walter AM et al (2003) A decision algorithm for patch spraying. Weed Res 43:276–284

Christensen S, Sogaard HT, Kudsk P et al (2009) Site-specific weed control technologies. Weed Res 49:233–241

de Castro AI, Jurado-Expósito M, Peña-Barragán JM et al (2012) Airborne multi-spectral imagery for mapping cruciferous weeds in cereal and legume crops. Precis Agric 13:302–321

de Castro AI, López Granados F, Gómez-Candón D et al (2013) In-season site-specific control of cruciferous weeds at broad-scale using quickbird imagery. in: 9th European Conference on Precision Agriculture (9th ECPA). Lleida, Spain, Universidad de Lleida

de Castro AI, Torres-Sánchez J, Peña JM et al (2018) An automatic random forest-OBIA algorithm for early weed mapping between and within crop rows using UAV imagery. Remote Sens 10:285

Dammer KH (2016) Real-time variable-rate herbicide application for weed control in carrots. Weed Res 56:237–246

Dammer KH, Wartenberg G (2007) Sensor-based weed detection and application of variable herbicide rates in real time. Crop Prot 26:270–277

Felton WL, Mccloy KR (1992) Spot spraying. Agric Eng 73:9–12

Fennimore SA, Slaughter DC, Siemens MC et al (2016) Technology for automation of weed control in speciality crops. Weed Technol 30:823–837

Fernandez-Quintanilla C, Dorado J, San Martin C et al (2011) A five-step approach for planning a robotic site-specific weed management program for winter wheat. In: Gonzalez-de-Santos P, Rabatel G (eds) RHEA-2011. Robotics and associated high-technologies and equipment for agriculture, Montpellier, pp 3–12

Fernandez-Quintanilla C, Dorado J, Andujar D et al (2018) Is the current state of the art of weed monitoring suitable for site-specific weed management in arable crops? Weed Res 58:259–272

Franco C, Pedersen SM, Papaharalampos H et al (2017) The value of precision for image-based decision support in weed management. Precis Agric 18:366–382

Gao J, Liao W, Nuyttens D et al (2018) Fusion of pixel and object-based features for weed mapping using unmanned aerial vehicle imagery. Int J Appl Earth Obs Geoinform 67:43–53

Gerhards R, Oebel H (2006) Practical experiences with a system for site-specific weed control in arable crops using real-time image analysis and GPS-controlled patch spraying. Weed Res 46:185–193

Gómez-Candón D, López-Granados F, Caballero-Novella J et al (2012a) Sectioning remote imagery for characterization of Avena sterilis infestations. Part A: Weed abundance. Precis Agric 13:322–336

Gómez-Candón D, López-Granados F, Caballero-Novella J et al (2012b) Sectioning remote imagery for characterization of Avena sterilis infestations. Part B: Efficiency and economics of control. Precis Agric 13:337–350

Gonzalez-de-Santos P, Ribeiro A, Fernandez-Quintanilla C et al (2016) Fleets of robots for environmentally-safe pest control in agriculture. Precis Agric:1–41

Gray CJ, Shaw DR, Gerard D et al (2008) Utility of multispectral imagery for soybean and weed species differentiation. Weed Technol 22:713–718

Griffin TW (2016) Adoption of precision agricultural technology in Kansas. Kansas State University. Department of Agricultural Economics Extension. Publication. KFMA Research Article KSU-AgEcon-TG-2016. https://www.agmanager.info/adoption-precision-agricultural-technology-kansas. Accessed 21 Jan 2017

Guo Q, Wu F, Pang S et al (2017) Crop 3D-a LiDAR Based Platform For 3D High-Throughput Crop Phenotyping. Sci China Life Sci

Gutiérrez PA, López-Granados F, Peña-Barragán JM et al (2008) Mapping sunflower yield as affected by Ridolfia segetum patches and elevation by applying evolutionary product unit neural networks to remote sensed data. Comput Elect Agric 60:122–132

Gutiérrez S, Tardaguila J, Fernández-Novales J, Diago MP (2016) Data Mining and NIR Spectroscopy in Viticulture: Applications for Plant Phenotyping under Field Conditions. Sensors 16:236

Gutjar C, Gerhards R (2010) Decision rules for site-specific weed management. In: Oerke EC et al (eds) Precision crop protection—the challenge and use of heterogeneity. Springer Science+Business Media, Berlin, pp 223–239

Hengl T (2006) Finding the right pixel size. Comput Geosci 32:1283–1298

Huang H, Deng J, Lan Y et al (2018) Accurate weed mapping and prescription map generation based on fully convolutional networks using UAV imagery. Sensors 18:3299. https://doi.org/10.3390/s18103299

Jiménez-Brenes FM, López-Granados F, Torres-Sánchez J et al (2019) Automatic UAV-based detection of Cynodon dactylon for site-specific vineyard management. PLoS One 14:e0218132. https://doi.org/10.1371/journal.pone.0218132

Jurado-Expósito M, de Castro AI, Torres-Sánchez J, Jiménez-Brenes FM, López-Granados F (2019) Papaver rhoeas L. mapping with cokriging using UAV imagery. Precis Agric. https://doi.org/10.1007/s11119-019-09635-z

Lamastus-Stanford FE, Shaw DR (2004) Evaluation of site-specific weed management implementing the herbicide application decision support system (HADSS). Precis Agric 5:411–426

Lambert JPT, Hicks HL, Childs DZ et al (2017) Evaluating the potential of unmanned aerial systems for mapping weeds at field scales: a case study with Alopecurus myosuroides. Weed Res 58(1):35–45. https://doi.org/10.1111/wre.12275

Liakos K, Busato P, Moshou D et al (2018) Machine learning in agriculture: a review. Sensors 18:2674

López-Granados F, Peña-Barragan JM, Jurado-Exposito M et al (2008) Multispectral classification of grass weeds and wheat (Triticum durum) using linear and nonparametric functional discriminant analysis and neural networks. Weed Res 48:28–37

López-Granados F, Torres-Sánchez J, Serrano-Pérez A et al (2016) Early season weed mapping in sunflower using UAV technology: variability of herbicide treatment maps against weed thresholds. Precis Agric 17:183–199

Louargant M, Jones G, Faroux R (2018) Unsupervised classification algorithm for early weed detection in row-crops by combining spatial and spectral information. Remote Sens 10:761. https://doi.org/10.3390/rs10050761

Ludovisi R, Tauro F, Salvati R et al (2017) UAV-based thermal imaging for high-throughput field phenotyping of black poplar response to drought. Front Plant Sci 8:1681. https://doi.org/10.3389/fpls.2017.01681

Lutman PJW, Miller PCH (2007) Spatially variable herbicide application technology; opportunities for herbicide minimisation and protection of beneficial weeds. Research Review No. 62, Home-Grown Cereals Authority (HGCA), UK

Neeser C, Dille JA, Krishnan G et al (2004) WeedSOFT®: a weed management decision support system. Weed Sci 52:115–122

Nguyen T, Slaughter D, Max N et al (2015) Structured Light-Based 3D Reconstruction System for Plants. Sensors (Basel, Switzerland). 15:18587–612. https://doi.org/10.3390/s150818587

Parsons DJ, Benjamin L, Clarke J et al (2009) Weed manager—a model-based decision support system for weed management in arable crops. Comput Elect Agric 65:155–167

Peña JM, Torres-Sánchez J, de Castro AI et al (2013) Weed mapping in early-season maize fields using object-based analysis of unmanned aerial vehicle (UAV) images. PLoS One 8:e77151. https://doi.org/10.1371/journal.pone.0077151

Peña-Barragán JM, López-Granados F, Jurado-Expósito M et al (2007) Mapping Ridolfia segetum patches in sunflower crop using remote sensing. Weed Res 47:164–172

Peña-Barragán JM, López-Granados F, Jurado-Expósito M et al (2010) Sunflower yield related to multi-temporal aerial photography, land elevation and weed infestation. Precis Agric 11:568–585

Pérez-Ortiz M, Peña JM, Gutiérrez PA et al (2015) A semi-supervised system for weed mapping in sunflower crops using unmanned aerial vehicles and a crop row detection method. Appl Soft Comput 37:533–544

Pérez-Ortiz M, Peña JM, Gutiérrez PA et al (2016) Selecting patterns and features for between- and within- crop-row weed mapping using UAV-imagery. Expert Syst Appl 47:85–94

Peteinatos G, Weis M, Andujar D et al (2014) Potential use of ground-based sensor technologies for weed detection. Pest Manag Sci 70:190–199

Pflanz M, Nordmeyer H, Schirrmann M (2018) Weed mapping with UAS imagery and a bag of visual words based image classifier. Remote Sens 10:1530. https://doi.org/10.3390/rs10101530

Rasmussen J, Nielsen J, Garcia-Ruiz F (2013) Potential uses of small unmanned aircraft systems (UAS) in weed research. Weed Res 53:242–248

Rydahl P, Berti A, Munier-Jolain N (2008) 24-Decision support systems (DSS) for weed control in Europe–state-of-the-art and identification of 'best parts' for unification on a European level. https://www.researchgate.net/publication/228431432_O_24. Accessed 7 Feb 2008

Sønderskov M, Rydahl P, Bøjer OM et al (2016) Crop protection online—weeds: a case study for agricultural decision support systems. In: Papathanasiou J, Ploskas N, Linden I (eds) Real-world decision support systems. Integrated Series in Information Systems, Springer International, Berlin, 37:303–320

Torres-Sánchez J, López-Granados F, de Castro AI et al (2013) Configuration and specifications of an unmanned aerial vehicle (UAV) for early site specific weed management. PLoS One 8:e58210. https://doi.org/10.1371/journal.pone.0058210

Weis M, Gerhards R (2007) Feature extraction for the identification of weed species in digital images for the purpose of site-specific weed control. In: Stafford JV (ed) Precision agriculture. Wageningen Academic, Wageningen, pp 537–545

Yost MA, Sudduth KA, Walthall CL et al (2019) Public–private collaboration toward research, education and innovation opportunities in precision agriculture. Precis Agric 20:4–18

Yu J, Sharpe SM, Schumann AW et al (2019) Deep learning for image-based weed detection in turfgrass. Eur J Agron 104:78–84

Zhang Y, Slaughter DC (2011) Hyperspectral species mapping for automatic weed control in tomato under thermal environmental stress. Comput Elect Agric 77:95–104

Zhang Y, Slaughter DC, Staab ES (2012a) Robust hyperspectral vision-based classification for multi-season weed mapping. J Photogram Remote Sens 69:65–73

Zhang Y, Staab ES, Slaughter DC et al (2012b) Automated weed control in organic row crops using hyperspectral species identification and thermal micro-dosing. Crop Prot 41:96–105

# Part III
# Environmental Risk Modelling

# Chapter 8
# Theory and Practice for Environmental Risk Assessment: Understanding the Trade-Off Between the Benefits and Risks Behind Herbicide Use as Tool for Designing Sustainable Weed Management Systems

**Felipe Ghersa, Sebastián Pessah, Alejandra C. Duarte Vera, and Diego O. Ferraro**

**Abstract** The adoption of herbicides as a weed control strategy has allowed farmers to reduce the short-term effects of biological adversities on crop yields. However, they have also jeopardized agroecosystem sustainability by causing negative alterations of social and environmental subsystems. The physicochemical properties of herbicides (volatility, adsorption, or water solubility) can make them persist in the soil, air, and water, changing the structure and function of key environmental compartments. The occurrence of herbicide-resistant weed populations has generated a positive feedback loop requiring the application of higher doses, aggravating negative externalities. Hence, the economic benefits of herbicides as a unique control strategy substantially decrease in time. In addition, the dependence of agricultural systems on external inputs generates an herbicidal "lock-in" process that hinders the transition towards more sustainable integrated management systems. Therefore, there is a pressing need to elucidate the principal aspects of environmental risk analysis of herbicide use in agroecosystems. The objectives of this chapter are: (1) to introduce key concepts related to the construction and application of environmental risk indicators with a focus on agricultural system risk assessment, (2) to list the potentially negative effects associated with the use of herbicides, (3) to understand the processes that regulate herbicides' fate and behavior in farming systems, (4) to highlight the importance of decision support systems (DSS) in reducing herbicide use in favor of integrated weed management (IWM), and (5) to understand the

F. Ghersa · S. Pessah · A. C. D. Vera · D. O. Ferraro (✉)
Departamento de Producción Vegetal, Cátedra de Cerealicultura, Facultad de Agronomía, CONICET, Instituto de Investigaciones Fisiológicas y Ecológicas Vinculadas a la Agricultura (IFEVA), Universidad de Buenos Aires (UBA), Buenos Aires, Argentina
e-mail: felipeghersa@agro.uba.ar; spessah@agro.uba.ar; vduarte@agro.uba.ar; ferraro@agro.uba.ar


G. R. Chantre, J. L. González-Andújar (eds.), *Decision Support Systems for Weed Management*, https://doi.org/10.1007/978-3-030-44402-0_8

decision-making logic behind the increasing adoption of chemical weed control despite its negative socio-environmental effects.

**Keywords** Risk modelling · Technology adoption · Pesticides · Ecotoxicity · Indicators · Decision making

## 8.1 Introduction

The growing demands for food and fiber products has brought about a substantial increase in the use of nonrenewable resources, such as soil and fossil fuels, to close the gap between attainable and actual yield levels (Van Ittersum and Rabbinge 1997). However, the high dependence on external inputs, due to the simplification of social and environmental subsystems, has altered the possibility of sustaining modern industrial agricultural systems in time (Hansen 1996; Pretty 2007). One of the principle determinants of this productive gap is the presence of biological adversities such as weeds, which have led to increasing use of chemical products.

The term "weed" is generally centered around the anthropocentric viewpoint of any plant species that interferes with productive activities and human well-being (Radosevich et al. 2007; Neve et al. 2009). There are no fixed biological characteristics that allow the identification of a particular plant as a weed, although some ideal traits have been identified (Baker 1974). Therefore, those plants that are considered weedy, aside from having no particular use, generally have negative effects on crop yields and crop product quality (Gibson et al. 2008). Through control strategies (chemical or mechanical), weed populations can be reduced or suppressed within a defined area; however, this seldom implies their complete eradication. The fact that weed species populations increase in number, and expand geographically, greatly influences agricultural decision-makers on the amount of resources they spend to inhibit their occurrence (Gould et al. 2018).

Despite the increase in herbicide use, weeds are still present as a productive adversity because of an increase in the abundance of individuals that elude herbicide action or that present evolutionary mechanisms towards the establishment of herbicide-resistant populations (Jasieniuk et al. 1996; Vidal et al. 2010). These processes generate positive feedback loops (Chapin III et al. 1996) that imply a burgeoning dependence on chemical applications in order to control weeds, with a consequent increase in weed resistance followed by a cyclic need to apply higher herbicide doses. Therefore, decisions on how to control weeds are principally influenced by species biology and the particular technological package applied (Radosevich et al. 1997). It becomes evident that these management strategies can influence agroecosystems' sustainability by affecting their principal components—i.e., water, soil, and atmosphere. Some of these effects include emissions of herbicides and their byproducts into the environment, decreases in native species populations, and negative impacts in animal welfare (Pimentel et al. 1992). In addition, these externalities generate additional costs, which producers do not usually include in their gross margin estimations (Wilson and Tisdell 2001). In order to quantify the effects of cultural practices on agroecosystem services (Costanza et al. 2014), both environmental risk indicators (ERI) and decision support systems (DSS) can be used. These provide a knowledge base regarding how technologies affect systems' properties, sustainability, and resource use (Ferraro et al. 2003), and allow for transitions from damaging positive feedback cycles to more stable and desirable states with lower environmental and social impacts (Swift and Anderson 1994).

In addition to these ecological considerations, social factors are a key determinant in appropriate design and implementation of integrated weed management (IWM) strategies. This requires a thorough understanding of farmers' behavioral patterns by analyzing their decision-making process using tools that capture the associated complexity and heterogeneity and integrates them with the ecological aspects of the agricultural system practices. Multi-agent models are a possible way to represent human behavior and link it with the physical functions of productive systems (Schreinemachers and Berger 2011). In these models the interaction between agents and their environment determines the possible functional and structural configurations that a system can take. In addition, multi-agent models could be coupled with optimization models or algorithms allowing more flexibility to decision-making by increasing the scope and precision of inputs used (Whittaker et al. 2017). This could allow more informed fact-based decisions that may be counter to established cultural knowledge regarding optimal presence of weed populations and trade-offs between increasing chemical control and accruing environmental and economic costs (Rossi et al. 2014).

Based on the previously depicted benefit-risk trade-off between herbicide use and weed control the objectives of this chapter are to: (1) introduce some key concepts related to the agricultural systems environmental risk assessment (ERA), (2) list the potentially negative effects associated with the use of herbicides, (3) clarify the endogenous and exogenous aspects of the herbicides that regulate their fate in the environment, (4) provide evidence for understanding the logic of increasing adoption of herbicides, despite their negative socio-environmental effects, and (5) review the role of DSS as useful tools in promoting better and informed decisions to reduce herbicide use in favor of more IWM strategies.

## 8.2 Environmental Risk Assessment in Agricultural Systems

Risk is defined as the probability of occurrence of an event and its associated consequences. Risk analysis gathers several statistical modeling and database management techniques that allow the prediction of future unknown events through the analysis of previous patterns and occurrences. These models are known as environmental risk indicators (or estimators) (ERI) (Bockstaller and Girardin 2003), and their construction requires knowledge about key system variables which determine the presence of risk. In this way, most analyses follow a basic structure in order to identify predictive variables (i.e., explanatory variables) and response variables (i.e., actual risk), as well as, the causal mechanisms that link them together (Jakeman et al. 2006).

As a result, using ERI to solve environmental, ecological, or agricultural-derived problems allows an in-depth systemic understanding (Scoullar et al. 2010), giving decision-makers the possibility of identifying low-risk management practices that comply with both productivity and sustainability goals. However, it is important to note that for ERI's adequate development and use, certain conditions must be met, such as, they: (1) should be updated and improved, as knowledge about the system structure and function increase; (2) must have a meaning and interpretability about the system beyond its objective value; (3) should respond to the user's necessities, (4) must be accessible and easily interpretable at a given temporal and spatial scale, (5) must be as objective as possible, and (6) should be easily applied by its "target population" (e.g., farmers and scientists) (Fernandes and Woodhouse 2008). Nevertheless, the choice and selection of ERI and their respective input variables depend ultimately on the problem to be solved and the particular analysis objective (Ghersa et al. 2000). In addition, ERI's predictive variables should also comply with certain criteria. Specifically, explanatory variables should: (1) be quantitative values (a numerical meaning must be assigned), (2) be sensitive to changes at a systemic level, (3) condense complex information in a simple and concise manner, and (4) be simple/easy to store and extract from database structures. It is worth noting that variables selection is an important step in measuring risk. For example, not considering or involuntarily omitting key ones, may lead to increase the probability of associated error (Von Wirén-Lehr 2001). Conversely, using too many variables (overparameterization) may hinder the development of models that are easy to manage, interpret, analyse, and/or update (Van Cauwenbergh et al. 2007). In addition, variables should be able to detect direct and indirect relationships between components of the system such as trade-offs or synergies (Bennett 2009). Trade-offs arise when a change in a variable generates an inverse effect on another variable, while synergies consist of situations in which at least two variables increase or decrease simultaneously generating a larger overall effect than the individual sum of effects (Chapin III et al. 1996; Raudsepp-Hearne et al. 2010; Luukkanen et al. 2012). In agricultural systems, knowledge about the interrelationship between variables determines which decisions are adequate to reduce the associated risk thus enabling prosperous management strategies (Chapin III et al. 1996).

Finally, ERA requires the consideration of the ecological effects of chemicals across different scales and levels of biological organization. This appears to be a daunting task, as Köhler and Triebskorn (2013) suggest that it is increasingly difficult to track the effects of pesticides beyond the population level, especially within a context of dynamic global change. In agreement with this argument, Rohr et al. (2016) point out that, in general, ERA shows a negative relationship between level of biological organization and ease of assessing cause–affect relationships when a high-throughput screening of a large numbers of chemicals is considered. In turn, two initial consequences arise from these assertions, (1) the difficulty of capturing ecological effects beyond toxicological laboratory tests and (2) a clear trade-off between precision of measurements and their final utility to assess the anthropic impact on biophysical environmental components. As noted by Jepson (1993), Freemark and Boutin (1995), and Topping et al. (2015) the integration of ecological theory with toxicological and environmental chemistry are key aspects to generate appropriate risk assessment frameworks. Particularly, the pattern and frequency of exposure to the toxic chemical and the landscape structure are important determinants of risk for pollutants with short persistence, in temporary habitats or those affecting dispersive invertebrates.

In sum, the choice and identification of ERI are of significant importance in order to evaluate possible anthropic effects on key ecosystem properties (Ghersa et al. 2000). In agricultural systems, ERA presents an alternative framework that encompasses the biophysical functionality with the productivist (utilitarian) dimension into a systemic viewpoint (Müller 2005). Also, they offer the possibility of evaluation of ecosystem change in relation to weed management strategies and technologies, and how to apply viable solutions to ameliorate environmental degradation without necessarily compromising economic output (Duarte Vera et al. 2015).

## 8.3 Fate of Herbicides in Agricultural Systems

Beyond their typification, in terms of the hazard that herbicides use may imply over nontarget populations, a large part of the environmental effects of these substances are associated with both their fate and behavior once applied. These two aspects are closely related to both (1) their physicochemical structure and (2) herbicides' transport processes in soil, water, and air, which are in turn influenced by the environmental conditions. Although herbicides represent the largest proportion of total pesticide use (Zhang 2018), physicochemical and environmental processes that determine their fate in the environment are generally applicable to the majority of pesticides.

### *8.3.1 Physicochemical Description of Herbicides*

The most common substances used as herbicides are organic chemicals, which are mainly composed of hydrogen and carbon atoms. In addition, few other elements compose the active ingredient molecules such as oxygen, nitrogen, sulfur, phosphorus, and the halogen group. The chemical structure of most herbicides is heterocyclic, and many contain a halogen within their structure. By way of chemical processes, these substances are altered into forms such as esters, alcohols, and organic acids which allow changes in physicochemical properties that are more effective for the intended use and allow easier application. In addition, these formulations greatly influence the fate of these chemicals in the environment due to their solubility in water, electrical charge, and volatility. The key chemical structure of an herbicide is its biological active portion, or *active ingredient* (a.i.) which determines its ability to kill "target" individuals. Small changes in the molecular composition and/or configuration can directly alter the a.i. biological activity and physical properties. Active ingredients can be altered by chemical processes like esterification or transformation into organic salts to enhance biological activity, change the method of application and finally alter the fate of the herbicide in the environment. It is often the case that active ingredients derived from alcohols, phenols, and organic acids are more soluble in water, while esters are more soluble in organic solvents and have a tendency to produce vapors (Leonard et al. 1995). In the case of esters, for example, the size of the molecule may also influence its degree of volatility, determining both the method of application as well as the interaction between the "target population" and the surrounding environment.

## 8.3.2 Herbicides Transport in the Environment

Ultimately, displacement amongst ecosystem compartments, soil adsorption, and physical, chemical, and/or biological degradation determine the fate of the herbicide in the environment. This makes knowledge about key ecosystem processes linked to persistence important in planning safe pest management strategies, performing damage control in highly contaminated areas and prevent increasing tradeoffs between high productivity, technology, and environmental health.

Figure 8.1 presents a flow diagram of herbicide molecules after spraying or ground application. The molecules that are not metabolized by target organisms flow through the environmental compartments (atmosphere, waterbodies, soil, and biota) until they are degraded into other compounds or removed by crop harvest.

### 8.3.2.1 Herbicide Soil Persistence

Half-life is a measure of persistence of an active ingredient (or formulation) in the environment. Specifically, half-life refers to the time required to degrade 50% of the applied dose of any substance. According to Kamrin (1997), a criteria to classify herbicide persistence is as follows: (1) nonpersistent or weakly persistent is <30 days, (2) moderately persistent is between 30 and 100 days, and (3) strongly persistent is >100 days. In addition to its chemical structure, the herbicide's persistence also depends on the environmental conditions, such as soil adsorption (attraction of ions or molecules to the soil surface, $K_{oc}$), soil temperature and moisture, vegetation, and soil microorganisms. These factors determine the susceptibility of a substance to both chemical and microbial degradation. As a result, the interaction between half-life and soil conditions will effectively determine herbicide's persistence. For example, an herbicide can have a long half-life but due to a low sorptive

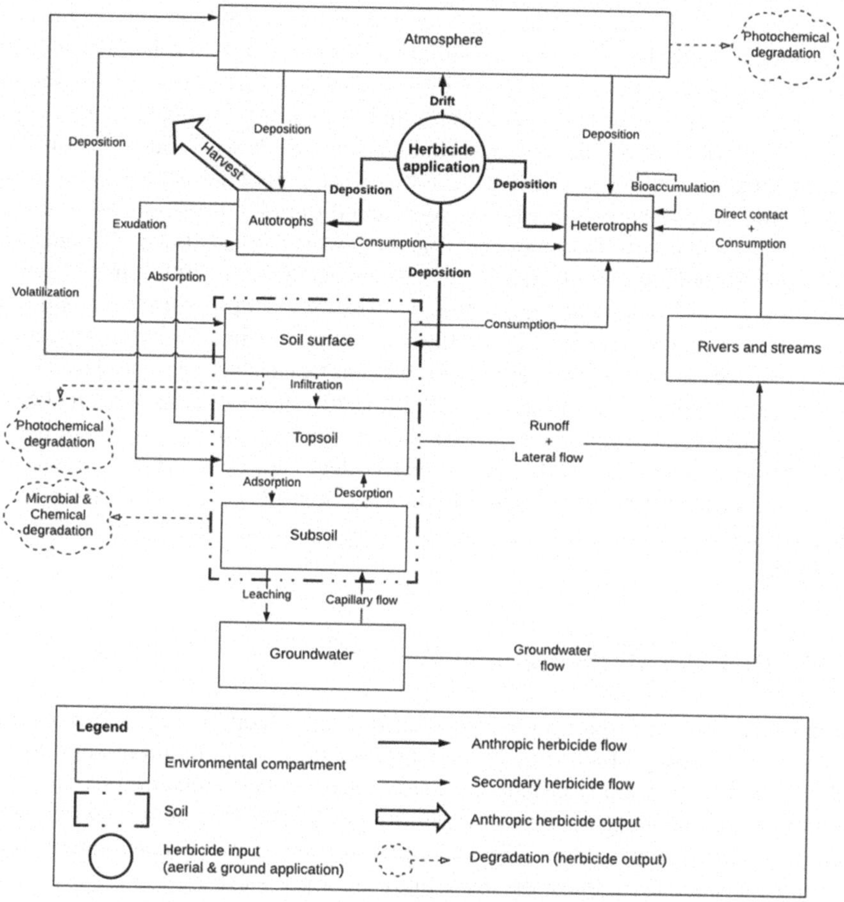


**Fig. 8.1** Fate of herbicides in the environment. Environmental compartments refer to the principal herbicide pools (atmosphere, waterbodies, soil, and biota). Anthropic herbicide flows are a direct consequence of application, while secondary flows refer to herbicide molecules that have gone unaltered through an environmental compartment. Anthropic herbicide output is the removal of biomass from the system by way of crop harvest. Dotted clouds that represent degradation processes are system outputs, since herbicide molecules are no longer in the system but have been altered into new compounds. These remain in the system, either dissolved in water or adsorbed to clay or organic matter

capacity it cannot exert toxic effects into the environment. Conversely, a substance can be virtually inactive to target organisms but affect nontarget organisms or biomagnify after a short period of time (Radosevich et al. 2007).

Herbicides reach the soil through various pathways (Fig. 8.1). Some herbicides are mixed or applied directly into the soil, while others reach the soil surface after canopy interception as runoff from treated foliage or when affected individuals decompose into the soil (Weber et al. 1989). Once they have reached the soil, herbicides interact with the solid, liquid, gaseous, and biological components. Some herbicides interact with the solid phase of soil; that is, they adsorb to soil colloids (e.g., negatively charged particles of clay and organic matter (OM)). These processes are a key factor affecting herbicide persistence and availability for plant uptake, since adsorption inhibits uptake, degradation, and leaching. Soils with course texture and low OM content have low adsorptive capacity favoring both herbicides degradation and leaching. Additionally, soil pH has been found to influence persistence of certain herbicides like sulfonylureas and triazines, which break down through acid hydrolysis (Hiltbold and Buchanan 1977; Sarmah and Sabadie 2002). In general, at pH levels above 6.8 chemical degradation by hydrolysis stops, while water solubility increases. As a consequence, under these conditions both sulfonylureas and triazines are more persistent and available for plant uptake causing potential carryover effects on succeeding crops (Bailey and White 1970). In acidic soils (i.e., pH below 7) adsorption to soil particles is increased, which may increase persistence of herbicides such as imidazolinones that are primarily degraded by microbial activity. In the case of sulfonylurea and triazine herbicides, low pH levels can render these products ineffective since they are both highly adsorbed and broken down by hydrolysis.

#### 8.3.2.2 Herbicide Air Displacement

Herbicides can move through air by way of drift or volatilization (Fig. 8.1). Volatility refers to the tendency of a chemical to change its phase from liquid or solid to gas (volatilize). This process is principally determined by their temperature and vapor pressure of the chemical, so all herbicides are potentially volatile with wide range of degrees. Herbicides that have low vapor pressure are relatively nonvolatile, while the opposite occurs with those that have high vapor pressure.

Drift refers to the movement of particles—solid or liquid—through air, away from the site of application. Since most herbicides are applied through air in the form of a suspension, it is the most serious concern since it can potentially damage nontarget organisms, generate residue in adjacent fields, or contaminate water and food sources. To reduce the risk of displacement through air, several methods can be used. When an herbicide is sprayed, the size of the droplet will likely influence the propensity to drift. In general, heavier droplets (i.e., more than 200 μm in diameter) reach the intended target, while smaller droplets are more likely to drift away (Matthews et al. 2014). Thus, using larger droplets and spraying the product downward can greatly reduce the risk that the substance will move through the air into unintended sites. Notwithstanding, for nonsystemic herbicides larger droplets may be less precise and provide worse control, since they do not cover as much surface area. Buffer zones are also a suitable method to reduce the effects of substances that drift from the site of application (de Snoo and de Wit 1998).

#### 8.3.2.3 Herbicide on Surface and Leaching Water

Water dynamics and soil physical chemistry greatly influence the movement of herbicides. Water moves through the soil profile and the molecules that are not adsorbed to soil colloids are leached. Therefore, herbicide movement depends on soil properties, herbicide solubility, and the amount of water that percolates into the soil. Most leaching involves vertical movements, although lateral or even upward leaching may take place (Fig. 8.1). In cases when herbicides do not degrade completely or leave residues and enough water falls onto the soil, herbicides have been found to reach groundwater supplies (Thurman et al. 1991; Johnson et al. 2001).

#### 8.3.2.4 Herbicide Degradation

Degradation of the original molecule into simpler chemical structures occurs because of exposure to sunlight (photochemical decomposition), soil chemical processes (chemical decomposition), and degradation by microorganism activity (microbial decomposition). These processes mostly occur in the soil, but may also take place in air, water, plants, microbes, and animals (Goring et al. 1975; Fenner et al. 2013). Persistence curves determine the amount of time that an herbicide is active in the environment and depends on soil factors and climatic factors such as pH, soil composition, moisture, temperature, and radiation, as well as the physicochemical properties of the applied substances (Curran 2016). Photochemical decomposition results from the photolysis process that breaks down molecules of herbicides that lie on the surface of leaves or topsoil. However, not all herbicides are degraded by solar radiation at the same rate such that photolysis is an herbicide-specific process. In addition, herbicides can undergo chemical decomposition in the soil through processes of oxidation-reduction, hydrolysis, and water-insoluble salts and chemical complex formations. Microbial decomposition accounts for a large proportion of herbicide degradation in the soil (Chapin III et al. 2011). Several factors influence these processes, such as soil moisture, temperature, pH, and OM content. Strong alterations in these factors can cause microbial activity to slow down and hinder decomposition (Curran 2016). Microorganisms decompose herbicides through the secretion of specific enzymes that break down complex organic molecules. These compounds are then used by microbes as a source of carbon, nutrients, and energy. According to Anderson (1996), microorganisms perform degradative reactions that alter the original herbicide molecule, for example, dehalogenation (removes chlorine, bromine, or other halogen atoms), dealkylation (removes organic side chains), hydrolysis (removes amides or esters), ring hydroxylation (adds hydroxyl (-OH) groups to aromatic ring), ring cleavage (breaks structure of aromatic ring), and reduction (addition of hydrogen to $NO_2$ groups under anaerobic conditions). Although they are a principal agent in herbicide degradation, soil macrofauna is also adversely affected by certain toxic components. Nitrogen-fixing bacteria and mycorrhizal fungi are two groups that are known to be negatively affected by certain pesticides (Trappe et al. 1984; Johnsen et al. 2001).

## 8.4 Potentially Negative Effects of Herbicides

Despite the irrefutable advantage of weed management practices for food and fibers production, the extended and continual use of chemical weed control strategies can have deleterious effects on the environment. Some of these consequences are: (1) food chain disruption, (2) increased risk of invasive species and weed resistance, (3) leaching of herbicide components and other agrochemicals to waterbodies, (4) direct and indirect effect on local fauna, such as short- and long-term toxicity as well as alteration of vegetative structure and cover for feeding and habitat, and (5) negative effects on soil physicochemical properties and microorganism communities (Levitan et al. 1995; Pimentel and Burgess 2014).

The physicochemical properties of herbicides such as volatility, adsorption, or solubility in water can make them persist in the environment in their original form or as toxic metabolites. In addition, some substances bind to body lipids in organisms making them likely to bioaccumulate (Fig. 8.1), which increases their level of toxicity across food chains (a process known as biomagnification).

The toxicological effects range from acute after immediate oral, dermal, or inhalation exposure to chronic such as cancer or organ failure because of long-term exposure or carcinogenicity of certain active ingredients or surfactants.

### *8.4.1 Ecotoxicity*

Adequate use, safety, and further development of commercial herbicide products require their classification into groups regarding similarities in chemical structure, use, and effect on plants (Briggs 1992). However, it has been recognized that toxicity and dose of application are major factors in determining the final effect of herbicides (Ferraro et al. 2003). Therefore, they can also be classified based on their toxicity and hazard level, providing a more thorough description of their effects on target population as well as possible environmental externalities.

Based on the premise that all chemicals are toxic at some dose (the Roman Paracelsus, known as the father of toxicology, expressed *as dosis sola facit venenum*), ecotoxicity refers to the chemical, biological, or physical stress that a substance can cause to the constituents of an ecosystem (such as animals, plants, and microorganisms) (Truhaut 1977). These effects can occur after: (1) immediate herbicide exposure (acute toxicity), such as eye and skin irritation or neurotoxicity; or (2) after long-term gradual exposure or accumulation (chronic or sub-chronic toxicity), such as impaired liver function, reproductive abnormalities, and cancer (Mansour 2004).

Toxicological evaluations are performed through experimental procedures on laboratory animal tests, which are exposed to different doses of herbicides on varying timescales from hours to years (Whitford 2002). Only those results that arise from experiments on species that have similar known reactions to humans can be extrapolated to generate precise warnings on human health. However, real human susceptibility can only be estimated when systematic exposure of a group of humans to a substance has produced consistent results over a reasonable time period (Mostafalou and Abdollahi 2013).

#### 8.4.1.1 Acute Toxicity

The effect of a substance after immediate exposure on an individual can be measured in relation to its body weight. As the dose applied on a test population increases, the response (such as death) on the individuals increases in a proportional manner. A common test to measure acute toxicity is Lethal Dose 50 ($LD_{50}$), or Lethal Concentration 50 ($LC_{50}$), for individuals exposed through air or water, respectively. For a group of individuals of a given species, the toxicity test measures the amount of a substance that is necessary to kill 50% of the evaluated population within a specified time. The test has been criticized for being a rough estimate of a substance's toxicity, which is affected by species tested, age, weight, sex, genetic strain, health, diet, temperature, housing conditions, season, and probably other environmental conditions at the time of the test (Briggs 1992). In addition, many ratings for substances have been done when requirements were less rigorous and could provide inadequate measures of toxicity (Zbinden and Flury-Roversi 1981). In general, toxicity ratings are represented by way of a sigmoidal dose-response curve. At the bottom of the curve, the is no significant effect on the affected population which is referred to as *no observed adverse effect level* (NOAEL). This is followed by an increased linear response which starts with the *lowest observed adverse effect level* (LOAEL), the lowest dose at which a significant effect on the test population is noticeable. Finally, the curve reaches a saturation plateau where 100% of the test population is killed, or the *maximum observed adverse effect level* (MOAEL).

Once a dose-response curve is found, a rating scale can be generated to determine the risk associated with amount and type of exposure. A commonly used rating scale was devised by the United States Environmental Protection Agency (USEPA), which distinguished three categories of exposure (Table 8.1): oral (ingestion risk), dermal (skin absorption), and inhalation (inhalatory risk). The fifth column in Table 8.1 details a dose that could be lethal for an average weight human, also known as *reference dose* (RfD) or *margin of exposure* (MOE). Considering that dose-response curves are found using test animals, a safety factor of 100 is used to find the RfD once the NOAEL is determined.

#### 8.4.1.2 Chronic Toxicity

Chronic and sub-chronic toxicity refer to long-term effects of exposure such as organ damage, failure, or cancer. These effects are measured on laboratory test animals over periods ranging from weeks to 1 or 2 years (Whitford 2002). $LD_{50}$ and RfD for chronic toxicity are obtained through the same procedures as for actuate

**Table 8.1** Required pesticide labels based on EPA rating for acute toxicity

| Required label | EPA rating | Type of exposure | Amount of exposure | Probable lethal dose for 150-pound human |
|---|---|---|---|---|
| "Danger" | I | Oral | 0–50 mg/kg | Teaspoon |
| | | Dermal | 0–200 mg/kg | |
| | | Inhalation | 0–0.2 mg/L | |
| "Warning" | II | Oral | 50–500 mg/kg | 1 teaspoon–1 ounce |
| | | Dermal | 200–2000 mg/kg | |
| | | Inhalation | 0.2–2 mg/L | |
| "Caution" | III | Oral | 500–5000 mg/kg | 1 ounce–1 pint (for 1 pound) |
| | | Dermal | 2000–20,000 mg/kg | |
| | | Inhalation | 2–20 mg/L | |
| No label | IV | Oral | Over 5000 mg/kg | Over 1 pint or pound |
| | | Dermal | Over 20,000 mg/kg | |
| | | Inhalation | Over 20 mg/L | |

Oral and dermal ratings are expressed in terms of LD50 and inhalation is expressed in terms of LC50. LD50 = lethal dose that kills 50% of test animals in a given time. LC50 = lethal concertation in air or water in which animals live that kills 50% in a given time mg/L = milligrams per liter. A milligram is 1/1000 of a gram. This measurement is comparable to parts per million (ppm). mg/kg = milligrams (of toxin) per kilogram (of body weight of animal). This measurement is comparable to parts per million (ppm). Adapted from Briggs (1992) and Radosevich et al. (2007)

toxicity. Briggs (1992) adequately points out that it is often difficult to estimate the true nature of chronic effects as long-term testing performed in a systematic basis would be necessary, which is clearly impractical on humans for ethical reasons.

Biomagnification is also a result of the cumulative character of some herbicides, such as organochlorines that bind to body lipids. It is the accumulation of a substance in a food chain, and it occurs when small amounts of a toxic substance in the soil, air, or water is taken up by plants, later eaten by small animals, and successively building-up at the top of the food chain. Predators at the top of food chains, such as humans, can present thousands of times the amount compared to the environment. To estimate the propensity of an herbicide to bioaccumulate the *biological concentration factor* (BCF) and $K_{OW}$ are used (Kamrin 1997). BCF measures the concentration of a substance in a living organism in relation to the concentration in the surrounding environment, while $K_{OW}$ is the octanol-water coefficient—(i.e., how well a chemical substance is distributed at equilibrium between octanol and water). Despite the fact that BCF is more sensitive to biological accumulation, $K_{OW}$ measure provides a reference value of 1000:1 octanol-water differential gradient when the $\ln(K_{OW})$ is 3 or more. This level indicates that a substance is very likely to concentrate at the top levels of a food chain (Chadwick and Shaw 2016).

In addition to causing organ damage, due to the metabolic and accumulative effects of toxic substances, certain components can also be carcinogenic. Carcinogenic substances are especially dangerous since no minimum threshold has been found, below which no effect is observed. Thus, in theory, a given molecule of a carcinogenic substance could affect a susceptible cell inducing abnormal growth. Chronic exposure to herbicides can also cause adverse reproductive effects such as increased tendency to abort, reduced offspring weight, malformations, birth defects, and behavioral and learning disorders in offspring (Prüss-Ustün et al. 2011). To determine mutagenic herbicide effects testing is performed on developmental and fertility processes, as well as in vivo and in vitro. This allows to identify if an herbicide causes damage on either chromosomes or genes and serves as a screen for suspect carcinogens since most mutagens may also cause cancer (Ames 1979; Cairns 1981). Finally, it has been noted that some herbicides may act as endocrine disruptors by mimicking hormones and disrupting processes such as metabolism, stress, development, and reproduction (McKinlay et al. 2008). However, Radosevich et al. (2007) state that the subject is still quite controversial and more evidence has to be collected to establish conclusive evidence.

## 8.4.2 *Herbicides and Human Health*

The bourgeoning use of herbicides to control adversities in cropping systems has increased the risk of exposure to both nontarget species and humans. Exposure can be both intentional (i.e., self-poisoning) or unintentional (i.e., occupational and take-home pathways, home and garden use, public health applications, and residues in food or water) (Hoppin and LePrevost 2017). The physiological response to exposure depends on both dosage and time (Fig. 8.2).

Although the widespread use of herbicides is an increasing risk factor in the lives of the general population, the most at-risk segment of society are agricultural workers. Occupational exposures are the most common in terms of concentration, frequency, and duration as well as the high levels of toxicity found in the formulations used in these settings (Aiassa et al. 2019). In addition, the general population is at risk from non-occupational and environmental (air, water, and soil) herbicide exposure through diet (e.g., chemicals or residues that have been applied to fruits and vegetables) and when herbicides are applied to homes, gardens, or drift from nearby agricultural fields (Krieger 2001; López et al. 2012; Hoppin and LePrevost 2017).

### 8.4.2.1 Chronic and Acute Toxicity in Humans

Herbicides and other pesticides pose a risk to human health due to acute and chronic effects. It is estimated that there are between 3 and 25 million reported cases of acute pesticide poisoning worldwide (Jeyaratnam 1990); although, the difficulty in measuring the effects and bad access to healthcare in rural areas makes most cases go underreported or undiagnosed. Ratings for acute toxicity are generally done by organizations such as the United States Environmental Protection Agency (USEPA 2003) or the World Health Organization (WHO 2010). Hoppin and LePrevost

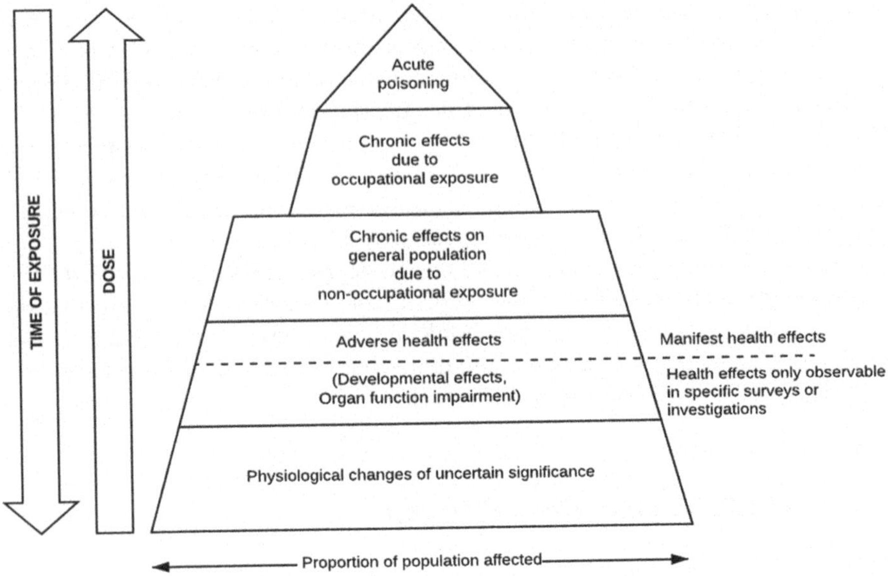


**Fig. 8.2** Human health effects of herbicides in response to time of exposure and the dose. Acute poisoning occurs at high doses and short-term exposure, while physiological changes of uncertain significance are a result of long-term exposure at low doses. Presence of an individual in one polygon does not exclude presence in the others. Adapted from Prüss-Ustün et al. (2011)

(2017) also note that many agricultural workers fail to report symptoms due to cultural factors and lack of awareness regarding health risks.

Symptoms for acute poisoning in humans range from a mild irritation in the skin, eyes or throat to possible death. This depends on several factors such as the specific herbicide, the amount of exposure and the age and size of the individual. Most symptoms are nonspecific and include headaches, dizziness, abdominal pain, nausea, diarrhea, and vomiting, which in turn makes it difficult to diagnose as poisoning and prescribe an adequate treatment.

The most difficult adverse effects to diagnose and detect result from long-term low-level exposure, which cause chronic conditions such as cancer, neurological consequences, respiratory outcomes, diabetes, birth defects, and cardiovascular diseases (Krieger 2001). However, it is often difficult to know which substances are the cause of these adverse effects, since not all of them are persistent or bioaccumulate making some chemicals untraceable. In the case of substances with low levels of persistence, such as organophosphates, chronic toxicity must be measured through alternative measures of exposure like surveys and questionnaires. This can lead to questionable data since it relies on the knowledge of the person being interviewed regarding information on active ingredient and doses of applied herbicide products. In addition, it is often costly and time-consuming to perform these measurements.

Most studies present an exposure–response relationship, such that longer exposure would mean higher and more significant risk levels (Hoppin and LePrevost

2017). Weichenthal et al. (2010) found relationships between specific pesticides and lung, pancreatic, colon, rectum, bladder, and brain cancers in a US Agricultural Health Study cohort. Nevertheless, evidence to support carcinogenic effects is often inconclusive, especially regarding childhood cancer (Hoppin and LePrevost 2017). In the case of adult cancers, evidence seems more conclusive, with data from occupational and population-based studies. In general, individuals with long-term exposures (over 10 years) presented increased risks of lymphohematopoietic cancers (Merhi et al. 2007). In addition, other effects of chronic effects on humans have been associated with damage of genetic material (Porcel de Peralta et al. 2011; López et al. 2012; Aiassa et al. 2019), neurological conditions (Kamel and Hoppin 2004), and respiratory ailments (Proskocil et al. 2008; Cho et al. 2008; Fukuyama et al. 2009).

#### 8.4.2.2 Occupational Hazard: The Case of Argentina

In Argentina, agricultural expansion and intensification associated to strong commodities markets and technological advances related to genetic modifications have engendered the increment of crop production from 8.8 million (in 1961) to more than 32 million harvested hectares in 2018 (Rolla et al. 2018). This almost fourfold increase in agricultural land use can be chiefly explained by the expansion of soybean production, which had the largest overall increase in harvested area especially since the late 1980s with the introduction of no-tillage farming (Grau et al. 2005). Soybean went from representing less than 1% of total harvested area in 1961 to approximately 60% in 2017 (54 million ton). The combined effect of intensification, expansion, and the wide transition towards no-tillage farming systems (which relies more heavily on chemical inputs to control adversities than conventional farming) entailed an increased risk of exposure to pesticides.

The expansion of soy production principally occurred in the Pampa region and its surrounding areas—Santa Fe, Entre Rios, and Cordoba provinces. Now, more than 50 years into the process of agricultural expansion and intensification, the effects of long-term exposure to pesticides are becoming evident. Therefore, chronic toxicity has recently become a relevant factor in ERA in addition to the risk of acute toxicity posed by chemical products and their derivatives to both rural and urban populations (López et al. 2012).

In Córdoba, where approximately 10% of the population live in towns with less than 2000 inhabitants and small settlements dispersed within the agricultural productive matrix, several studies have begun to trace the carcinogenic and genotoxic effects in adult and children populations.

Recent findings show a positive correlation between years of exposure and genotoxicity indicators (i.e., frequency of micronucleus, frequency of chromosome aberrations, DNA fragmentation, and plasma cholinesterase levels) (Martínez-Valenzuela and Gómez-Arroyo 2007; Do Carmo and Alvarez 2009; López et al. 2012; Aiassa et al. 2019). It is worth noting that most studies on long-term exposure cannot isolate the effect of a single product or class of pesticide, rather they evaluate the effect of a technological package, which generally includes a combination of agrochemicals (herbicides, insecticides, fungicides, fertilizers).

Figure 8.3 shows Aiassa et al. (2019) results on genotoxic effects of long-term occupational pesticide exposure on a rural population of male workers in Cordoba, Argentina. Trial sample population were exposed individuals who were selected considering labor activity, number of applications sprayed in the area per year (≥3), age (18–65 years), and exposure time ≥3 years. The control (reference) group consisted of inhabitants whose residences were ≥1000 m away from areas sprayed with agrochemicals (without any contact with pesticides), age 18–65 years, and a similar lifestyle habit to that of the exposed sample. Biomarkers of exposure analyzed by Aiassa et al. (2019) were chromosomal aberrations, micronuclei, and DNA fragmentation through the comet assay and as a biomarker of effect on plasma cholinesterase enzyme. Correlation analysis did not isolate a single risk factor (i.e., herbicide product) but rather a technological package generally consisting of several of pesticides (i.e., glyphosate, cypermethrin (CY), 2,4-D, endosulfan, atrazine, and chlorpyrifos).

In particular, higher levels of chromosome aberration and micronuclei are indicative of genetic damage and instability that is correlated to several neoplastic diseases and increased risk of cancer (Gollapudi and Krishna 2000; López et al. 2012). Several other studies from Cordoba (Mañas et al. 2009; Peralta et al. 2011) and Santa Fe (Simoniello et al. 2010) provinces, Paraguay (Benítez-Leite et al. 2010) and Mato Grosso in Brazil (Do Carmo and Alvarez 2009) reached similar results in populations of long-term occupational exposure. It is worth noting that none of these studies were able to isolate a specific product as a cause of increased cancer

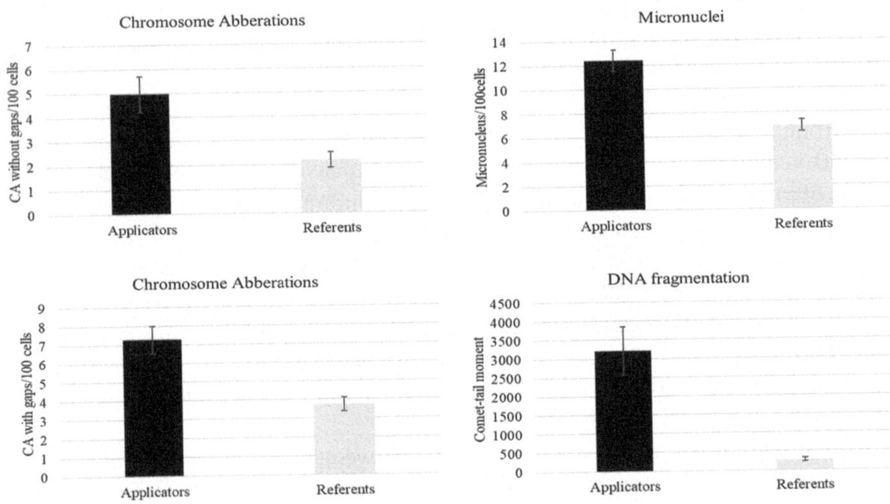


**Fig. 8.3** Genotoxic effects of occupational pesticide exposure in Córdoba, Argentina. Study conducted on 30 pesticide applicators from the province of Cordoba, Argentina. Difference in effects between applicators and referents (control) were significant for all indicators ($p < 0.05$). Extracted from Aiassa et al. (2019)

risk, rather several pesticides that are sequentially applied on a regular basis are probably involved.

Non-occupational exposure can also generate negative health effects, although responsible chemicals are even more difficult to trace (Whitmore et al. 1994; Damalas and Eleftherohorinos 2011). The principal risk arises due to ingestion of food or contaminated water. Lepori et al. (2013) reviewed Argentinian studies on pesticide contamination in food products and waterways. Their results suggest and increasing level of pesticides on household food products and waterways in the main productive areas of Argentina. Although more information is needed to accurately assess the indirect health risk effects that these patterns could pose on the general population (especially organochlorine pesticides, which tend to bioaccumulate) there seems to be enough evidence to warn both agricultural workers and the general public regarding the possible risks associated with long-term direct and indirect pesticide exposure.

## 8.5 Why Farmers Continue to Use Herbicides?

Environmental and human health deleterious effects of pesticides (including herbicides) are evident. Operational advantages of chemical control over alternative weed management methods have generated an almost exclusive dependence on this technology on a global scale (Walsh et al. 2013). Although stern regulations exist at both national and international levels, the dominance of both transgenic and tolerant crops has brought about the overuse of chemicals (Green 2014) with the consequent risk increase on both society and the environment (Pimentel et al. 1992; Pimentel and Greiner 1997). Despite their relevance and recurrent warnings about the externalities of herbicide use, these aspects are seldom taken into account for weed management decision-making (Wilson and Tisdell 2001). In addition to effectively implement sustainable agricultural technologies it is necessary to adequately measure risk indicators related to toxicology, pollution, or other forms of contamination, as well as the responsible application of tools and technologies by decision-makers (i.e., understand the use and potential damage that agricultural technologies can cause). Farmers are ultimately the ones responsible for the decision of which weed management strategies they adopt, so understanding the decision-making process can be helpful to promote a more sustainable agricultural.

### *8.5.1 Externalities and Agroecosystem Sustainability*

Environmental hazard depends on both potential risk of chemicals and other technologies, as well as the capacity that biophysical systems have of processing the external inputs (Mejer and Jørgensen 1979).

The costs associated with herbicide externalities expand beyond the agricultural sector, and they not only affect present producers but also jeopardize the future productive capacity of ecosystems (Carlson 1989; Strange and Scott 2005; Damalas 2009; Beketov et al. 2013; Malaj et al. 2014; Stehle and Schulz 2015). The inherent complexity of environmental and social problems hinders their evaluation and quantification, which in turn makes them rarely appear within producers' cost matrix generating market failures or externalities (Devine and Furlong 2007; Leach and Mumford 2008; Waterfield and Zilberman 2012).

Negative externalities produced can be grouped into three categories: (1) damage to human health, (2) impacts on system productivity, and (3) generation of herbicide-resistant weeds (Bullock et al. 2018). The costs incurred by the effects of herbicide exposition to human health (both short- and long-term) are often ignored by producers and actors associated with agricultural activities. This is because of incorrect diagnoses of observed ailments, scarce medical resources (especially in developing countries), and general lack of certainty regarding the temporal relationship between herbicide application and human health effects. Regarding functional systemic alterations, changes in biodiversity can cause a steep drop in yields due to loss of soil fertility, ecosystem services, and modifications in functional relations between pest and beneficial species. Lastly, the appearance of herbicide-resistant populations can be understood from *common pool resources* management perspective. Given that a producers' actions can have repercussions on both spatial (e.g., dispersion of weed seeds and reproductive materials) and temporal scales (e.g., future weed populations), pest management strategies should incorporate factors that affect pests' population dynamics. A clear compromise exists between controlling present weed populations and preserving herbicide susceptibility of future populations (Ambec and Desquilbet 2012).

The cause of negative externalities can be understood through the prisoner's dilemma, a classical game theory situation (Fig. 8.4). The short-term economic gains that chemical control can offer over other management strategies has made it a dominant and unsustainable agricultural practice. As it is depicted in Fig. 8.4a, chemical control offers a small advantage over Integrated Weed Management (IWM). According to Kogan (1998), IWM is centered upon the premise of optimal crop growth with the least possible alteration of agroecosystem functions. This may entail forgoing maximum profits through total resource exploitation, which explains why chemical control offers higher socio-environmental returns before the threshold is crossed.

However, continued herbicide use as a principal strategy would make the relative economic benefits eventually decrease as resources are depleted and environmental compartments become polluted (i.e., less yields and stability). Once the initial threshold is crossed, IWM becomes the better alternative, providing more stable and higher socio-environmental welfare. In addition, the rise in crop supply would mean a consequent decrease in their market price. Thus, the use of herbicides could turn into an inescapable need for those producers that did not use them previously, to reach a satisfactory economic performance. This attractor point can be observed in quadrants II and III of Fig. 8.4b. When a farmer decides to use IWM practices he/

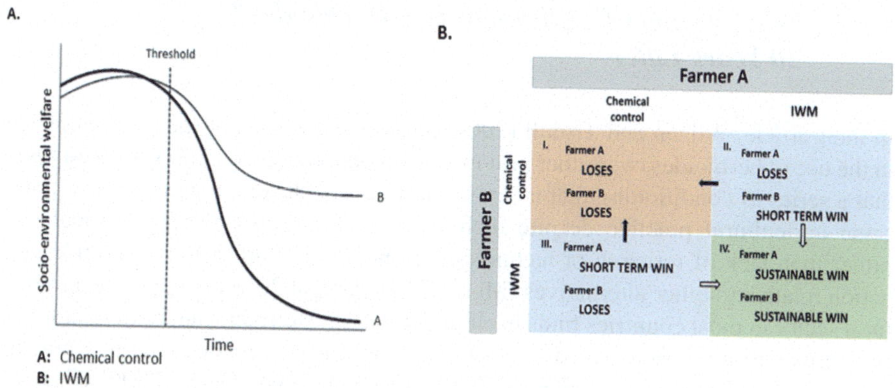


**Fig. 8.4** Prisoner's dilemma for weed management strategies. (**a**) Socio-environmental welfare (i.e., economic benefits and ecosystem services) because of either IWM or chemical control, (**b**) prisoner's dilemma outcomes for multiple farmers under different combinations of technological decisions

she may be affected negatively by the externalities produced by those who implement chemical control. In addition, in this context the farmer that chose IWM also has an economic disadvantage, increasing his opportunity cost even more. This scenario forces the farmers that chose IWM to move towards the more profitable chemical control methods that in the long term will lead towards quadrant I. When both farmers chose the chemical control methods the system may become highly unsustainable making socio-environmental welfare drop in the mid- to long-term (Fig. 8.4a). Therefore, strong incentive systems for collaborative networks, science-based modeling such as DSS to incorporate environmental data and functions into decision-making, and sociopolitical institutions are necessary to aid farmers in modifying weed management decisions towards integrated practices that allow for optimal equilibrium between economic return and socio-environmental welfare (Ostrom 2009).

The perception that the short-term costs of externalities are lower than ordinary costs of production, in addition to lack of individual incentives that push producers to use alternative management methods, places the benefits that society reaps from controlling weeds into a suboptimal equilibrium (Fig. 8.4b). That is, society would be better off if individual actors incorporated collective benefits into their decision paradigm, rather than simply short-term gains. This situation generates a strong need for policies and legislation that promote a more responsible use of herbicides that minimize the negative effects on the environment and human health. Possible alternatives are integrated, systemic and science-based pest management strategies that would greatly reduce chemical control and the negative side effects.

### 8.5.2 *Why Farmers Continue to Use Herbicides? 20 Years Later*

In their article, Wilson and Tisdell (2001) analyze the causes of the predominance in the use of herbicides over other pest management strategies. Their results suggest that a series of conditioning factors exist that force producers to continue using the same agricultural practice despite their negative long-term effects. Particularly, rather than lack of research or technology, it has to do with farmers' attitude and action relative to other alternatives (Moss 2019). Almost 20 years later, the agricultural sector in most countries finds itself under the same circumstances in relation to problems that arise from weed control (e.g., yield loss, dependence on synthetic herbicides, resistance to herbicides) (Chauvel et al. 2001; Oerke 2006; Chauvel et al. 2012) and new growing environmental and social pressures (Lechenet et al. 2014; Petit et al. 2015).

According to Tisdell (2005) there is a strong relationship between the adoption of an agricultural practice or technology and its short-term economic outcome, although its future costs may increase by way of market mechanisms or social and environmental externalities. This situation makes productive systems oscillate around inefficient attractor points generating economic barriers which hinder the adoption of more sustainable alternatives. For the weed management case once a control strategy is adopted it often becomes a dominant strategy (Cowan and Gunby 1996). This "lock-in" phenomenon was initially observed by Arthur (1989). If a technology or practice has a competitive advantage over others it will end up dominating the market, even though it is not the most efficient. When resistant crop hybrids started to be commercialized, most of these products offered higher yields than IWM strategies; however, most farmers and decision-makers did not account for the long-term externalities that were associated with the technology.

### 8.5.3 *Factors That Generate the "Lock-In" Effect*

Economic costs for changing towards a different technology are among the factors that strengthen the feedback loops that lock-in producers into certain control practices. Not only are new technologies' learning curves generally steep, but many are also usually incompatible with IWM methods. In addition, the size of productive fields used under chemical control schemes make a steady transition towards IPM more difficult. Regarding crop rotations, the pressure to produce certain "cash crops," such as agricultural commodities for food and biofuels, has generated a simplification of rotation strategies leading to more favorable environments for both resistant and tolerant weed species. Finally, there exists a research bias towards chemical control—i.e., integrated weed management has not been a central part of the research agenda.

## 8.5.4 *Adoption Process for New Agricultural Practices*

A thorough review of factors that affect adoption rates for new technology allowed the identification of those that have a positive impact on farmers' decisions. These factors can be classified into three main groups:

### 8.5.4.1 Social Factors

A study by Prokopy et al. (2008) suggests that there are multiple social factors that affect a farmers decisions on the adoption of a new technology. The authors find that age, level of education, perceived earnings, disposable income and capital, interaction with other farmers, farm size, and the diversity of agricultural activities are key determinants.

### 8.5.4.2 Perception of Profitability

The notion that economic variables play a central role in the likelihood of technology adoption was first established during the 1950s (Griliches 1957, 1960; Mansfield 1961).Those farmers that perceive that a productive practice may be more profitable are more likely than others to adopt it. Therefore, long-term studies on the profitability of IPM could be a critical factor in increasing its rate of adoption. Tisdell (2005) points out that a strong limiting factor in the decision-making process regarding weed control strategies is the lack for information about economic outcomes of each practice (also known as *bounded rationality*). By having limited knowledge about the effects and cost-benefit relationship of alternative control methods, farmers are usually incapable of taking optimal decisions from an economic standpoint and they are easily influenced by marketing campaigns and commercial information.

### 8.5.4.3 Attitude Towards Risk

Risk aversion has been frequently associated with a reduction in the adoption rate of new technological practices (Lindner et al. 1982; Lindner 1987; Tsur et al. 1990; Leathers and Smale 1991; Feder and Umali 1993). Studies by Finnoff et al. (2005, 2007) show that higher risk aversion can lead to higher rates of reactive behavior (weed control or eradication) compared to proactive alternatives (i.e., preventive measures) . Given that chemical control strategies are more likely to generate negative externalities, prevention measures can provide higher social and environmental benefits, despite being associated with higher risk-taking decision-makers.

## 8.6 Decision Support Systems: A Tool Towards Integrated Weed Management

Decision support systems (DSS) can play an important role in improving weed management strategies, as well as communication, collaboration, and engagement between the scientific community, farmers, and other agricultural decision-makers (Montull et al. 2014). These tools generally consist of a framework of computer models or software packages that connect on-site data at different scales (i.e., weed composition and community structure, climate and soil conditions, crop genetics and ecophysiology, and management strategies) with ERI, agronomic and ecological models. (see DSS in Section I and examples in Section IV). Their main purpose is to simulate environmental stocks and flows within agricultural systems, such as nutrients, energy, or species composition, to understand the relationship between human action (i.e., management strategies) and ecosystem structure and function (Blackshaw et al. 2006). As a result, weed management DSS are instrumental in shedding light on optimal control and management strategies in order to reduce environmental risk while maintaining yield levels. By highlighting trade-offs and synergies between herbicide use, tillage or other weed control methods and environmental outputs, farmers can use these results to expand their decision-making horizon beyond a short-term chemical-based viewpoint (Colbach et al. 2017).

However, it is often the case that farmers are also reluctant to incorporate these tools into their decision-making due to low herbicide cost, variability in chemical weed control outcomes, and lack of interest or resources to collect necessary input data to run the models (Olson and Eidman 1992; Rossi et al. 2014). In consequence, crop protection products are generally applied implying worst-case scenario of adversities, irrespective of heterogeneity in field conditions regarding weed composition, crop development, soil, and climate (Montull et al. 2014). Breaking these barriers to increase DSS adoption amongst decision-makers will require close-knit collaboration between the scientific community, extensionists, and farmers to reduce possible learning curves, facilitate data collection, and incorporate long-term ecosystem dynamics (Swanton and Weise 1991; Buhler 2002). It is the case that IWM will not necessarily increase average income by way of higher yields, but rather by lowering both social and environmental burdens that result from indiscriminate use of chemical products (Berti et al. 2003).

## 8.7 Conclusions

Farmers continue to use herbicides despite clear warnings from the scientific community since the 1990s about the detrimental effects on the environment, human health, and questionable economic benefits. In this chapter the causes for a positive feedback loop of chemical weed control which hinders the adoption of alternative technologies and management strategies are analyzed. Lack of available information for decision-makers regarding the complete costs of chemical weed control, in addition to scarce knowledge about agroecosystem function under different weed management practices, makes indicators that reflect environmental and human health under alternative strategies necessary. ERA could provide information to characterize potential negative effects of herbicide use, evaluate the costs and risks associated with alternative management strategies, and generate tools and channels to allow information regarding environmental and human health risks to reach herbicide users and the general public. However, such broad frameworks present trade-offs between precision and utility of models used, especially when considering different spatial and temporal scales. These challenges expose the need for transdisciplinary research efforts to integrate toxicological and environmental chemistry with ecological theory and agent-based and optimization models that make possible the generation of predictive frameworks in relation to both ecological and socioeconomic effects of herbicide use.

Nevertheless, in order to close the gap between knowledge generation and its adequate application, close-knit feedback networks must be created between researchers, extensionists, land manager, and communities within the context of improved productive system development (Freebairn and King 2003). Altogether, IWM strategies require a balanced value system that incorporates the possibility of increasing material well-being without compromising essential biophysical structures and functions, and ecosystems services. In addition, it should incorporate the idea of justice as fairness such that all peoples, both present and future, have the same right to usufruct natural resources (Radosevich et al. 2007). Difficulties to reach unifying consensus about the value systems require strong social and governmental institutions as well as precise science-based information, such as DSS, to provide adequate incentives and knowledge on how to avoid tragedy of the commons situations (Hardin 1968; Ostrom 2009). Ultimately, reducing environmental risk in agriculture depends on strong value systems, social and political consensus, and tools that allow the generation of precise information and decision support systems on both biophysical and socioeconomic variables.

## References

Aiassa DE, Mañas FJ, Gentile NE, Bosch B, Salinero MC, Gorla NB (2019) Evaluation of genetic damage in pesticides applicators from the province of Córdoba, Argentina. Environ Sci Pollut Res 26:20981–20988

Ambec S, Desquilbet M (2012) Regulation of a spatial externality: refuges versus tax for managing pest resistance. Environ Resource Econ 51(1):79–104

Ames BN (1979) Identifying environmental chemicals causing mutations and cancer. Science 204(4393):587–593

Anderson W (1996) Weed science: principles and applications. West Publishing Company, St. Paul, MN

Arthur WB (1989) Competing technologies, increasing returns, and lock-in by historical events. Econ J 99(394):116–131

Bailey GW, White JL (1970) Factors influencing the adsorption, desorption, and movement of pesticides in soil. In: Single pesticide volume: the Triazine herbicides. Springer, New York, pp 29–92
Baker HG (1974) The evolution of weeds. Annu Rev Ecol Syst 5(1):1–24
Beketov MA, Kefford BJ, Schäfer RB, Liess M (2013) Pesticides reduce regional biodiversity of stream invertebrates. Proc Natl Acad Sci 110(27):11039–11043
Benítez-Leite S, Macchi M, Fernández V, Franco D, Ferro E, Mojoli A, Cuevas F, Alfonso J, Sales L (2010) Daño celular en una población infantil potencialmente expuesta a pesticidas. Pediatría (Asunción) 37(2):97–106
Bennett J (2009) Vibrant matter: a political ecology of things. Duke University Press, Durham
Berti A, Bravin F, Zanin G (2003) Application of decision-support software for postemergence weed control. Weed Sci 51(4):618–627
Blackshaw RE, O'DONOVAN JT, Harker KN, Clayton GW, Stougaard RN (2006) Reduced herbicide doses in field crops: a review. Weed Biol Manage 6(1):10–17
Bockstaller C, Girardin P (2003) How to validate environmental indicators. Agr Syst 76(2):639–653
Briggs SA (1992) Basic guide to pesticides: their characteristics and hazards. Taylor & Francis, Washington, DC
Buhler DD (2002) 50th anniversary—invited article: challenges and opportunities for integrated weed management. Weed Sci 50(3):273–280
Bullock DS, D'Arcangelo FM, Desquilbet M (2018) A discussion of the market and policy failures associated with the adoption of herbicide-tolerant crops. TSE working papers 18-959, Toulouse School of Economics (TSE)
Cairns J (1981) The origin of human cancers. Nature 289(5796):353
Carlson GA (1989) Externalities and research priorities in agricultural pest control. Am J Agric Econ 71(2):453–457
Chadwick J, Shaw I (2016) Principles of environmental toxicology. CRC, London
Chapin FS III, Torn MS, Tateno M (1996) Principles of ecosystem sustainability. Am Nat 148(6):1016–1037
Chapin FS III, Matson PA, Vitousek P (2011) Principles of terrestrial ecosystem ecology. Springer Science & Business Media, New York
Chauvel B, Guillemin J, Colbach N, Gasquez J (2001) Evaluation of cropping systems for management of herbicide-resistant populations of blackgrass (Alopecurus myosuroides Huds.). Crop Prot 20(2):127–137
Chauvel B, Guillemin J-P, Gasquez J, Gauvrit C (2012) History of chemical weeding from 1944 to 2011 in France: changes and evolution of herbicide molecules. Crop Prot 42:320–326
Cho YS, Oh SY, Zhu Z (2008) Tyrosine phosphatase SHP-1 in oxidative stress and development of allergic airway inflammation. Am J Respir Cell Mol Biol 39(4):412–419
Colbach N, Colas F, Pointurier O, Queyrel W, Villerd J (2017) A methodology for multi-objective cropping system design based on simulations. Application to weed management. Eur J Agron 87:59–73
Costanza R, de Groot R, Sutton P, Van der Ploeg S, Anderson SJ, Kubiszewski I, Farber S, Turner RK (2014) Changes in the global value of ecosystem services. Glob Environ Chang 26:152–158
Cowan R, Gunby P (1996) Sprayed to death: path dependence, lock-in and pest control strategies. Econ J 106(436):521–542
Curran WS (2016) Persistence of herbicides in soil. Crops Soils 49(5):16–21
Damalas CA (2009) Understanding benefits and risks of pesticide use. Sci Res Essays 4(10):945–949
Damalas CA, Eleftherohorinos IG (2011) Pesticide exposure, safety issues, and risk assessment indicators. Int J Environ Res Public Health 8(5):1402–1419
de Snoo GR, de Wit PJ (1998) Buffer zones for reducing pesticide drift to ditches and risks to aquatic organisms. Ecotoxicol Environ Saf 41(1):112–118
Devine GJ, Furlong MJ (2007) Insecticide use: contexts and ecological consequences. Agric Hum Values 24(3):281–306

Do Carmo R, Alvarez MF (2009) Expansión del cultivo de soja, salud y medio ambiente. Situación en Córdoba (Argentina) y Mato Grosso (Brasil). In: Población y medio ambiente en Latinoamérica y el Caribe: Cuestiones recientes y desafíos para el futuro. Asociación Latinoamericana de Población (ALAP), Rio de Janeiro, pp 183–200

Duarte Vera AC, Batlla D, Ghersa CM, Ferraro DO (2015) Cuando la clave es integrar: introducción al desarrollo de grupos de comportamiento de dormición de malezas en cultivos agrícolas. Rev Agron Ambiente 35(2):153–169

Feder G, Umali DL (1993) The adoption of agricultural innovations: a review. Technol Forecast Soc Chang 43(3–4):215–239

Fenner K, Canonica S, Wackett LP, Elsner M (2013) Evaluating pesticide degradation in the environment: blind spots and emerging opportunities. Science 341(6147):752–758

Fernandes LA, Woodhouse PJ (2008) Family farm sustainability in southern Brazil: an application of agri-environmental indicators. Ecol Econ 66(2–3):243–257

Ferraro DO, Ghersa CM, Sznaider GA (2003) Evaluation of environmental impact indicators using fuzzy logic to assess the mixed cropping systems of the Inland Pampa, Argentina. Agric Ecosyst Environ 96(1–3):1–18

Finnoff D, Shogren JF, Leung B, Lodge D (2005) Risk and nonindigenous species management. Rev Agric Econ 27(3):475–482

Finnoff D, Shogren JF, Leung B, Lodge D (2007) Take a risk: preferring prevention over control of biological invaders. Ecol Econ 62(2):216–222

Freebairn D, King C (2003) Reflections on collectively working toward sustainability: indicators for indicators! Aust J Exp Agric 43(3):223–238

Freemark K, Boutin C (1995) Impacts of agricultural herbicide use on terrestrial wildlife in temperate landscapes: a review with special reference to North America. Agric Ecosyst Environ 52(2–3):67–91

Fukuyama T, Tajima Y, Ueda H, Hayashi K, Shutoh Y, Harada T, Kosaka T (2009) Allergic reaction induced by dermal and/or respiratory exposure to low-dose phenoxyacetic acid, organophosphorus, and carbamate pesticides. Toxicology 261(3):152–161

Ghersa C, Omacini M, Ferraro D, Martinez-Gherza M, Perelman S, Satorre E, Soriano A (2000) Estimación de indicadores de sustentabilidad de los sistemas mixtos de producción en la Pampa interior. Rev Argentina Prod Anim 2(1):49–66

Gibson L, Kingwell R, Doole G (2008) The role and value of eastern star clover in managing herbicide-resistant crop weeds: a whole-farm analysis. Agr Syst 98(3):199–207

Gollapudi BB, Krishna G (2000) Practical aspects of mutagenicity testing strategy: an industrial perspective. Mutat Res Fundam Mol Mech Mutagen 455(1–2):21–28

Goring C, Laskowski D, Hamaker J, Meikle R (1975) Principles of pesticide degradation in soil. In: Environmental dynamics of pesticides. Springer, Boston, pp 135–172

Gould F, Brown ZS, Kuzma J (2018) Wicked evolution: can we address the sociobiological dilemma of pesticide resistance? Science 360(6390):728–732

Grau HR, Aide TM, Gasparri NI (2005) Globalization and soybean expansion into semiarid ecosystems of Argentina. AMBIO J Hum Environ 34(3):265–267

Green JM (2014) Current state of herbicides in herbicide-resistant crops. Pest Manag Sci 70(9):1351–1357

Griliches Z (1957) Hybrid corn: an exploration in the economics of technological change. Econometrica 25:501–522

Griliches Z (1960) Hybrid corn and the economics of innovation. Science 132(3422):275–280

Hansen J (1996) Is agricultural sustainability a useful concept? Agr Syst 50(2):117–143

Hardin G (1968) The tragedy of the commons. Science 162(3859):1243–1248

Hiltbold A, Buchanan G (1977) Influence of soil pH on persistence of atrazine in the field. Weed Sci 25(6):515–520

Hoppin JA, LePrevost CE (2017) Pesticides and human health. In: Environmental pest management: challenges for agronomists, ecologists, economists and policymakers. Wiley, Oxford, pp 251–276

Jakeman AJ, Letcher RA, Norton JP (2006) Ten iterative steps in development and evaluation of environmental models. Environ Model Software 21(5):602–614
Jasieniuk M, Brûlé-Babel AL, Morrison IN (1996) The evolution and genetics of herbicide resistance in weeds. Weed Sci 44:176–193
Jepson PC (1993) Ecological insights into risk analysis: the side-effects of pesticides as a case study. Sci Total Environ 134:1547–1566
Jeyaratnam J (1990) Acute pesticide poisoning: a major global health problem. World Health Stat Q 43(3):139–144
Johnsen K, Jacobsen CS, Torsvik V, Sørensen J (2001) Pesticide effects on bacterial diversity in agricultural soils—a review. Biol Fertil Soils 33(6):443–453
Johnson AC, Besien TJ, Bhardwaj CL, Dixon A, Gooddy DC, Haria AH, White C (2001) Penetration of herbicides to groundwater in an unconfined chalk aquifer following normal soil applications. J Contam Hydrol 53(1–2):101–117
Kamel F, Hoppin JA (2004) Association of pesticide exposure with neurologic dysfunction and disease. Environ Health Perspect 112(9):950
Kamrin MA (1997) Pesticide profiles: toxicity, environmental impact, and fate. CRC, New York
Kogan M (1998) Integrated pest management: historical perspectives and contemporary developments. Annu Rev Entomol 43(1):243–270
Köhler H-R, Triebskorn R (2013) Wildlife ecotoxicology of pesticides: can we track effects to the population level and beyond? Science 341(6147):759–765
Krieger R (2001) Handbook of pesticide toxicology, two-volume set: principles and agents, vol 1. Academic press, New York
Leach A, Mumford J (2008) Pesticide environmental accounting: a method for assessing the external costs of individual pesticide applications. Environ Pollut 151(1):139–147
Leathers HD, Smale M (1991) A Bayesian approach to explaining sequential adoption of components of a technological package. Am J Agric Econ 73(3):734–742
Lechenet M, Bretagnolle V, Bockstaller C, Boissinot F, Petit M-S, Petit S, Munier-Jolain NM (2014) Reconciling pesticide reduction with economic and environmental sustainability in arable farming. PLoS One 9(6):e97922
Leonard R, Knisel W, Davis F (1995) Modelling pesticide fate with GLEAMS. Eur J Agron 4(4):485–490
Lepori ECV, Mitre GB, Nassetta M (2013) Situación actual de la contaminación por plaguicidas en Argentina. Rev Int Contam Ambiental 29:25–43
Levitan L, Merwin I, Kovach J (1995) Assessing the relative environmental impacts of agricultural pesticides: the quest for a holistic method. Agric Ecosyst Environ 55(3):153–168
Lindner RK (1987) Adoption and diffusion of technology: an overview. ACIAR Proc Ser:144–152
Lindner RK, Pardey PG, Jarrett FG (1982) Distance to information source and the time lag to early adoption of trace element fertilisers. Aust J Agric Econ 26(2):98–113
López SL, Aiassa D, Benítez-Leite S, Lajmanovich R, Manas F, Poletta G, Sánchez N, Simoniello MF, Carrasco AE (2012) Pesticides used in south American GMO-based agriculture: a review of their effects on humans and animal models. In: Advances in molecular toxicology, vol 6. Elsevier, Amsterdam, pp 41–75
Luukkanen J, Vehmas J, Panula-Ontto J, Allievi F, Kaivo-oja J, Pasanen T, Auffermann B (2012) Synergies or trade-offs? A new method to quantify synergy between different dimensions of sustainability. Environ Policy Gov 22(5):337–349
Malaj E, Peter C, Grote M, Kühne R, Mondy CP, Usseglio-Polatera P, Brack W, Schäfer RB (2014) Organic chemicals jeopardize the health of freshwater ecosystems on the continental scale. Proc Natl Acad Sci 111(26):9549–9554
Mañas F, Peralta L, Gorla N, Bosh B, Aiassa D (2009) Aberraciones cromosómicas en trabajadores rurales de la Provincia de Córdoba expuestos a plaguicidas. BAG J Basic Appl Genet 20(1):9–13
Mansfield E (1961) Technical change and the rate of imitation. Econometrica 29:741–766
Mansour SA (2004) Pesticide exposure—Egyptian scene. Toxicology 198(1–3):91–115

Martínez-Valenzuela C, Gómez-Arroyo S (2007) Riesgo genotóxico por exposición a plaguicidas en trabajadores agrícolas. Rev Int Contam Ambiental 23(4):185–200

Matthews G, Bateman R, Miller P (2014) Spray of droplets. In: Pesticide application methods, 4th edn. Wiley-Blackwell, Oxford, pp 91–124

McKinlay R, Plant J, Bell J, Voulvoulis N (2008) Endocrine disrupting pesticides: implications for risk assessment. Environ Int 34(2):168–183

Mejer H, Jørgensen SE (1979) Exergy and ecological buffer capacity. In: State-of-the-art in ecological modelling. Elsevier, Amsterdam, pp 829–846

Merhi M, Raynal H, Cahuzac E, Vinson F, Cravedi J, Gamet-Payrastre L (2007) Occupational exposure to pesticides and risk of hematopoietic cancers: meta-analysis of case–control studies. Cancer Causes Control 18(10):1209–1226

Montull J, Soenderskov M, Rydahl P, Boejer O, Taberner A (2014) Four years validation of decision support optimising herbicide dose in cereals under Spanish conditions. Crop Prot 64:110–114

Moss S (2019) Integrated weed management (IWM): why are farmers reluctant to adopt non-chemical alternatives to herbicides? Pest Manag Sci 75(5):1205–1211

Mostafalou S, Abdollahi M (2013) Pesticides and human chronic diseases: evidences, mechanisms, and perspectives. Toxicol Appl Pharmacol 268(2):157–177

Müller F (2005) Indicating ecosystem and landscape organisation. Ecol Indic 5(4):280–294

Neve P, Vila-Aiub M, Roux F (2009) Evolutionary-thinking in agricultural weed management. New Phytol 184(4):783–793

Oerke E-C (2006) Crop losses to pests. J Agric Sci 144(1):31–43

Olson KD, Eidman VR (1992) A farmer's choice of weed control method and the impacts of policy and risk. Rev Agric Econ 14(1):125–137

Ostrom E (2009) A general framework for analyzing sustainability of social-ecological systems. Science 325(5939):419–422

Peralta P, Mañas F, Gentile N, Bosch B, Méndez A, Aiassa D (2011) Evaluación del daño genético en pobladores de Marcos Juárez expuestos a plaguicidas: estudio de un caso en Córdoba, Argentina. Rev Diálogos 2(1):7–26

Petit S, Munier-Jolain N, Bretagnolle V, Bockstaller C, Gaba S, Cordeau S, Lechenet M, Mézière D, Colbach N (2015) Ecological intensification through pesticide reduction: weed control, weed biodiversity and sustainability in arable farming. Environ Manag 56(5):1078–1090

Pimentel D, Burgess M (2014) Environmental and economic costs of the application of pesticides primarily in the United States. In: Integrated pest management. Springer, Berlin, pp 47–71

Pimentel D, Greiner A (1997) Environmental and socio-economic costs of pesticide use. In: Pimentel D (ed) Techniques for reducing pesticide use: environmental an economics benefits. Wiley, Chichester, pp 51–78

Pimentel D, Acquay H, Biltonen M, Rice P, Silva M, Nelson J, Lipner V, Giordano S, Horowitz A, D'amore M (1992) Environmental and economic costs of pesticide use. BioScience 42(10):750–760

Porcel de Peralta M, Scagnetti J, Grigolato R, Sylvestre J, Kleinsorge E, Simoniello M (2011) Evaluación del daño oxidativo al ADN y efecto de la susceptibilidad genética enuna población laboral y ambientalmente expuesta a mezclas de plaguicidas. FABICIB 15:119–129

Pretty J (2007) Agricultural sustainability: concepts, principles and evidence. Philos Trans R Soc B Biol Sci 363(1491):447–465

Prokopy LS, Floress K, Klotthor-Weinkauf D, Baumgart-Getz A (2008) Determinants of agricultural best management practice adoption: evidence from the literature. J Soil Water Conserv 63(5):300–311

Proskocil BJ, Bruun DA, Lorton JK, Blensly KC, Jacoby DB, Lein PJ, Fryer AD (2008) Antigen sensitization influences organophosphorus pesticide–induced airway hyperreactivity. Environ Health Perspect 116(3):381

Prüss-Ustün A, Vickers C, Haefliger P, Bertollini R (2011) Knowns and unknowns on burden of disease due to chemicals: a systematic review. Environ Health 10(1):9

Radosevich SR, Holt JS, Ghersa C (1997) Weed ecology: implications for management. Wiley, New York

Radosevich SR, Holt JS, Ghersa CM (2007) Ecology of weeds and invasive plants: relationship to agriculture and natural resource management. Wiley, New York

Raudsepp-Hearne C, Peterson GD, Bennett E (2010) Ecosystem service bundles for analyzing tradeoffs in diverse landscapes. Proc Natl Acad Sci 107(11):5242–5247

Rohr JR, Salice CJ, Nisbet RM (2016) The pros and cons of ecological risk assessment based on data from different levels of biological organization. Crit Rev Toxicol 46(9):756–784

Rolla AL, Nuñez MN, Guevara ER, Meira SG, Rodriguez GR, de Zárate MIO (2018) Climate impacts on crop yields in Central Argentina. Adaptation strategies. Agr Syst 160:44–59

Rossi V, Salinari F, Poni S, Caffi T, Bettati T (2014) Addressing the implementation problem in agricultural decision support systems: the example of vite.net®. Comput Electron Agric 100:88–99

Sarmah AK, Sabadie J (2002) Hydrolysis of sulfonylurea herbicides in soils and aqueous solutions: a review. J Agric Food Chem 50(22):6253–6265

Schreinemachers P, Berger T (2011) An agent-based simulation model of human–environment interactions in agricultural systems. Environ Model Software 26(7):845–859

Scoullar K, Seely B, Welham C, Kimmins H, Blanco JA (2010) Forecasting forest futures: a hybrid modelling approach to the assessment of sustainability of forest ecosystems and their values. Routledge, London

Simoniello MF, Kleinsorge EC, Scagnetti JA, Mastandrea C, Grigolato RA, Paonessa AM, Carballo MA (2010) Biomarkers of cellular reaction to pesticide exposure in a rural population. Biomarkers 15(1):52–60

Stehle S, Schulz R (2015) Agricultural insecticides threaten surface waters at the global scale. Proc Natl Acad Sci 112(18):5750–5755

Strange RN, Scott PR (2005) Plant disease: a threat to global food security. Annu Rev Phytopathol 43:83–116

Swanton CJ, Weise SF (1991) Integrated weed management: the rationale and approach. Weed Technol 5(3):657–663

Swift M, Anderson J (1994) Biodiversity and ecosystem function in agricultural systems. In: Biodiversity and ecosystem function. Springer, Berlin, pp 15–41

Thurman EM, Goolsby D, Meyer M, Kolpin D (1991) Herbicides in surface waters of the midwestern United States: the effect of spring flush. Environ Sci Technol 25(10):1794–1796

Tisdell CA (2005) Economics of environmental conservation, vol 1. Edward Elgar, Cheltenham

Topping CJ, Craig PS, de Jong F, Klein M, Laskowski R, Manachini B, Pieper S, Smith R, Sousa JP, Streissl F (2015) Towards a landscape scale management of pesticides: ERA using changes in modelled occupancy and abundance to assess long-term population impacts of pesticides. Sci Total Environ 537:159–169

Trappe JM, Molina R, Castellano M (1984) Reactions of mycorrhizal fungi and mycorrhiza formation to pesticides. Annu Rev Phytopathol 22(1):331–359

Truhaut R (1977) Ecotoxicology: objectives, principles and perspectives. Ecotoxicol Environ Saf 1(2):151–173

Tsur Y, Sternberg M, Hochman E (1990) Dynamic modelling of innovation process adoption with risk aversion and learning. Oxf Econ Pap 42(2):336–355

USEPA (2003) Series 870—health effects test guidelines. United States Environmental Protection Agency, Washington, DC

Van Cauwenbergh N, Biala K, Bielders C, Brouckaert V, Franchois L, Cidad VG, Hermy M, Mathijs E, Muys B, Reijnders J (2007) SAFE—a hierarchical framework for assessing the sustainability of agricultural systems. Agric Ecosyst Environ 120(2–4):229–242

Van Ittersum M, Rabbinge R (1997) Concepts in production ecology for analysis and quantification of agricultural input-output combinations. Field Crop Res 52(3):197–208

Vidal RA, Rainero HP, Kalsing K, Trezzi MM (2010) Prospección de las combinaciones de herbicidas para prevenir malezas tolerantes y resistentes al glifosato. Planta Daninha 28(1):159–165

Von Wirén-Lehr S (2001) Sustainability in agriculture—an evaluation of principal goal-oriented concepts to close the gap between theory and practice. Agric Ecosyst Environ 84(2):115–129
Walsh M, Newman P, Powles S (2013) Targeting weed seeds in-crop: a new weed control paradigm for global agriculture. Weed Technol 27(3):431–436
Waterfield G, Zilberman D (2012) Pest management in food systems: an economic perspective. Annu Rev Env Resour 37:223–245
Weber J, Miller C, Sawhney B, Brown K (1989) Reactions and movement of organic chemicals in soils. Soil Sci Soc Am Spl Publ 22:305–333
Weichenthal S, Moase C, Chan P (2010) A review of pesticide exposure and cancer incidence in the agricultural health study cohort. Environ Health Perspect 118(8):1117–1125
Whitford F (2002) The complete book of pesticide management. In: Science, regulation, stewardship, and communication. Wiley-Interscience, New York
Whitmore R, Immerman F, Camann D, Bond A, Lewis R, Schaum J (1994) Non-occupational exposures to pesticides for residents of two US cities. Arch Environ Contam Toxicol 26(1):47–59
Whittaker G, Färe R, Grosskopf S, Barnhart B, Bostian M, Mueller-Warrant G, Griffith S (2017) Spatial targeting of agri-environmental policy using bilevel evolutionary optimization. Omega 66:15–27
WHO (2010) The WHO recommended classification of pesticides by hazard and guidelines to classification 2009. World Health Organization, Geneva
Wilson C, Tisdell C (2001) Why farmers continue to use pesticides despite environmental, health and sustainability costs. Ecol Econ 39(3):449–462
Zbinden G, Flury-Roversi M (1981) Significance of the LD 50-test for the toxicological evaluation of chemical substances. Arch Toxicol 47(2):77–99
Zhang W (2018) Global pesticide use: profile, trend, cost/benefit and more. Proc Int Acad Ecol Environ Sci 8(1):1

# Chapter 9
# Environmental Risk Indicators for Weed Management: A Case Study of Ecotoxicity Assessment Using Fuzzy Logic

**Diego O. Ferraro, Alejandra C. Duarte Vera, Sebastián Pessah, and Felipe Ghersa**

**Abstract** Herbicide use is a key element in the current intensification of agricultural production systems that usually leads to increases in crop yield. However, development of theoretical frameworks and tools is necessary to allow for environmental assessment of herbicides. In this chapter, we present a series of elements that should be considered for designing these types of tools. In addition, we describe the structure of RIPEST, a simple model based on fuzzy logic that evaluates the ecotoxicological hazard of pesticides (herbicides, fungicides and insecticides). RIPEST was run using a time series of pesticide use and actual soybean yields from Argentina. Results from this cropping system assessment allows for discussion of the ecotoxicological risk of herbicide use, in particular, and pesticides, in general.

**Keywords** Sustainability · Risk modelling · Pesticides · Indicators · Decision rules · Eco-efficiency

## 9.1 Introduction

Intensification in modern agriculture includes the use of genetically modified crops, the expansion of agricultural frontiers and the increased use of inputs (Ferraro and Benzi 2015; Qaim and Traxler 2005). These changes positively impacted cropping systems with a significant increase in yields (Foley et al. 2011). However, the potential environmental costs of this intensification process have become a cause for concern (Bommarco et al. 2013). Thus, current agricultural intensification highlighted

D. O. Ferraro (✉) · A. C. Duarte Vera · S. Pessah · F. Ghersa
Facultad de Agronomía, Cátedra de Cerealicultura, Instituto de Investigaciones Fisiológicas y Ecológicas Vinculadas a la Agricultura (IFEVA), UBA-CONICET, Universidad de Buenos Aires (UBA), Buenos Aires, Argentina
e-mail: ferraro@agro.uba.ar; vduarte@agro.uba.ar; spessah@agro.uba.ar; felipeghersa@agro.uba.ar


G. R. Chantre, J. L. González-Andújar (eds.), *Decision Support Systems for Weed Management*, https://doi.org/10.1007/978-3-030-44402-0_9

the importance of assessing potential environmental impacts on agroecosystems. Particularly, rising pesticide use (herbicides, insecticides and fungicides) has been related to both human health and environmental degradation processes (Kim et al. 2017; Arias-Estévez et al. 2008; Imfeld and Vuilleumier 2012). Herbicides account for approximately 50% of all pesticides applied for pest management purposes in the USA and approximately 40% globally (Prosser et al. 2016). The global increase of herbicide-resistant weeds jeopardizes these figures by inferring a potential rise in herbicide dosage required for future weed management. Thus, an understanding and a practical assessment of the impact of agrochemical inputs are essential goals for designing sustainable cropping systems (Pretty 2008). Based on these antecedents, the aims of this chapter are (1) to highlight some key aspects for developing risk assessment frameworks, (2) to show the potential of fuzzy logic for risk modelling, (3) to illustrate the use of fuzzy logic in the development of a hazard assessment model: RIPEST (Risk of PESTicides) and (4) to assess the long-term dynamics (1986–2018) of ecotoxicological hazard due to both pesticide use, in general, and herbicide use, in particular, in one of the main Argentinean cropping systems.

## 9.2 Key Aspects for Risk Modelling

In this section, it is presented a brief synthesis of three key aspects related to the requirements and potential constraints for designing an environmental risk assessment process due to chemical inputs (e.g. herbicides). First, emphasis is placed on the need to define, unequivocally, the idea of risk and the laboratory-scale procedures that are followed to obtain parameters that allow to define risk levels on organisms. Then, the limitations imposed by uncertainty are raised, both in terms of parameters obtained and also regarding the different organization levels involved that imply several interactions between organisms and that will define the environmental risk value in relevant scales (e.g. populations, ecosystems). Finally, the decision-making component that should embraced the risk modelling approaches, highlighting aspects like the simplicity and clarity when defining the risk indicators, and also the ability to communicate conclusions about the environmental risk that may improve the design of agricultural practices involving the use of pesticides.

## 9.3 A Comprehensive Risk Modelling Approach

By definition, risk is a combination of the probability of occurrence of a dangerous event and the severity of the damage or problems that may be caused for that event (Kaplan and Garrick 1981). It follows from this definition that both probability and severity are two key aspects for developing a risk model that allows addressing issues related to exposure (i.e. probability) and hazard (i.e. severity). For example, herbicides are assessed through ecological risk assessments in order to show that its use will cause no unacceptable effects (Thorbek et al. 2009) on non-target organisms (i.e. beneficial arthropods, native flora, local biota). Hazard identification is the process of identifying the effects that are considered to be adverse, and is the first step in ecological risk assessments, especially when dealing with chemical compounds (Renwick 2002). When considering contaminants, hazard is related with preliminary toxicity test, epidemiological data, of adverse effects records (Renwick 2002; Hayes et al. 2004). Once, the hazard is identified, a subsequent process of hazard characterization should be carried out. During this hazard characterization process, in vivo dose–response curves are required for assessing the toxicological properties related to specific compounds. This quantitative step is the basis of standard calculations for hazard assessment such as the no-observed-adverse-effect-level (NOAEL), the lowest-observed adverse-effect-level (LOAEL), the acceptable daily intake (ADI), the reference dose (RfD) or the most widely end point used for pesticide evaluation: the lethal dose 50 (LD50) if exposure occurs through the oral or dermal route, and the lethal concentration 50 (LC50) if the exposure occurs through inhalation (Lewis et al. 2016). A subsequent step in a comprehensive risk modelling approach must include an exposure assessment to estimate the contact of a chemical, physical or biological agent with the outer boundary of an organism (WHO 2002). Finally, the overall risk model is usually based on ratios between predicted environmental concentrations (i.e. exposure) and predicted no-effect concentrations (i.e. no hazard level). Although this process of risk assessment has generated much of the information and knowledge for developing chemical regulations, its application is not exempt from difficulties and limitations (Kramer et al. 2011), which must be considered when defining a final approach for risk assessment for defining, for example, sustainable weed management practices in agricultural systems.

## 9.4 Dealing with Uncertainty

The very definition of risk imposes the notion of uncertainty (Kaplan and Garrick 1981). There are several uncertainty sources but a simple way to group them is into (1) model structure uncertainty, which is the uncertainty about the form of the model itself (including the behavior of the system, parameter estimation uncertainty and the interrelationships among the system elements) and (2) model technical uncertainty, which is uncertainty arising from the computer implementation of the model. The former uncertainty group is crucial for understanding the modelling output because it includes the imperfection of our knowledge and the inherent variability of the phenomena being described. When dealing with environmental risk, much of the uncertainty comes from the difficulty for understanding and parameterizing the changes from small and controllable scales based on few interactions (i.e. an herbicide dose–survival response curve) to larger and more complex scales (i.e. beneficial population size change). Thus, an important aspect of ecological risk assessment is linked to the scales of space and time (Bradbury et al. 2004).

Much of the possibility of understanding the functioning of a system (and thus diagnose the risk associated with a change) has to do with being able to understand how the processes studied at different scales are coupled (Levin 1992). Large spatial and temporal phenomena (such as climate patterns) may generate restrictions that operate on a smaller scale (top-down effects). An example of this may be the modelling the effects of humid climate cycles on crop yields, starting from inferring a cause–effect ratio of the major scale to the minor (Jones 2010). Following this example, it is also possible to consider the effects of scales lower-to-high (bottom-up effects), such as farmers' decision-making or crop genetic improvement, to cope with larger-scale constraints. As far as pesticides are concerned, this integration between the top-down and bottom-up effects is highly relevant, since (as explained in the previous section) ecotoxicology has traditionally based its risk measurement approach through extrapolation from organisms to ecosystems and from small-scale to large-scale systems. This approach is reductionist because it is unable to capture the complex and variable nature of the interaction between biota that occurs in real-world systems at large spatial scales (Beketov and Liess 2012). However, while the quantification of the individual effects is more accessible, the effects on a population or a community imply indirect effects such as the interaction between populations, the possible effects of mitigation in a trophic network or other emergent properties that derive from the interaction between different life forms (Odum 1994).

## 9.5 Focus on Decision-Making Process

The last aspect to consider in relation to risk assessment is its connection to the decision-making process in the use of natural resources. It is known that environmental risk assessments are framed in specific regulations, but they still need to be adapted to the decision-making needs, the quality of the available data and the legally recognized protection objectives (Forbes et al. 2009). However, environmental decisions are generally complex and, in addition to presenting the multiplicity of aspects presented above, they imply a combination of agents with different decision logic (Le et al. 2008). This subjectivity implies that the tools (models) designed to assess environmental risk should be clear enough for the user to evaluate the costs and benefits of a decision. This clarity is based on (1) the explanation of each of the principles used to assess environmental risk, (2) the trade-off assessment among risk components (e.g. herbicide risk on different organisms) or even between legal or economic aspects and (3) a visualization of results according to the logic of the users which make the decisions. For environmental risk assessment to be effective, it is therefore necessary to integrate, in the model structure, some results that may be useful in the user decision-making process. Although, for the purposes of discussions on conservation or ecological processes, pesticide risk models have an important consideration, their real adoption by other agents (farmers, technicians, extensionists) is still limited (Nienstedt et al. 2012). Two key issues emerge that would hinder the adoption of risk models by these agents: (1) the need for modelling outcome as guide towards the adoption of Good Agricultural Practices (GAP) (Hobbs 2003), as well as (2) the scarcity of case studies that can be used to explore the added value of ecological models for risk assessment (Forbes et al. 2009; Schmolke et al. 2010).

A key aspect in relation to the design of environmental risk assessment tools has to do with the way in which the results are transferred. The information provided must have some link to the objectives pursued by the end user. In other words, in an agricultural system, where the objective is to maximize the economic benefit by increasing crop yields; decision makers must be able to evaluate the actions that lead to the risk that has been identified, with a similar logic to the one used to make productive decisions. For example, if a set of herbicides at a specific dose generates a particular environmental impact, the risk model must provide the information of those doses and of these products so that the user (technician, agronomist, producer) can evaluate how their input-cost matrix will change, using other weed management that may eventually improve the environmental performance of the cropping system. This approach may clearly highlight the potential trade-offs between economic results (agricultural income) and environmental and health outcomes (environmental pollution or health risks for both non-target organism and rural populations), and environmental risk models would increase adoption chances to the extent that they show compromised solutions for a given set of resources and technology (Lahr and Kooistra 2010).

## 9.6 A Simple and Clear Risk Assessment Framework: The Fuzzy Logic Approach

The need to include the aforementioned aspects in the environmental assessment process (management of uncertainty, clarity in the assumptions adopted and efficiency in communicating the results) implies having a framework of modelling appropriate to these objectives. Moreover, this analytical framework should be quantitative and be able to integrate different types of information, which are not always expressed in the form of functional relationships based on empirical evidence but may also represent a desirable state in terms of acceptance (or not) of a hazard level.

Fuzzy logic has been used meaningfully in knowledge-based systems for both the knowledge representation and inference mechanisms (Zimmermann 1996; Tan 2005). It is a very flexible framework that allows integration of different types of information to formalize conclusions, and has already been applied to ecosystem assessment in agriculture (Metternicht and Gonzalez 2005; Ferraro 2009); forestry (Iliadis 2005) and social aspects of environmental management (Zhang et al. 2013). Fuzzy logic is capable of handling the goal's ambiguity and is well-suited for eliciting expert knowledge when data is lacking or there is no full agreement in the desirable level of the system key attributes (Fleming et al. 2007).

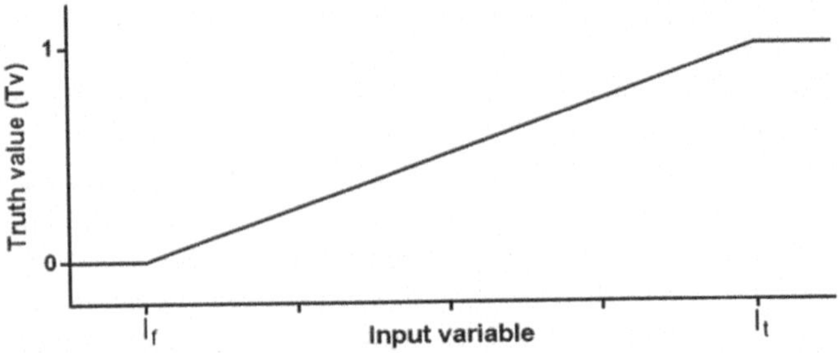


**Fig. 9.1** Generic linear membership function for assessing the truth value of a fuzzy proposition. The function represents the gradual change from the totally false (Tv = 0) to totally true (Tv = 1) over the variable domain. $I_f$ and $I_t$ are the values where the fuzzy proposition becomes 100% false or 100% true, respectively

The structure of a fuzzy model comprises three main elements: (1) the input variables that are related to measurable conditions, (2) the membership functions that functionally relate input variables and system processes or attributes (i.e. fuzzy propositions) and (3) the logical nodes that combine and weight the different processes for assessing the system performance. A membership function (Fig. 9.1) defines the fuzzy nature of the indicators describing the degree to which an event occurs, but not whether it effectively occurs. Membership functions may take any value from the interval [1, 0] representing the full (i.e. "true") or null (i.e. "false") degree to which each proposition occurs, respectively. The contribution of an input variable to the final truth value of the proposition (Fig. 9.1) is expressed in terms of "linguistic variables". Linguistic variables take linguistic values, such as "true" or "false" and "sustainable" or "unsustainable". A linguistic value therefore is a fuzzy subset of a fuzzy proposition; and a membership function defines each linguistic value by determining to what degree (i.e. truth value) an input variable is "true" or "false" (Cornelissen et al. 2001). Therefore, membership functions reflect the knowledge available in the literature and the authors' perception about effect of system input variables on key system processes. Linear membership functions, with two thresholds assigning each parameter value to a "True" and/or "False" class, are more often chosen due to their simplicity for showing continuous changes of the variable under study, from a minimum to a maximum level.

The outputs from fuzzy propositions (i.e. the truth value) are combined by using either logical operators (i.e. logical node) or a set of rules (i.e. rule node) for assessing both the intermediate and the final system performance. A logical node comprise the use of a logical connective in order to evaluate a proposition in terms of the strength of evidence provided by analysis of its subordinate propositions (i.e. its antecedents) (Reynolds 1999). A fuzzy network may use different logical connectives: (1) the (U) union, (2) the OR and (3) the AND operator. The U connective is used to specify that low strength of evidence for one topic can be compensated by strong evidence from another. Arithmetically, the U connective computes the average of its arguments (Marchini 2011). Eventually, when some specific criteria of relative importance need to be represented, the U connective computes the weighted average of its antecedents. The relative importance value can range from 1 (equal importance) to 0 (null importance of the antecedent in the logical node). The OR connective uses the logical max-operator, and implies that the maximum value of the subordinate proposition determines the final value of the proposition evaluated. Oppositely, the AND connective uses the min-operator which implies that final value of the subordinate proposition depends on the minimal value of the antecedents, and represents a precautionary view when assessing ecosystem integrity.

Conversely, the rule nodes are represented by fuzzy rules of the form IF (antecedent)–THEN (consequent). One fuzzy rule consists of K arguments in the form of fuzzy sets Ai,j with membership functions μAj and one response variable r with a numerical conclusion wi:

$$\mathrm{Ri)\,IF\left(x1\,is\,Ai,1\right)AND\left(x2\,is\,Ai,2\right)THEN\,r\,is\,wi} \tag{9.1}$$

where Ri is the i rule, Ai,j are the fuzzy subsets corresponding to a partitioned domain of the input variable xj (j = 1, …, K), r is an output variable and wi is the numerical conclusion of the i rule. For example, in the case that the IF–THEN rule includes the logical connective AND among the antecedents, the intersection of both fuzzy sets is represented through the min-operator:

$$\mathrm{mRi}\left(\mathrm{x1;x2;xK}\right)=\min\left(\mu Ai,;,;,1,;,;,\mu Ai,;,;,2,;,;,\mu Ai,;,;,\mathrm{K}\right) \tag{9.2}$$

where mRi (x1;x2;xK) is the final membership value of the i rule and μAi,1 to K are the observed membership value of 1 to K fuzzy subsets. This final value depends on the minimal value of the antecedents and represents a precautionary view when assessing ecosystem integrity.

In fuzzy logic, the conclusion of each one of the rules has usually expresses using linguistic values (i.e. very high risk or moderate risk). Nevertheless, the logical inference allows for the replacing of these conclusions by numerical values in the interval [0, 1] (Zadeh 1965). Consequently, in all fuzzy rules used, the numerical conclusion is graded between 1 (100% true) and 0 (100% false). Then, fuzzy rules are combined in each module through a rule node. A rule system consists of i = 1 to M rules and can be represented in the form of a matrix with positive integer values Ai,j and wi:

$$\mathrm{R}=\begin{bmatrix} A_{1,1} & \dots & A_{1,K} & w_1 \\ .. & \dots & \dots & \dots \\ A_{M,1} & \dots & A_{M,K} & w_K \end{bmatrix} \tag{9.3}$$

Finally, rules are aggregated (i.e. defuzzification process) with the purpose of transforming the set of numeric conclusions from each module into a single value using different defuzzification methods. There are several defuzzification methods, such as the weighted-average, maximum membership, average maximum membership and centre of gravity (Takagi and Sugeno 1985).

## 9.7 RIPEST: An Example of Pesticide Assessment Model

RIPEST is a simple fuzzy-based model to estimate the ecotoxicological hazard of pesticides in agricultural systems. Its approach is based on the link between the toxicity of the different plant protection products (herbicides, insecticides or fungicides) with its dose used to estimate an environmental potential harmful value. Users can access the RIPEST site to register and create their own information repository. RIPEST can be used free of charge from the website of the Faculty of Agronomy, University of Buenos Aires (http://malezas.agro.uba.ar/ripest/). The use of RIPEST allows to assess the ecotoxicological hazard for (1) insects, (2) mammals and (3) the joint hazard of both impacts. RIPEST is built from ecotoxicological information of 3054 formulations registered in the Argentinean National Service for Sanitary and Quality of Agriculture and Food (SENASA), for extensive grain crops (corn, wheat, sunflower, soybeans, cotton, rice, oats, barley, rye, rapeseed, flax, peanuts and sorghum) (SENASA 2018). In the next sections, the elements of the RIPEST model are described, following the generic structure of a fuzzy logic model described previously: (1) the input variables, (2) the membership functions and (3) the decision rules.

## 9.8 Input Variables

The indicator of pesticide impact used three input variables that describe the toxicity and the amount of active ingredients utilized in each field: (1) oral acute lethal dose 50 for rats, (2) contact acute LD50 for bees and (3) the dose applied for each pesticide application. Therefore, each active ingredient was characterized by means of two different toxicity values: (1) mammal toxicity and (2) insect toxicity. In order to assess the magnitude of the impact of each application, the values of mammal and insect toxicity were measured using the concept of Toxic Units (TU) (Newman 2010).

$$\text{Tmam}\left[\text{TUm}\right] = \text{D} / \text{LD50r} \tag{9.4}$$

$$\text{Tins}\left[\text{TUi}\right] = \text{D} / \text{LD50b} \tag{9.5}$$

where Tmam is the mammal toxicity of each pesticide application, Tins the insect toxicity of each pesticide application, D the dose applied (g formulated product/ha), LD50r the oral acute lethal dose 50 for rats (mg formulated product/1000 g rat weight), LD50b the contact acute lethal dose 50 for bees (μg formulated product/bee) and TUi and TUm the toxic units for insects and mammals, respectively.

When the formulated product includes a single active ingredient the LD50 value is calculated by considering the concentration of the active ingredient in the final formulation. In the case of mixtures, RIPEST applies the principle of additivity for calculating the LD50 value (Altenburger et al. 2003). According to this principle, after the LD50 affect the concentration of each active component of the mixture, the concentrations of all toxic can be summed to obtain a value that can be used to predict toxicity. According to the additive model, the total concentration of a mixture which is expected an effect can be calculated using the following equation (Faust et al. 2000)

$$\frac{100}{TE_{mix}} = \sum_{n} \frac{C_i}{TE_i} \tag{9.6}$$

where TE = Toxicity estimator (in the case of RIPEST, LD50 of rats and bees); Ci = Concentration of component i of the mixture, i = component of the mixture and n = number of ingredients.

After calculating LD50 of a single active ingredient in formulations and mixtures, RIPEST uses the sum of the toxic units (TU) of all the pesticides applied in each field order to calculate the overall toxicity value (Newman 2010; Rose 1998):

$$\text{Sum Tmam}\left[\text{TUm}\right] = \sum_{i}^{n} \text{T mam} \tag{9.7}$$

$$\text{Sum Tins}\left[\text{TUi}\right] = \sum_{i}^{n} \text{T ins} \tag{9.8}$$

where Sum Tmam is the mammal toxicity of all the pesticides applied, Sum Tins the insect toxicity of all the pesticides applied and n the number of pesticide applications on each field, during a single cropping cycle.

## 9.9 Membership Functions

Once defined the input variables, it is necessary to move forward from measurement to assessment. In fuzzy logic, membership functions are primarily involved in this process by defining the fuzzy subsets and the interval shape. In RIPEST, the overall toxicity values (i.e. Sum Tmam and Sum Tins) were used to calculate two different indexes: (1) mammal index (M) and (2) insect index (I) (Fig. 9.2).

These indexes show the level of achievement of the adopted criteria for assessing the pesticide use impact. These criteria are characterized by the fuzzy subsets of "accepted" and "unaccepted" values of Sum Tmam and Sum Tins and the membership function shape. In RIPEST, the criteria adopted are as follows:

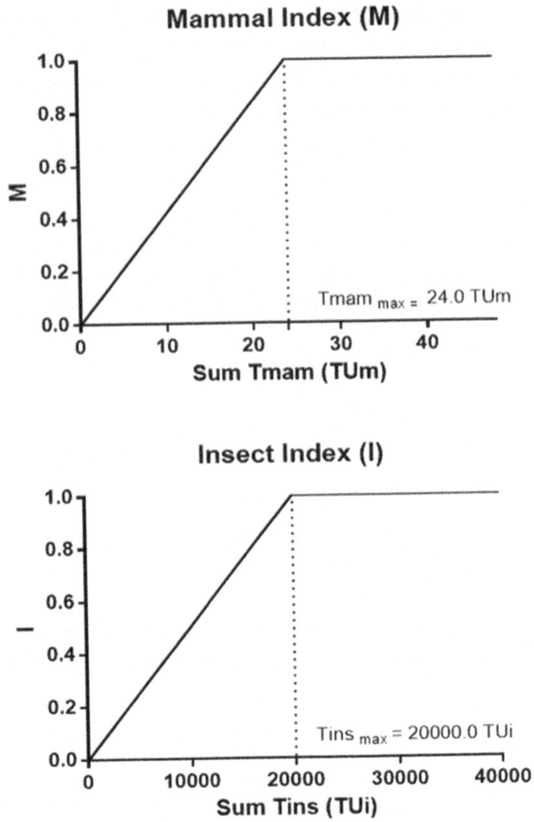


**Fig. 9.2** Membership function for calculating both the I and the M index. Sum Tmam and Sum Tins are the overall mammal and insect toxicity of all the pesticides applied, respectively. $Tmam_{max}$ and $Tins_{max}$ represent the numerical value (in toxic unit) for defining the full membership to the "unacceptable" fuzzy set of each index. Data sources for $Tmam_{max}$ and $Tins_{max}$ are provided in Tables 9.1 and 9.2, respectively

(a) M = 0 and I = 0, correspond to a pesticide use scenario without any pesticide applied (i.e. the "acceptable" fuzzy subsets).
(b) M = 1 (i.e. the "unacceptable" fuzzy subset for mammal toxicity) corresponds to a value of Sum Tmam = $Tmam_{max}$.
(c) I = 1, (i.e. the "unacceptable" fuzzy subset for insect toxicity) corresponds to a value of Sum Tins = $Tins_{max}$.

The full memberships to the fuzzy subsets "unacceptable" (i.e. $Tmam_{max}$ and $Tins_{max}$) are part of the assessment criteria of RIPEST, and they are calculated using both ecotoxicological data and the pesticide maximum dose (Dose max) (Tables 9.1 and 9.2). Value of Dose max are from pesticides registered in the Argentinean National Service for Sanitary and Quality of Agriculture and Food (SENASA 2018). As RIPEST is focused on cropping system assessment, the $Tmam_{max}$ and $Tins_{max}$ values were set up from pesticides registered for the following extensive crops: wheat, barley, rye, oats, corn, sunflower, soybean and cotton. From these pesticide profile, RIPEST selected the pesticides that, applied at its maximum recommended dose, result more toxic (i.e. the maximum TU value) for both mammals (Table 9.1) and insects (Table 9.2).

**Table 9.1** Pesticide data for calculating the unacceptable fuzzy set value of the M index ($Tmam_{max}$)

| Pesticide | Crop | Dose max (g/ha) | LD50r (mg/kg) | $Tmam_{max}$ (TUm) |
|---|---|---|---|---|
| Methidathion 0.4 | Cotton | 1500 | 62.5 | 24.0 |

Dose max: the highest pesticide dose registered in the Argentinean National Service for Sanitary and Quality of Agriculture and Food (SENASA 2018); LD50b: pesticide oral acute lethal dose for bees; TUm: Toxic units for mammals

**Table 9.2** Pesticide data for calculating the unacceptable fuzzy set value of the I index ($Tins_{max}$)

| Active ingredients | Crop | Dose max (g/ha) | LD50b (μg/bee) | $Tins_{max}$ (TUi) |
|---|---|---|---|---|
| Zeta-cypermethrin 0.2 | Corn | 200 | 0.01 | 20,000.0 |

Dose max: the highest pesticide dose registered in the Argentinean National Service for Sanitary and Quality of Agriculture and Food (SENASA 2018); LD50r: pesticide oral acute lethal dose 50 for rats; TUi: Toxic units for insects

## 9.10 Decision Rules

Finally, in order to calculate the overall pesticide impact of pesticides, the (M) and (I) indexes are integrated by two fuzzy rules of the form IF (antecedent)–THEN (consequent) to assemble the pesticide index (P) which indicates the overall impact of pesticide on each analyzed field. All indexes range from 0 (totally unacceptable use of pesticides) to 1 (totally acceptable use of pesticides). In RIPEST the rule node is calculated following Eqs. (9.1), (9.2) and (9.3):

$$\text{R1) IF}\left(\text{M is 1}\right)\text{AND}\left(\text{I is 1}\right)\text{THEN P is } w_1 = 1.0$$

$$\text{R2) IF}\left(\text{M is 1}\right)\text{AND}\left(\text{I is 0}\right)\text{THEN P is } w_2 = 0.1$$

$$\text{R3) IF}\left(\text{M is 0}\right)\text{AND}\left(\text{I is 1}\right)\text{THEN P is } w_3 = 0.1$$

$$\text{R4) IF}\left(\text{M is 0}\right)\text{AND}\left(\text{I is 0}\right)\text{THEN P is } w_4 = 0.0$$

where R1 to 4 are fuzzy rules, M is the mammal index, I is the insect index, P is the Pesticide index, 1 is the full membership condition to the fuzzy subset “unacceptable” of each fuzzy variable, 0 is the full membership condition to the fuzzy subset “acceptable” of each fuzzy variable and $w_1$ to $w_4$ are the numerical conclusions of each rule. The logical connective AND among the antecedents defines the intersection of both fuzzy sets through the min-operator following Eq. (9.2):

$$\text{mR1} = \min\left(\mu M,;,1,;,\mu I,;,1\right)$$

$$\text{mR2} = \min\left(\mu M,;,1,;,\mu I,;,0\right)$$

$$\text{mR3} = \min\left(\mu M,;,0,;,\mu I,;,1\right)$$

$$\text{mR4} = \min\left(\mu M,;,0,;,\mu I,;,0\right)$$

where mR1 to mR4 are the membership value each rule, µM,1 is the observed membership value to the fuzzy subset "unacceptable" of the M index, µM,0 is the observed membership value to the fuzzy subset "acceptable" of the M index. µI,1 is the observed membership value to the fuzzy subset "unacceptable" of the I index and µI,0 is the observed membership value to the fuzzy subset "acceptable" of the I index. Finally, the final membership values of all rules are integrated in a single crisp value by defuzzification process (Takagi and Sugeno 1985). RIPEST use the weighted average method, which can be used only for symmetrical output membership functions. The crisp value according to this method is as follows:

$$P = \frac{\sum_{i=1}^{n} mRi\,wi}{\sum_{i=1}^{n} mi} \tag{9.9}$$

where P is the final crisp value of the fuzzy rule node (i.e. the P index value), mRi is the membership value the rule i and wi is the numerical conclusions of the i rule.

## 9.11 Illustrative Examples

For illustrative purpose, RIPEST was used for building a time series (1986–2018) of ecotoxicological hazard in the main cropping area of Argentina (Rolling Pampa, Pergamino, Buenos Aires). The Rolling Pampa is the subregion of the Río de la Plata grasslands with a cropping history of more than 100 years (Soriano et al. 1991). Traditionally a mixed grazing-crop area, the spread of no-tillage in the early 1990s as well as the wheat–soybean double cropping and the lower cost of inputs (fertilizers, pesticides) lead to a rapid expansion and intensification of agricultural production (Manuel-Navarrete et al. 2009), mainly by the increase in the soybean production area (MinAgri 2018). The change from the conventional tillage system to no-tillage system has also led to a strategy shift for weed control, from a scheme based on tillage to a pesticide-based management strategy. Pesticide time series was built using the annual profile of pesticides used in the soybean crop. Annual pesticide profiles were extracted from the Argentine trade magazine *Márgenes Agropecuarios* (http://www.margenes.com). We also used the district average yield (MinAgri 2018) for this time period in order to link the dynamics of pesticide hazard and crop yield during the period analyzed.

Temporal dynamics of both I and M indexes as well as the overall P index revealed different time trends during the analyzed period (Fig. 9.3).

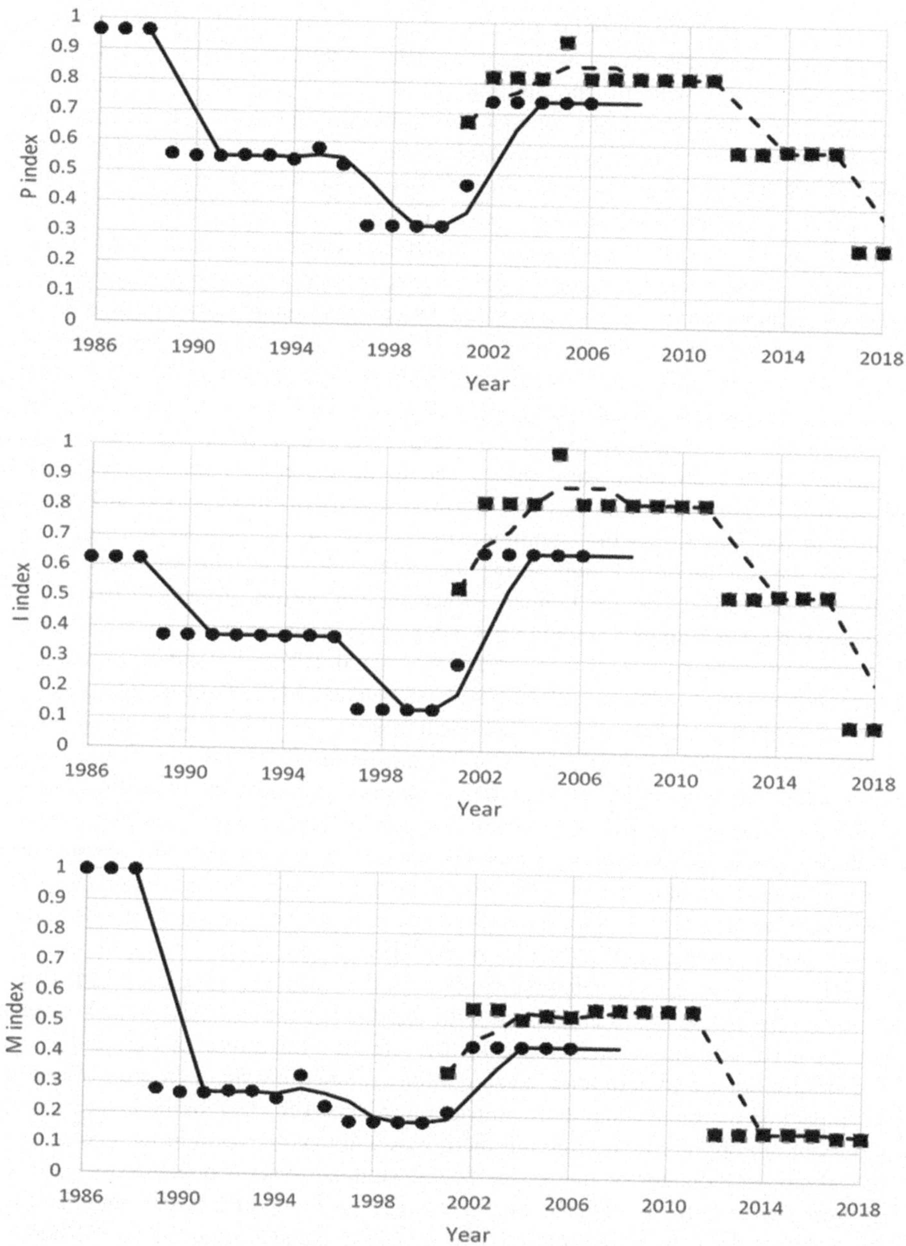


**Fig. 9.3** Time series plots of P, M and I indexes for soybean crop in the main cropping area of Argentina (Pergamino) from 1986 to 2018 presented as annual values (points), and 3-year moving averages (lines). Broken and full lines indicate soybean under conventional and no-tillage systems, respectively

All indexes exhibited a similar dynamic, characterized by high ecotoxicological hazard values at the beginning of the period (1986–1988) followed by a drop in all indexes associated with the prohibition of highly toxic products (Heptachlor, Parathion). The assessment of ecotoxicological hazard on insects (I index) showed a further improvement in the ecotoxicological profile in the mid-1990s, while the M index was stabilized until the beginning of 2000s (Fig. 9.3). By this time, the no-tillage system began to spread in the studied area, by coexisting a few years with conventional tillage (Fig. 9.3). Results from RIPEST showed that the ecotoxicological hazard soared from this technological change, not only in the no-tillage system (which was stabilized at a higher risk value than conventional tillage) but also in the initial system of conventional tillage (Fig. 9.3). The technological changes in the study area were not only due to the tillage system change of farming system but also due to the appearance of RR soybean (resistant to glyphosate herbicide) in 1996. This technological change of higher input levels for weed control, together with an economic context of high grain prices, led to an increase in the pesticide use (CASAFE 2012). Data from RIPEST showed a consequent increased in P index values in the 2000s, reaching similar values to those of mid-1980s (Fig. 9.3). RIPEST indexes remained high during the 2000s, although with relatively lower values for M index than for I index. However, the current decade (2010–2018) shows a remarkable improvement in the indexes analogous to those observed at the beginning of 1990s and towards the end of the period. RIPEST results exhibited the lowest index values in the period analyzed, particularly in the I index (Fig. 9.3).

One remaining aspect in ecotoxicological assessment is the relative impact due to herbicides among all pesticides. Current agricultural systems are experiencing remarkable rise in herbicide use due to the continual difficulty in controlling weeds (Benbrook 2012; Green 2014). Moreover, the rise of herbicide resistance has forced producers to use active ingredients that have ceased to be used, either by the appearance of more efficient products or because of an unacceptable environmental impact (Westwood et al. 2018). Data analysis from the illustrative example is able to show the relative contribution of herbicides in the RIPEST indexes (Fig. 9.4). In this figure, RIPEST indexes were calculated only with the herbicides (Ph, Ih and Mh) and related to the index values calculated with the whole pesticides (P, I and M; from Fig. 9.3). Results show an increase in the relative importance of herbicides for determining the ecotoxicological hazard values, something that is more evident in the impact on mammals than in the impact on insects (Fig. 9.4). In the latter case, it is possible to observe that, although there was a noticeable drop in the ecotoxicological hazard indicator on mammals (Fig. 9.3), this impact is now almost entirely attributed to the lethality of herbicides. These results would then be defining that the system studied has increased the relative impact associated with the use of herbicides, most likely derived from the weed management challenges described above.

As mentioned earlier, the importance of an environmental indicator resides not only in the system assessment but also in the possibility of achieving a rational choice towards more desirable states. The success of this improvement is partly affected by the way the results are reported and it is important to do so using elements used in the process of decision-making by farmers. One example is the

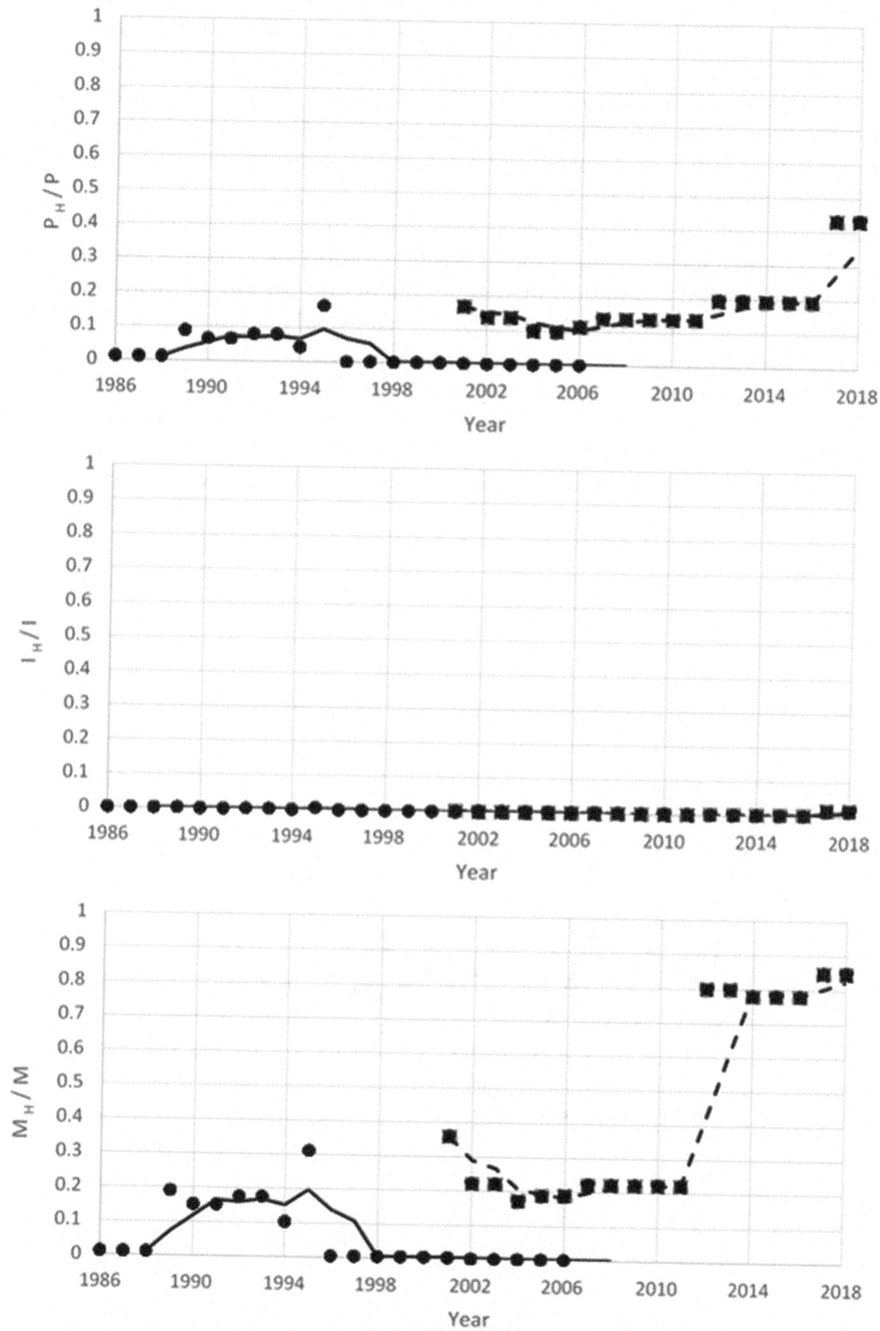


**Fig. 9.4** Time series plots of the ratio between P, M and I indexes calculated using only the herbicides and using all the pesticides ($P_H/P$; $M_H/M$ and $I_H/I$, respectively). Data, points and line pattern description are the same as that in Fig. 9.3

assessment of the environmental impact per unit of economic activity, a very frequent approach in the life cycle analysis (Bockstaller et al. 2008). For an ecotoxicological assessment, this approach can be followed by using the ratio between final crop yield and the ecotoxicological hazard associated to the pesticide profile used. In RIPEST terms, this ratio (P index/yield) should be able to assess the dynamic of a proxy indicator for the environmental efficiency of the crop system analyzed, regarding pesticide use (Fig. 9.5).

RIPEST results first showed a decreasing phase of pesticide hazard per unit of crop yield, which ended by the time of no-tillage adoption. The following period (2000–2005) showed a sustained increase of the hazard per unit of yield obtained, stabilizing P index values of 0.2 per ton of soybean yield until the end of the 2000s (Fig. 9.5). This result supports the idea of Fig. 9.3 about a relative worsening in the soybean cropping system analyzed in terms of the ecotoxicity of the pesticides used during this sub-period. However, the stabilization values achieved towards the end of the 2000s represent half of the observed impacts by the beginnings of the time series analyzed (Fig. 9.3). Finally, by the end of the 2000s there was a decay period

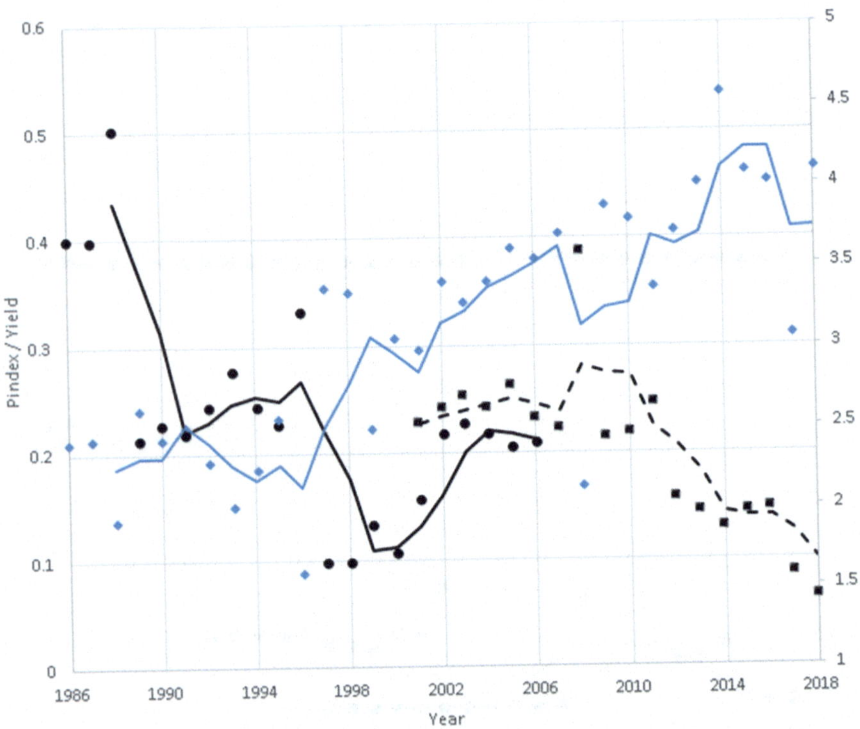


**Fig. 9.5** Time series plots of yield (in blue, right axis) and P index/yield ratio (in black, left axis) for soybean crop in the main cropping area of Argentina (Pergamino) from 1986 to 2018 presented as annual values (points), and 3-year moving averages (lines). Black broken and full lines indicate soybean under conventional and no-tillage systems, respectively

of hazard per unit of yield which implied not only a positive improvement of the ecotoxicological hazard in this last decade (Fig. 9.3) but also a non-appreciable trade-off between this improvement and the observed yields which consistently grow during this period, improving environmental efficiency in the use of pesticides (Fig. 9.5).

## 9.12 Conclusions

The intensification of crop production systems has been achieved, among other elements, by the intensive use of pesticides, with a considerable contribution from herbicides. The need for assessing the environmental consequences of these changes entails also the need for the development of clear indicators, with biological significance, and that express changes that can be incorporated into the farmers' decision-making process. In this chapter we presented the development of an environmental hazard indicator based on fuzzy logic (RIPEST). The simple and explicit characteristic of its structure provides the elements for designing a robust tool for assessing the environmental consequences of the use of pesticides. Illustratively, the use of RIPEST in a time series of pesticide use and soybean yield in the main cropping area of Argentina showed the RIPEST aptitude for identifying trajectories for improvement and deterioration in terms of ecotoxicity of production systems, both in absolute terms of ecotoxicity and in its relation to the obtained yield. Beyond the conclusions that derive from this type of RIPEST applications, it is important to note that risk is a variable that includes not only the hazard but also the severity of the damage and mainly the level of exposure to chemical compounds. Therefore, all these sources of variation should be included in a full risk assessment of pesticide use in agricultural systems. Moreover, within the framework of the decision support systems (DSS), an analytical tool such as RIPEST can play an important role in improving weed management strategies through assessment of environmental damage of potential integrated management decisions.

## References

Altenburger R, Nendza M, Schüürmann G (2003) Mixture toxicity and its modeling by quantitative structure-activity relationships. Environ Toxicol Chem 22(8):1900–1915

Arias-Estévez M, López-Periago E, Martínez-Carballo E, Simal-Gándara J, Mejuto J-C, García-Río L (2008) The mobility and degradation of pesticides in soils and the pollution of groundwater resources. Agric Ecosyst Environ 123(4):247–260

Beketov MA, Liess M (2012) Ecotoxicology and macroecology – time for integration. Environ Pollut 162:247–254. https://doi.org/10.1016/j.envpol.2011.11.011

Benbrook CM (2012) Impacts of genetically engineered crops on pesticide use in the US—the first sixteen years. Environ Sci Eur 24(1):24

Bockstaller C, Guichard L, Makowski D, Aveline A, Girardin P, Plantureux S (2008) Agri-environmental indicators to assess cropping and farming systems. a review. Agron Sustain Dev 28(1):139–149
Bommarco R, Kleijn D, Potts SG (2013) Ecological intensification: harnessing ecosystem services for food security. Trends Ecol Evol 28(4):230–238. https://doi.org/10.1016/j.tree.2012.10.012
Bradbury SP, Feijtel TC, Leeuwen CJV (2004) Peer reviewed: meeting the scientific needs of ecological risk assessment in a regulatory context. Environ Sci Technol 38(23):463A–470A
CASAFE (2012) Cámara Argentina de Sanidad Agripecuaria y Fertilizantes. Evolución del mercado de herbicidas en Argentina (online). www.casafe.org. www.casafe.org
Cornelissen AMG, van der Berg J, Koops WJ, Grossman M, Udo HMJ (2001) Assessment of the contribution of sustainability indicators to sustainable development: a novel approach using fuzzy set theory. Agric Ecosyst Environ 86:173–185
Faust M, Altenburger R, Backhaus T, Bödeker W, Scholze M, Grimme L (2000) Predictive assessment of the aquatic toxicity of multiple chemical mixtures. J Environ Qual 29(4):1063–1068
Ferraro DO (2009) Fuzzy knowledge-based model for soil condition assessment in Argentinean cropping systems. Environ Model Softw 24(3):359–370
Ferraro DO, Benzi P (2015) A long-term sustainability assessment of an Argentinian agricultural system based on emergy synthesis. Ecol Model 306:121–129. https://doi.org/10.1016/j.ecolmodel.2014.06.016
Fleming G, Merwe M, McFerren G (2007) Fuzzy expert systems and GIS for cholera health risk prediction in southern Africa. Environ Model Softw 22(4):442
Foley JA, Ramankutty N, Brauman KA, Cassidy ES, Gerber JS, Johnston M, Mueller ND, O'Connell C, Ray DK, West PC, Balzer C, Bennett EM, Carpenter SR, Hill J, Monfreda C, Polasky S, Rockström J, Sheehan J, Siebert S, Tilman D, Zaks DPM (2011) Solutions for a cultivated planet. Nature 478:337
Forbes VE, Hommen U, Thorbek P, Heimbach F, Van den Brink PJ, Wogram J, Thulke HH, Grimm V (2009) Ecological models in support of regulatory risk assessments of pesticides: developing a strategy for the future. Integr Environ Assess Manag 5(1):167–172
Green JM (2014) Current state of herbicides in herbicide-resistant crops. Pest Manag Sci 70(9):1351–1357. https://doi.org/10.1002/ps.3727
Hayes KR, Gregg PC, Gupta VVSR, Jessop R, Lonsdale WM, Sindel B, Stanley J, Williams CK (2004) Identifying hazards in complex ecological systems. Part 3: Hierarchical Holographic Model for herbicide tolerant oilseed rape. Environ Biosaf Res 3(2):109–128. https://doi.org/10.1051/ebr:2004012
Hobbs JE (2003) Incentives for the Adoption of Good Agricultural Practices (GAPs). Food and Agriculture Organization, Rome
Iliadis LS (2005) A decision support system applying an integrated fuzzy model for long-term forest fire risk estimation. Environ Model Softw 20(5):613
Imfeld G, Vuilleumier S (2012) Measuring the effects of pesticides on bacterial communities in soil: a critical review. Eur J Soil Biol 49:22–30
Jones R (2010) A risk management approach to climate change adaptation. In: RAC N, Wratt DS, Bornman JF, Jones K (eds) Climate change adaptation in New Zealand: future scenarios and some sectoral perspectives. New Zealand Climate Change Centre, New Zealand, pp 10–25
Kaplan S, Garrick BJ (1981) On the quantitative definition of risk. Risk Anal 1(1):11–27
Kim K-H, Kabir E, Jahan SA (2017) Exposure to pesticides and the associated human health effects. Sci Total Environ 575:525–535
Kramer VJ, Etterson MA, Hecker M, Murphy CA, Roesijadi G, Spade DJ, Spromberg JA, Wang M, Ankley GT (2011) Adverse outcome pathways and ecological risk assessment: Bridging to population-level effects. Environ Toxicol Chem 30(1):64–76. https://doi.org/10.1002/etc.375
Lahr J, Kooistra L (2010) Environmental risk mapping of pollutants: state of the art and communication aspects. Sci Total Environ 408:3899–3907
Le QB, Park SJ, Vlek PLG, Cremers AB (2008) Land-Use Dynamic Simulator (LUDAS): A multi-agent system model for simulating spatio-temporal dynamics of coupled human–landscape system. I. Structure and theoretical specification. Eco Informs 3:135–153

Levin SA (1992) The problem of pattern and scale in ecology: the Robert H. MacArthur award lecture. Ecology 73(6):1943–1967

Lewis KA, Tzilivakis J, Warner DJ, Green A (2016) An international database for pesticide risk assessments and management. Hum Ecol Risk Assess Int J 22(4):1050–1064. https://doi.org/10.1080/10807039.2015.1133242

Manuel-Navarrete D, Gallopín G, Blanco M, Díaz-Zorita M, Ferraro D, Herzer H, Laterra P, Murmis M, Podestá G, Rabinovich J, Satorre E, Torres F, Viglizzo E (2009) Multi-causal and integrated assessment of sustainability: the case of agriculturization in the Argentine Pampas. Environ Dev Sustain 11:612–638

Marchini A (2011) Modelling ecological processes with fuzzy logic approaches. In: Modelling complex ecological dynamics. Springer, Berlin, Heidelberg, pp 133–145

Metternicht G, Gonzalez S (2005) FUERO: foundations of a fuzzy exploratory model for soil erosion hazard prediction. Environ Model Softw 20(6):715

MinAgri (2018) Estimaciones agrícolas (Series of agricultural statistics by crop, year, province and department of the Argentine Republic)

Newman M (2010) Acute and cronic lethal effects to individuals. In: Newman MC (ed) Fundamentals of ecotoxicology. Ann Arbor Press, Chelsea, MI, pp 247–272

Nienstedt KM, Brock TC, van Wensem J, Montforts M, Hart A, Aagaard A, Alix A, Boesten J, Bopp SK, Brown C (2012) Development of a framework based on an ecosystem services approach for deriving specific protection goals for environmental risk assessment of pesticides. Sci Total Environ 415:31–38

Odum HT (1994) Ecological and general systems: an introduction to systems ecology. University Press of Colorado, Louisville, CO

Pretty J (2008) Agricultural sustainability: concepts, principles and evidence. Philos Trans R Soc Lond B Biol Sci 363(1491):447–465. https://doi.org/10.1098/rstb.2007.2163

Prosser RS, Anderson JC, Hanson ML, Solomon KR, Sibley PK (2016) Indirect effects of herbicides on biota in terrestrial edge-of-field habitats: a critical review of the literature. Agric Ecosyst Environ 232:59–72. https://doi.org/10.1016/j.agee.2016.07.009

Qaim M, Traxler G (2005) Roundup Ready soybeans in Argentina: farm level and aggregate welfare effects. Agric Econ 32(1):73–86

Renwick AG (2002) Pesticide residue analysis and its relationship to hazard characterisation (ADI/ARfD) and intake estimations (NEDI/NESTI). Pest Manag Sci 58(10):1073–1082. https://doi.org/10.1002/ps.544

Reynolds K (1999) NetWeaver for EMDS version 2.0 user guide: a knowledge base development system. US Department of Agriculture, Forest Service, Pacific Northwest Research Station, Portland, OR

Rose DG (1998) Environmental toxicology: current developments. Gordon and Breach Science Publishers, Amsterdam, The Netherlands

Registro nacional de terpeutica vegetal (National registry of vegetal therapeutics) (2018). http://www.senasa.gob.ar/sites/default/files/ARBOL_SENASA/INFORMACION/PROD%20VETE%20FITO%20Y%20FERTILI/PROD%20FITO%20Y%20FERTILIZANTES/REG%20NAC%20TERAPEUTICA%20VEGETAL/PRODUCTOS%20INSCRIPTOS/formulados_web_ago_2018_2.xls. Accessed Aug 2018

Schmolke A, Thorbek P, DeAngelis DL, Grimm V (2010) Ecological models supporting environmental decision making: a strategy for the future. Trends Ecol Evol 25(8):479–486

Soriano A, León RJC, Sala OE, Lavado RS, Deregibus VA, Cahuépé MA, Scaglia OA, Velázquez CA, Lemcoff JH (1991) Río de la Plata grasslands. In: Coupland RT (ed) Ecosystems of the world 8A. Natural grasslands. Introduction and western hemisphere, vol 19. Elsevier, New York, pp 367–407

Takagi T, Sugeno M (1985) Fuzzy identification of systems and its applications to modeling and control. IEEE Trans Syst Man Cybern 15(1):116

Tan RR (2005) Application of symmetric fuzzy linear programming in life cycle assessment. Environ Model Softw 20(10):1343

Thorbek P, Forbes VE, Heimbach F, Hommen U, Thulke H-H, van den Brink P, Wogram J, Grimm V (2009) Ecological models for regulatory risk assessments of pesticides: developing a strategy for the future. CRC press, Boca Raton, FL
Westwood JH, Charudattan R, Duke SO, Fennimore SA, Marrone P, Slaughter DC, Swanton C, Zollinger R (2018) Weed management in 2050: perspectives on the future of weed science. Weed Sci 66(3):275–285. https://doi.org/10.1017/wsc.2017.78
WHO (2002) The world health report 2002: reducing risks, promoting healthy life. World Health Organization, Geneva, Switzerland
Zadeh LA (1965) Fuzzy sets. Inf Control 8(3):338
Zhang H, Song J, Su C, He M (2013) Human attitudes in environmental management: Fuzzy Cognitive Maps and policy option simulations analysis for a coal-mine ecosystem in China. J Environ Manag 115:227–234. https://doi.org/10.1016/j.jenvman.2012.09.032
Zimmermann HJ (1996) Fuzzy set theory and its applications. Kluwer Academic Publishers, Boston

# Chapter 10
# DRASTIC and PIRI GIS-Based Indexes: Assessing the Vulnerability and Risk of Groundwater Pollution

**Jorgelina C. Montoya, Carolina Porfiri, Pablo M. Vazquez, Sharon Papiernik, Pamela Azcarate, and Zinda Roberto**

**Abstract** Groundwater resources in semiarid lands of central Argentina are currently threatened by contamination from agricultural pesticides. This chapter addresses the vulnerability and risk assessment of the quaternary aquifer system as regards pesticide pollution. We used pesticide DRASTIC Index and PIRI GIS-based models to assess the groundwater vulnerability and leaching potential of commonly used herbicides. DRASTIC and PIRI are two indices that provide complementary information. Incorporating them into a decision support system would help policy makers to identify areas most vulnerable to groundwater pollution and develop herbicide usage guidelines that protect groundwater resources.

**Keywords** Groundwater pollution · Pesticides · Aquifer vulnerability · Leaching potential · Risk assessment · Environmental tools · Risk indexes · GIS

## 10.1 Introduction

Agriculture is one of the main non-point sources of groundwater pollution. The intensive use of pesticides in modern agriculture has had an impact on groundwater quality worldwide (Parris 2011). Therefore, the protection of groundwater quality has become a global concern. It is important to supply local authorities and farmers

J. C. Montoya (✉) · C. Porfiri · P. M. Vazquez · P. Azcarate · Z. Roberto
Estación Experimental Agropecuaria Anguil "Ing. Agr. Guillermo Covas", Instituto Nacional de Tecnología Agropecuaria (EEA Anguil, INTA), Anguil, La Pampa, Argentina
e-mail: montoya.jorgelina@inta.gob.ar; porfiri.carolina@inta.gob.ar; vazquez.pablo@inta.gob.ar; azcarate.pamela@inta.gob.ar; roberto.zinda@inta.gob.ar

S. Papiernik
North Central Agricultural Research Laboratory, USDA Agricultural Research Service, Brookings, SD, USA
e-mail: Sharon.Papiernik@ars.usda.gov


G. R. Chantre, J. L. González-Andújar (eds.), *Decision Support Systems for Weed Management*, https://doi.org/10.1007/978-3-030-44402-0_10

with innovative and practical tools to support sustainable agricultural production. Groundwater pollution risk assessment is an important tool in protecting groundwater from pollution (Zhang et al. 2013). Integrating risk assessments into decision support systems (DSS) can aid in the development of environmental and agricultural policies that help protect the environment and ensure the safety of our food and water supplies. In areas where pesticide contamination of groundwater is a concern, it is critical to define the soils that are vulnerable to pesticide leaching, and to understand pesticide properties and use patterns that may affect their transport to groundwater.

In practice, sampling and monitoring can contribute to the assessment of the environmental impact of pesticides, but this is very costly and will only detect contamination after it happens. The risk of groundwater pollution by pesticides can be estimated based on the vulnerability of groundwater and the leaching potential of pesticides. Knowledge of the regional groundwater hydrogeological conditions and the type of agricultural activities (type and application rate of pesticides, frequency of use, etc.) can be used to assess potential groundwater contamination. Currently, overlay and index methods are among the most widely used tools for groundwater pollution risk assessment. These methods are considered as simple and practical tools that can assist in decision-making and policy formulation to minimize or avoid the negative impacts of pesticides on the natural quality of groundwater (Finizio and Villa 2002; Kookana et al. 2007).

Two of the most widely used models to assess groundwater vulnerability and leaching potential are DRASTIC (Aller et al. 1987) and PIRI (Kookana et al. 2005), respectively. In general, the DRASTIC index estimates groundwater pollution risk based on hydrogeological setting and climatic factors of the region of interest, without taking into account the pollutant characteristics. The pesticide impact rating index (PIRI) can be used to rank pesticides in terms of their mobility.

We used these two approaches to assess the potential risk of groundwater contamination: (1) by evaluating the groundwater vulnerability using DRASTIC, identifying if the hydrogeological system is under threat with respect to pesticides in general, and (2) by using PIRI to rank the capacity of six herbicides to leach to groundwater. Both the DRASTIC and PIRI indices made use of data extracted from national and international public databases and several regional studies.

The assessments were focused on the potential groundwater risk to contamination by herbicides from agricultural lands in the semiarid region of central Argentina. The semiarid lands of central Argentina have poor surface water resources; therefore, groundwater is the primary drinking water source for both urban and rural populations. These lands are an important productive region of Argentina characterized by intensive agriculture highly dependent on herbicides (Viglizzo et al. 2011), so the natural quality of groundwater is projected to be under substantial threat.

## 10.2 General Features of the Study Area

The study area of 27,612 km$^2$ is in the northeast of La Pampa province, displayed (Fig. 10.1) between 63°–64°15′ W and 35°–37°15′ S coordinates, within the semiarid lands of central Argentina. The climate of the study region is semiarid temperate. Most rainfall occurs between the spring and fall seasons (October–March), and average precipitation decreases from 800 to 600 mm per year along a NE-to-SW axis.

The study region includes (1) the sandy plain (SP) situated in the east of the region, and (2) the calcrete plain (CP) in the west of the area. The parent materials of soils are mainly eolic sediments with a low clay and high silt content. Sandy soils are mostly found in the SP sub-region where rainfall is higher and soil depth is not

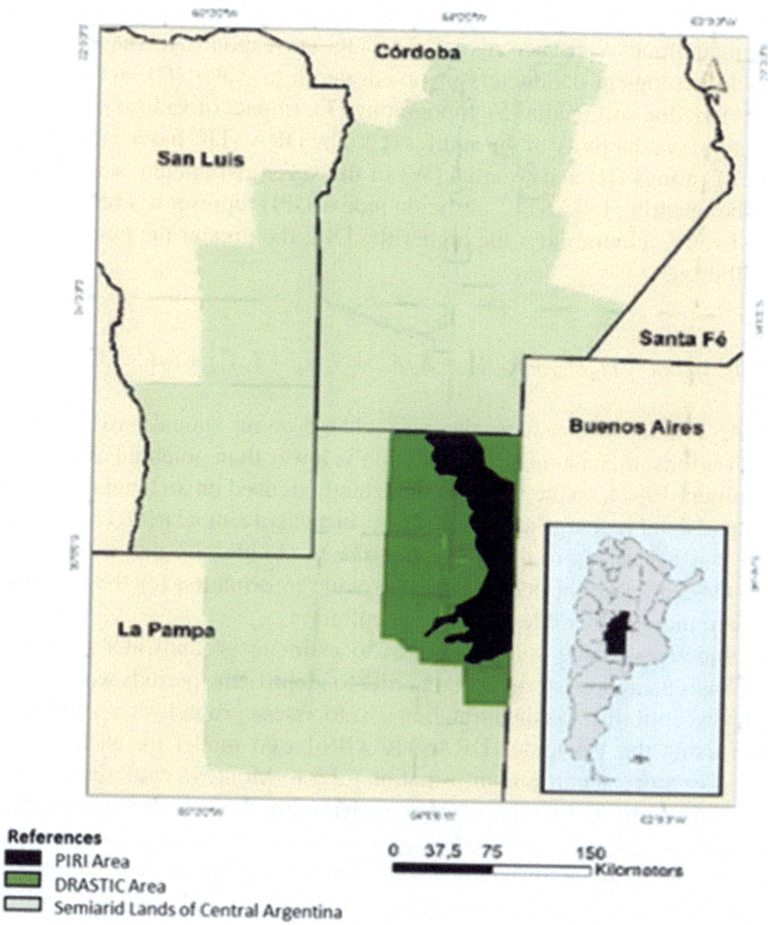


**Fig. 10.1** Location of the study area

limited by a calcrete layer, whereas CP soils have finer loamy textures but are shallow due to the presence of calcrete. The thickness of the upper sand layer in the landscape ranges between 15 m in dunes and 2–3 m in depressions (Malán 1983). In the western sector there is a calcrete/rocky layer close to the surface, underlying the loess deposits. Calcrete formation is characterized by a low permeability, poorly drained layer; a perched water table can be present in the calcrete formation. Calcrete fissures and fractures allow for water percolation to the water table (Giai et al. 2002).

## 10.3 DRASTIC Approach

The DRASTIC index was developed by the US Environmental Protection Agency (Aller et al. 1987) and is one of the most widely used groundwater vulnerability assessment methods. The acronym DRASTIC corresponds to the initials of the seven hydrogeological parameters involved: depth to water (*D*), net recharge (*R*), aquifer (*A*) media, soil media (*S*), topography (*T*), Impact of vadose zone media (*I*), and hydraulic conductivity of the aquifer (*C*). The DRASTIC index is the sum of the products of ratings (R) and weights (W) of the seven parameters according to Eq. (10.1). The resulting DRASTIC pesticide index (DPI) represents a relative measure of groundwater vulnerability: the higher the DPI, the greater the potential groundwater pollution.

$$\text{DRASTIC Index} = D_R D_W + R_R R_W + A_R A_W + S_R S_W + T_R T_W + I_R I_W + C_R C_W \quad (10.1)$$

The DRASTIC Index considers the net recharge on an annual basis. However, in semiarid regions, mean annual precipitation is lower than potential evapotranspiration (Simmers 1997), so the water balance method, used on an annual basis, results in no groundwater recharge at all. Therefore, instead of annual net recharge, monthly recharge was calculated in this study in order to (1) identify the times of the year when water excess might occur and (2) provide information for those times when aquifers are more vulnerable to pesticide pollution.

The objectives of the study were (a) to estimate groundwater recharge on a monthly basis using long-term data, in order to identify the periods when the aquifer has high susceptibility to contamination, (b) to assess groundwater vulnerability to pollution using the Pesticides DRASTIC GIS-based model for each month with recharge. For more details and information refer to Montoya et al. (2018).

Our results indicate that the aquifers under study are susceptible to pollution in March, April and November (Fig. 10.2). According to the results of the groundwater vulnerability assessment, the study area can be divided in two zones: moderate vulnerability (DPI interval 100–159) and high vulnerability (DPI interval 160–266). In these months, low vulnerability zones accounted for <2% of the study area.

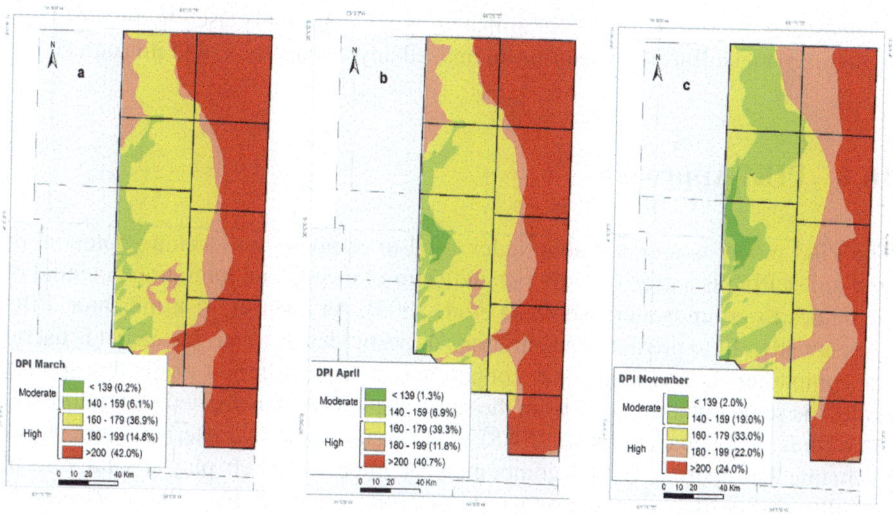


**Fig. 10.2** Maps of the pesticides aquifer vulnerability (DPI) for (**a**) March, (**b**) April, and (**c**) November

The maps revealed that 93.7 and 91% of the area have high risk of pollution for March and April, respectively, and 79% of the area shows high vulnerability in November. For all other months, the lack of recharge placed the entire region in the low vulnerability classification.

Based on the hydrogeological conditions, the three vulnerability maps of the present study showed that eastern lands are highly vulnerable, while central and western lands have moderate vulnerability. The high vulnerability of eastern lands can be explained by a combination of shallow water tables, highly permeable soils, a predomination of sandy components in the vadose zone, and positive water balances in March, April, and November.

The DRASTIC Index may be used as a pollution prevention tool in land use planning at the regional scale through the prioritization of areas where groundwater protection is a critical issue. While the original version of DRASTIC described the spatial vulnerability of lands, we incorporated a temporal scale in our study in order to identify periods of higher vulnerability. The temporal analysis also aids in implementing precise management strategies for dominant crops such as soybean, maize and sunflower. Based on recharge, November is a highly vulnerable month that corresponds to the sowing date of summer crops, so agrochemicals applied at sowing should be managed to prevent leaching. The index indicates that the aquifer is less vulnerable in spring than in March and April. At this moment the environmental and productivity factors indicate a yellow light related to the risk of groundwater contamination by pesticides in this region.

The vulnerability zonification provided by DRASTIC allows the user to focus the assessment of potential herbicide leaching to the most vulnerable scenarios. In this way, we modeled the potential of the SP, being the area most prone to groundwater pollution, to be contaminated by six commonly used herbicides (sulfometuron, metsulfuron, nicosulfuron, imazethapyr, imazamox, and imazapyr).

## 10.4 PIRI Approach

The PIRI model is a quantitative index used to estimate the leaching potential of pesticides. PIRI is a user-friendly risk indicator of the off-site migration potential of pesticides to groundwater (Kookana et al. 2005). As a simple risk indicator, PIRI does not attempt to predict concentrations of pesticides in water. Instead, it is useful for comparing the risks of water contamination from different pesticides applied under the same conditions, or from the same pesticide when applied under different conditions (Kookana and Oliver 2018). It is based on soil characteristics and physicochemical properties of the compounds of interest, which play a vital role in mobility and consequent contamination of groundwater. Indices like PIRI help to identify which pesticides have a relatively high risk to contaminate groundwater. Potential off-site pesticide movement is estimated using pesticide characteristics (amount used, sorption, and persistence in soil), soil physicochemical properties (organic carbon content, bulk density, and soil moisture), and other site conditions (recharge rate, soil layers/horizons, water table depth, etc.) (Kookana et al. 2005). Sorption coefficients are among the most sensitive input parameters in pesticide fate models, and accounting for the variation in sorption coefficients within soil landscapes could reduce uncertainties in pesticide environmental risk assessments when scaling up from the landscape to regional and national scales (Farenhorst et al. 2001).

The sulfonylurea (SU) herbicides are used to control broadleaf weeds and some grasses in a variety of crops. Sulfonylureas Tolerant Soybean (STS®) was launched to the Argentinian market in 2012 under the brand Ligate® (Sulfometuron methyl 15% + Clorimuron ethyl 20%, DuPont). SU herbicides have been considered reduced risk pesticides, in part because of their low application rates (4–50 g ia ha$^{-1}$) (Beyer et al. 1988; Hay 1990). However, there are several environmental concerns. Even at low application rates, these herbicides can persist in the soil throughout more than one growing season and may injure rotational crops (Bedmar et al. 2006). They also have a potential for off-site transport. Therefore, it is important to characterize their mobility in soil to better assess their impact on the environment.

Imidazolinones (IMI) represent a relatively new class of herbicides that can be used either pre- or post-emergence for the control of a wide range of weeds in broadleaf and cereal crops, and in non-crop situations. In addition, IMIs are used with imidazolinone-tolerant (Clearfield®) crops (Tan et al. 2005). In Argentina, much of the soybean crop is treated with imazethapyr. During the years 2005 to 2009, imazethapyr ranked as the 17th to 30th most used pesticide according to the Argentinian pesticide market reports (CASAFE 2019). In 2003, BASF SA registered the herbicide imazapyr for sunflower under the brand CLEARSOL® (24%), which was later available as CLEARSOL® DF (80%). During 2003 and 2004, imazapyr was the 24th most used herbicide in Argentina (CASAFE 2019). In 2010, the

Clearfield system launched to the market a new herbicide called CLEARSOL® Plus (BASF SA) which contains both imazamox (3.3% W/V) and imazapyr (1.5% W/V).

The goal of this study was to evaluate the groundwater contamination potential of three SU and three IMI herbicides in the SP sub-region. This work integrated the PIRI Index with a GIS into a spatial domain and included factors relating to the frequency with which each pesticide was used in the typical cropping sequence and the projected land area to which each herbicide was applied.

PIRI uses a modified version of the attenuation factor index (AF) and also incorporates the retardation factor (RF), both developed by Rao et al. (1985), for assessment of pesticide leaching in groundwater. AF serves as an index for pesticide leaching from the vadose zone, and its value ranges between 0 and 1 (Eq. 10.2). An AF value of 1 indicates that all of the surface-applied pesticide is likely to leach to the groundwater, whereas a value of 0 suggests that none of the applied pesticide will reach groundwater.

$$\mathrm{AF}_{\mathrm{GW}} = \exp\left[\frac{-\left(0.693 D\theta_{\mathrm{FC}}\mathrm{RF}\right)}{q t_{1/2}}\right] \tag{10.2}$$

where $D(m)$ is the depth to the groundwater or the depth ($m$) at which AF is to be calculated, $\theta_{FC}$ ($m^3\ m^{-3}$) is the volumetric water content at field capacity of soil, RF is the dimensionless retardation factor, $q$ is the groundwater recharge (m year$^{-1}$), and $t_{1/2}$ (year) is the pesticide degradation half-life (year).

RF is the retardation factor, which represents the retardation of pesticide leaching through soil due to partitioning of the pesticide between the sorbed and liquid phases (Eq. 10.3)

$$\mathrm{RF} = 1 + \left(K_{\mathrm{oc}} f_{\mathrm{oc}} \rho_{\mathrm{b}} / \theta_{\mathrm{FC}}\right) \tag{10.3}$$

where $\rho_b$ (kg m$^{-3}$) is the bulk density of the soil (kg m$^{-3}$), $f_{oc}$ (kg kg$^{-1}$) is the organic carbon content of the soil, and $K_{oc}$ ($m^3$ kg$^{-1}$) is the organic carbon normalized sorption coefficient of the pesticide (Table 10.1).

**Table 10.1** Recommended annual rates for each crop expressed as formulated products (fp), half-life ($t_{½}$), range of the normalized organic carbon sorption coefficient ($K_{oc}$), and retardation factor (RF) obtained for the studied soils profiles (0 to 100 cm depth)

| Herbicides | Crops | Rate fp kg ha$^{-1}$ | $t_{½}$ Days | $K_{oc}$ L kg$^{-1}$ | RF |
|---|---|---|---|---|---|
| Metsulfuron | Wheat | 0.008 | 30 | 0–42.6 | 1.00–2.13 |
| Sulfometuron | STS® soybean | 0.100 | 28 | 3.2–100.1 | 1.05–1.93 |
| Nicosulfuron | Maize | 0.070 | 21 | 32.3–363.4 | 1.36–9.16 |
| Imazamox | Clearfield® sunflower | 1.200 | 30 | 13.7–1166 | 1.42–9.82 |
| Imazethapyr | Soybean | 0.114 | 90 | 16.6–2332 | 1.43–11.13 |
| Imazapyr | Clearfield® sunflower | 0.100 | 142 | 22–2469 | 1.43–14.97 |

The expected impact on the groundwater is the product of the loading factor ($L_i$) (Eq. 10.4) and AF of each pesticide by year.

$$L_i = f_i\, d_i\, a_i\, p_i \tag{10.4}$$

where $f_i$ is the frequency of application, $d_i$ is the application rate (kg m$^{-2}$), $ai_i$ is the proportion of active ingredient in the product (kg kg$^{-1}$), and $p_i$ is the proportion of the study area that receives the pesticide (m$^2$ m$^{-2}$).

The total load of a pesticide that is likely to reach groundwater at a site is given by Eq. (10.5).

$$L_{\mathrm{GW}} = \Sigma L_i\, \mathrm{AF}_{\mathrm{GW}i} \tag{10.5}$$

It was assumed that the concentration of a pesticide in the groundwater is a function of the mixing of the pesticide residue in certain thickness of the aquifer and aquifer porosity ($n$). Assuming the top 1 m is the mixing zone (MZ) in the aquifer (a conservative estimate), the predicted concentration ($C_{\mathrm{GW}i}$ in kg m$^{-3}$) of pesticide in the top 1 m of the saturated zone is (Eq. 10.6) as follows:

$$C_{\mathrm{GW}i} = \left[ L_i\, \mathrm{AF}_{\mathrm{GW}i} \frac{1}{n} \right] \mathrm{MZ} \tag{10.6}$$

This concentration (Eq. 10.7) can be related to the detection limit of the method of analysis in a monitoring program. It can also be compared to the acceptable concentration of a pesticide in groundwater. Since the WHO Guidelines for Drinking-Water Quality values for SU and IMI have not been established, for the calculation of the groundwater risk index we considered that the acceptable concentration would not exceed the EU-maximum concentration limit (MCL) of 0.1 μg L$^{-1}$.

$$\mathrm{GRIndex} = \frac{C_{\mathrm{GW}}}{\mathrm{MCL}} \tag{10.7}$$

### *10.4.1 Spatiotemporal Integration of the Results with a GIS*

In order to assess the regional impact of non-point source herbicides pollution on groundwater, the PIRI Index was integrated with a GIS (Pollock et al. 2005). The annual $C_{\mathrm{GW}i}$ was calculated by PIRI for individual soil profiles that represent shoulder, middle, and foot slope positions. These values were integrated with the digital soil map in a GIS. To estimate the cumulative concentration of a pesticide, it was necessary to know the proportion that each crop is sowed during the 10-year period ($R$) and the proportion area occupied by crops (PC) in this region. $R$ was estimated by county using data provided by the Government of La Pampa province, Statistics

**Table 10.2** Mean annual sowed area expressed by county in hectares with the different crops during the studied period 2005–2015

| County | Total | Sunflower | Maize | Soybean | Others summer crops | Wheat | Others winter crops |
|---|---|---|---|---|---|---|---|
| Agustoni | 18,319 | 4072 | 4013 | 6975 | 2259 | 543 | 457 |
| Alta Italia | 12,855 | 3375 | 1917 | 2793 | 1892 | 2309 | 569 |
| Anguil | 25,824 | 8427 | 5611 | 1622 | 2670 | 4626 | 2868 |
| Catrilo | 28,500 | 10,562 | 5012 | 6642 | 3993 | 1476 | 815 |
| Ceballos | 22,867 | 3086 | 4461 | 10,973 | 3023 | 1016 | 308 |
| Colonia Barón | 21,696 | 5903 | 4219 | 3925 | 3156 | 3086 | 1407 |
| Coronel Hilario Lagos | 9513 | 781 | 2102 | 4549 | 1473 | 476 | 132 |
| Dorila | 14,249 | 4231 | 2601 | 4151 | 2303 | 741 | 222 |
| Falucho | 7101 | 1138 | 1236 | 3217 | 867 | 345 | 298 |
| General Pico | 60,487 | 11,212 | 12,254 | 22,391 | 8140 | 3679 | 2811 |
| Intendente Alvear | 37,145 | 3198 | 7347 | 19,841 | 4568 | 1250 | 941 |
| Lonquimay | 21,694 | 7595 | 3427 | 5240 | 2731 | 1884 | 817 |
| Miguel Cané | 10,517 | 4146 | 1685 | 2997 | 1129 | 223 | 337 |
| Quemú | 37,198 | 10,086 | 6791 | 12,402 | 4687 | 1714 | 1518 |
| Relmo | 3330 | 1400 | 688 | 900 | 314 | 0 | 28 |
| Sarah | 8192 | 596 | 1261 | 4796 | 949 | 580 | 10 |
| Speluzzi | 6671 | 2134 | 1385 | 1440 | 1052 | 260 | 400 |
| Uriburu | 24,965 | 9117 | 4512 | 3581 | 2490 | 3912 | 1353 |
| Vertiz | 12,885 | 4324 | 1786 | 4366 | 1578 | 549 | 282 |
| **Total** | **384,008** | **95,383** | **72,308** | **122,801** | **49,274** | **28,669** | **15,573** |

and Census Bureau: 2005–2015 (Table 10.2). PC was estimated by analyzing satellite image time series (Modis TERRA MOD13Q1 with spatial resolution = 250 m) during the period 2005–2015 (Vázquez et al. 2013) (Fig. 10.3). Values were summed to estimate the cumulative concentration of pesticides in the aquifer after 10 years of agricultural activity (CC) (Eq. 10.8).

$$CC_{GW10} = \left( \sum_{10}^{i=1} C_{GWi}\, R_i \right) PC_i \quad (10.8)$$

In Argentina, the information regarding the use of herbicides and technological adoption is scant, and this makes it difficult to define scenarios for the use of pesticides. In this study, it was assumed that 100% of wheat area was treated with metsulfuron; and 100% of soybean area was treated with imazethapyr (Table 10.3). Although STS® Soybean did not completely displace non-STS soybeans, given the lack of information it was assumed that 100% of soybeans were treated with sulfometuron. The proportion of sunflowers with Clearfield technology increased by 6% of the area during 2005–2015, reaching 75% of sunflowers planted in the southern region of the country. We assumed an average value of 50% of sunflowers with

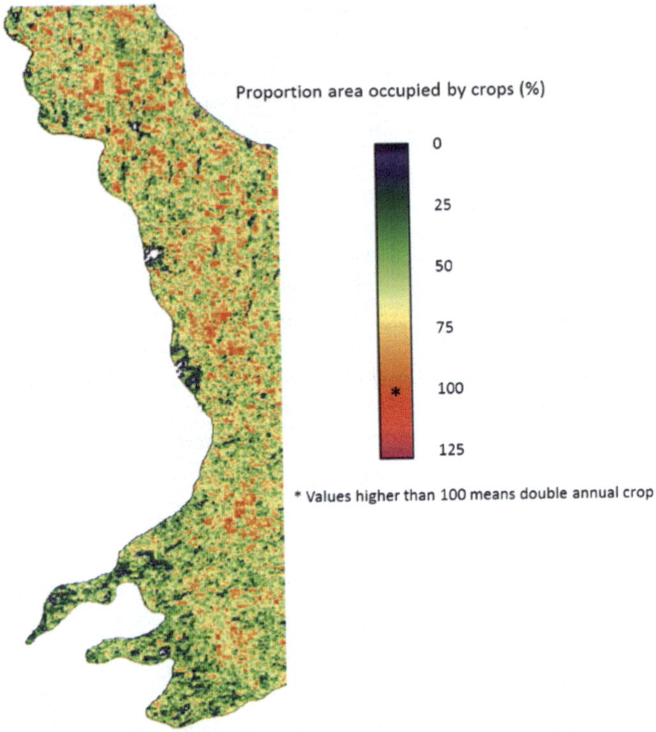


Fig. 10.3 Proportion area occupied by crops

Table 10.3 Proportion (%) of the area of crops treated with herbicides

| Herbicides | Sunflower | Maize | Soybean | Wheat |
|---|---|---|---|---|
| Metsulfuron | – | – | – | 100 |
| Sulfometuron | – | – | 100 | – |
| Nicosulfuron | – | 100 | – | – |
| Imazamox | 50 | – | – | – |
| Imazethapyr | – | – | 100 | – |
| Imazapyr | 50 | – | – | – |

this technology. In the early 2000s nicosulfuron was used massively (CASAFE 2019). Nowadays not much information is available about its use; therefore, it was assumed that 100% of maize crop was treated with nicosulfuron.

From the spatiotemporal integration of 10 years' crop sequence, we found that the $CC_{GW10}$ varied by herbicide and soil properties. PIRI estimated cumulative concentrations of imazethapyr and imazapyr in groundwater that exceed the EU's MCL of 0.1 μg $L^{-1}$. Low concentrations of imazamox and the SU herbicides were estimated (Fig. 10.4). Both SUs and IMIs have very low sorption to these soils, and they are used with low application rates. Imazapyr has long persistence in soil with

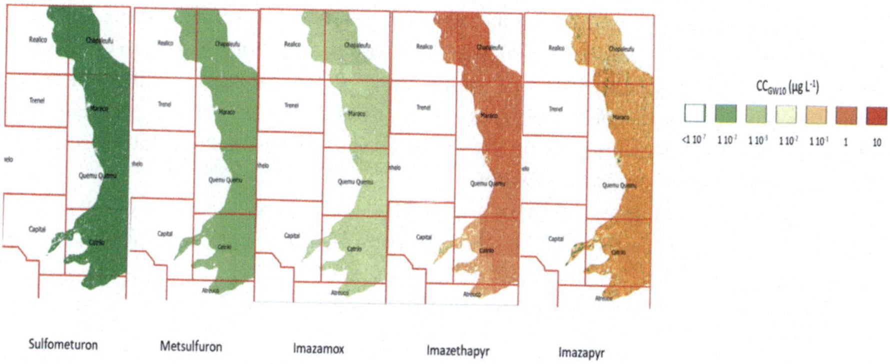


**Fig. 10.4** Estimation of the cumulative concentrations of pesticides in the groundwater ($CC_{GW10}$) for the herbicides

**Table 10.4** Spatiotemporal integration of GRIndex for each herbicide

| Herbicide | GRIndex | Classification |
|---|---|---|
| Nicosulfuron | <<<1 × $10^{-6}$ | Very low |
| Sulfometuron | <1 × $10^{-6}$ | Very low |
| Metsulfuron | 1 × $10^{-6}$ to 1 × $10^{-2}$ | Low |
| Imazamox | 1 × $10^{-1}$ to 1 × $10^{-1}$ | Low |
| Imazapyr | 1–10 | High |
| Imazethapyr | 1–10 and 10–100 | High to very high |

half-life of 142 days. Imazethapyr has intermediate persistence, and imazamox and the other tested herbicides have shorter half-lives, 30 days or less (Table 10.4). Imazethapyr has shorter half-life than imazapyr; however, we assumed 122,801 ha year$^{-1}$ was treated with imazethapyr, compared with 47,691 ha year$^{-1}$ treated with imazapyr. For herbicides like imazapyr and imazethapyr with relatively long half-life and low sorption to soils, the treated area becomes important for the prediction of $CC_{GW10.}$ Regional groundwater monitoring studies confirm these results. Porfiri et al. (2017) analyzed 758 groundwater samples from a rural area for atrazine and imazapyr. Of the 159 samples that contained imazapyr, 65% contained <5 µg L$^{-1}$, 22% contained 5–20 µg L$^{-1}$, 7.5% contained 20–40 µg L$^{-1}$, and 5% contained >40 µg L$^{-1}$ imazapyr. High concentrations of imazapyr suggested non-point and point source contamination. Imazapyr has been detected in 3.1% of groundwater samples in Argentina at concentrations ranging between 0.2 and 6.4 µg L$^{-1}$ (Montoya et al. 2018).

For this simulation, we assumed that 100% of soybean land was treated with imazethapyr and sulfometuron. Results for the spatial PIRI Index showed greater estimated concentrations of imazethapyr than of sulfometuron in the top 1 m of the superficial aquifer because of the longer persistence and higher application rate of imazethapyr than sulfometuron. Metsulfuron showed low risk of groundwater pollution, in part because it is only used on wheat (28,669 ha year$^{-1}$), it has short persistence, and the application rate is very low. Nicosulfuron has a short half-life (21 days) and was not predicted to reach groundwater (map not shown), even though it was modeled as applied to 100% of maize land, 72,300 ha year$^{-1}$. Although these herbicides have low sorption capacity to the studied soils, their usage, persistence, and application rates become important in the prediction of the cumulative concentration of pesticides in the groundwater.

The spatiotemporal integration of the GRIndex showed that imazethapyr and imazapyr have risk of leaching and polluting groundwater at concentrations exceeding the MCL established by the EU (Table 10.4).

## 10.5 Conclusions

In summary, the DRASTIC vulnerability maps revealed that groundwater in the northeast portion of La Pampa province is under high-to-moderate vulnerability to pollution during the months that aquifer recharge occurs. The SP of the study area is dominated by high pollution vulnerability class, and this is very strongly related to shallow groundwater systems, highly permeable sediments, and periodic positive net recharge.

The PIRI Index was calculated to estimate the relative risk of groundwater contamination of six commonly used herbicides. The PIRI Index offers a powerful tool to estimate groundwater contamination potential by herbicides. Because PIRI takes into account herbicide persistence, use patterns, and hydrogeological factors in addition to herbicide sorption data, it may provide a more accurate prediction of the risk of off-site transport of herbicides compared with an RF model or other simple models. The inclusion of the frequency and regional extent of herbicide usage produce a more accurate prediction of groundwater contamination by pesticides.

In agricultural soils of the semiarid region of Argentina, SU herbicides are applied at a low rate and typically once a year; they have relatively low sorption and low persistence. For SUs, the PIRI-GIS predicted a low risk of groundwater contamination. On the other hand, IMIs have similar application patterns, low sorption, and higher persistence in these soils; PIRI-GIS indicated a substantial risk that imazethapyr and imazapyr may leach to groundwater. The results of this study indicate the need to minimize the use of persistent herbicides and promote IWM practices to prevent the detrimental environmental effects of herbicide use on Argentina's water quality.

The DRASTIC and PIRI indices provide complementary information. Embedding them in a DSS would allow policy makers to identify the area most vulnerable to groundwater pollution and define effective scenarios of herbicide management to support the sustainable use of agrochemicals. Integrating these tools into a GIS framework provides a graphic demonstration of the projected consequences of herbicide usage. These tools must be coupled with good alternative pest management practices to minimize environmental risk.

**Acknowledgments** Reprinted by permission from Springer Nature: Springer, Sustainable Water Resources Management. Assessing the vulnerability of groundwater resources in semiarid lands of central Argentina. Montoya JC, Porfiri C, Roberto ZE, Viglizzo EF. © Springer International Publishing AG, part of Springer Nature (2018).

## References

Aller L, Bennet T, Leher JH et al. (1987) DRASTIC A standardized system for evaluating groundwater pollution potential using hydrogeologic settings. In: US Environmental Protection Agency report (EPA/600/2–87/035). p 622

Bedmar F, Perdigon JA, Monterubbianesi MG (2006) Residual phytotoxicity and persistence of chlorimuron and metsulfuron in soils of Argentina. J Environ Biol 27:175–179

Beyer EM, Duffy MJ, Hay JV et al (1988) Sulfonylureas. In: Kearney PC, Kaufman DD (eds) Herbicides: chemistry, degradation and mode of action. Marcel Dekker, New York, pp 117–189

CASAFE (2019) Datos del Mercado Argentino de fitosanitarios. https://www.casafe.org/publicaciones/datos-del-mercado-argentino-de-fitosanitarios/. Accessed 02 May 2019

Farenhorst A, Muc D, Monreal C, Florinski I (2001) Sorption of herbicides in relation to soil variability and landscape position. J Environ Sci Health 36:379–387

Finizio A, Villa S (2002) Environmental risk assessment for pesticides: a tool for decision making. Environ Impact Asses Rev 22:235–248

Giai S, Visconti G, Peinetti H (2002) La formación de falsas capas freáticas en Eduardo Castex, Provincia de La Pampa. EdUNLPam, Gil N° 353, Tercer piso, Santa Rosa, La Pampa (CP 3600), Argentina. Huellas 7:66–82

Hay JV (1990) Chemistry of sulfonylurea herbicides. Pest Sci 29:247–261

Kookana RS, Oliver DP (2018) Environmental risk indicators: their potential utility in pesticide risk management and communication. In: Maestroni B, Cannavan A (eds) Integrated analytical approaches for pesticide management. Academic Press, New York, pp 197–206

Kookana RS, Correll RI, Miller RB (2005) Pesticide impact rating index—a pesticide risk indicator for water quality. Water Air Soil Pollut 5:45–65. https://doi.org/10.1007/s11267-005-7397-7

Kookana RS, Kumar A, Oliver DP et al (2007) Pesticide risk indicators: their role in minimizing off-site impacts of pesticides on water quality. In: Kennedy IR, Solomon KR, Gee SJ et al (eds) Rational environmental management of agrochemicals. Risk assessment, monitoring, and remedial action, vol 966. American Chemical Society, Washington, DC, pp 37–52

Malán JM (1983) Estudio hidrogeológico para el abastecimiento de agua potable a la localidad de General Pico, departamento Maracó, provincia de La Pampa. In: Informe preliminar. Hidrología de las grandes llanuras: Actas del Coloquio de Olavarria, 1983 (1ª ed.). UNESCO/Comité Nacional Argentino para el Programa Hidrológico Internacional, Buenos Aires, pp 1449–1457

Montoya JC, Porfiri C, Roberto ZE et al (2018) Assessing the vulnerability of groundwater resources in semiarid lands of Central Argentina. Sustain Water Resour Manag:1–16. https://doi.org/10.1007/s40899-018-0246-4

Parris K (2011) Impact of Agriculture on Water Pollution in OECD Countries: Recent Trends and Future Prospects. International Journal of Water Resources Development 27(1):33–52

Pollock DW, Kookana R, Correll RL (2005) Integration of the pesticide impact rating index with a geographic information system for the assessment of pesticide impact on water quality. Water Air Soil Pollut 5:67–68

Porfiri C, Montoya JC, Azcarate P (2017) Distribución espacial de las concentraciones de los herbicidas imazapyr y atrazina a escala de estableciendo en un acuífero del NE de La Pampa. Resúmenes de Trabajos y Actas completas. VI CONGRESO BIANUAL PROIMCA—IV CONGRESO BIANUAL PRODECA. 6, 7 y 8 de septiembre de 2017 en la Facultad Regional

Bahía Blanca. Universidad Tecnológica Nacional. https://www.frbb.utn.edu.ar/prodeca-proimca/. Accessed 2 May 2019
Rao PSC, Hornsby AG, Jessup RE (1985) Indices for ranking the potential for pesticides contamination of groundwater. Soil Crop Sci Soc FL Proc 44:1–8
Simmers I (1997) Groundwater recharge principles, problems and developments. In: Simmers I (ed) Recharge of phreatic aquifers in (semi-)arid areas. International Association of Hydrogeologists/AA Balkema, Rotterdam/Brookfield, pp 1–18
Tan S, Evans RR, Dahmer ML et al (2005) Imidazolinone-tolerant crops: history, current status and future. Pest Manag Sci 61:246–257
Vázquez P, Adema E, Fernández B (2013) Dinámica de la fenología de la vegetación a partir de series temporales de NDVI de largo plazo en la provincia de La Pampa. Ecol Austral 23:77–86
Viglizzo EF, Frank FC, Carreño LV et al (2011) Ecological and environmental footprint of 50 years of agricultural expansion in Argentina. Glob Chang Biol 17:959–973
Zhang Q, Yang X, Zhang Y, Zhong M (2013) Risk assessment of groundwater contamination: a multilevel fuzzy comprehensive evaluation approach based on DRASTIC model. Sci World J 22:610390. https://doi.org/10.1155/2013/610390

# Part IV
# Weed Management Decision Support Systems: Study Cases

# Chapter 11
# How to Use a "Virtual Field" to Evaluate and Design Integrated Weed Management Strategies at Different Spatial and Temporal Scales

**Nathalie Colbach**

**Abstract** Switching from intensive herbicide-based to agroecological weed management needs models to explore the vast range of possible combinations of cropping techniques, to assess long-term effects and weed (dis)services. This chapter presents the mechanistic FLORSYS model, a "virtual field" simulating daily weed and crop growth and reproduction over the years, on which arable cropping systems can be experimented in temperate climates. The model inputs include a detailed description of the cropping system, soil characteristics, weather and the regional weed species pool. A detailed life cycle predicts daily state variables describing weeds, crops and soil conditions depending on inputs, with a 3D individual-based representation of the multispecies crop–weed canopy. Effects on a given plant or seed depend on weather and soil conditions, management operations, biophysical environment as well as species, plant morphology and stage. To simplify the addition of new species, difficult-to-measure model parameters are estimated with functional relationships from easily measured species traits, trait databases and expert opinion. To simplify the comparison of cropping systems, the detailed daily and 3D outputs are translated into indicators assessing crop production and weed (dis)services. A series of case studies illustrates how the model is used to (1) optimise individual cropping techniques with frequency analyses, (2) run multicriteria evaluations of existing and prospective cropping systems at the field and landscape scales, (3) identify the cropping techniques and species traits that drive crop production and weed (dis)services and (4) design innovative cropping systems and to promote integrated weed management in participatory workshops with farmers.

**Keywords** Cropping system design · 3D individual-based model · Mechanistic model · Agroecology · Integrated weed management · Decision support · Multicriteria evaluation · In silico experiment · Crop ideotype · Participatory farmer workshop

---

N. Colbach (✉)
Agroécologie, Subgroup Dijon, INRAE, Univ. Bourgogne, Univ. Bourgogne Franche-Comté, Dijon, France
e-mail: Nathalie.Colbach@inra.fr



G. R. Chantre, J. L. González-Andújar (eds.), *Decision Support Systems for Weed Management*, https://doi.org/10.1007/978-3-030-44402-0_11

## 11.1 Introduction

Because of environmental and health issues and the resulting changes in agricultural policies, weed management strategies must be rethought from scratch to rely little or not at all on herbicides. The switch from a single highly efficacious technique, that is, herbicides, to a combination of partially efficient preventive and curative techniques (Liebman and Dyck 1993; Liebman and Gallandt 1997) needs models to explore the vast range of possible combinations, to assess long-term effects and the many services (e.g. trophic resources for pollinators and pest enemies) and disservices (e.g. competition for soil resources, host for crop pests) depending on weeds.

In order to understand and predict the variability in effects observed for the different cropping techniques in a large range of situations without reparameterisation, mechanistic models are best. Such models decompose the life cycle of weeds and crops into elementary processes depending on biophysical effects of cropping systems, in interaction with biophysical variables (Colbach et al. 2005; Colbach 2010). Indeed, it is not sufficient to quantify the average effects of techniques; farmers also need to know the probability of success of a given management strategy and the risk of obtaining the opposite effect of the one they were originally aiming at (Colbach et al. 2014a).

For that purpose, the required model needs to consider most of the cropping system components, even if they do not directly target weeds as do herbicides or mechanical weeding. Indeed, any effect on the crop or the environmental conditions will also affect weeds. Model inputs must also include pedoclimatic conditions to take account of regional differences and, most importantly, to integrate interactions between cropping systems and environmental conditions (Colbach 2010). As weed seed banks persist for several years in the soil (Lewis 1973), a comprehensive model must allow for simulations over several years or even decades to assess how today's decisions could affect weed flora and crop production during the years to come (Colbach et al. 2014a). This model needs a daily time step to be consistent with the temporal scale of farming operations and the interactions with pedoclimate. The model should also be multispecies, both in terms of weeds and crops. Indeed, arable fields include several dozens or even hundreds of different weed species (Fried et al. 2008), and crop diversification is an important lever of integrated weed management (Liebman and Dyck 1993; Liebman and Gallandt 1997).

This chapter will present such a model and describe how it is used to design agroecological weed management strategies. FLORSYS (Colbach et al. 2014b, c, 2017c; Gardarin et al. 2012; Mézière et al. 2015; Munier-Jolain et al. 2013, 2014) is a "virtual field" (in silico) approach which allows for the simulation of weed and crop growth and reproduction on a daily basis over the years, on which cropping systems can be experimented and a large range of crop, weed and environmental measures estimated (Fig. 11.1).

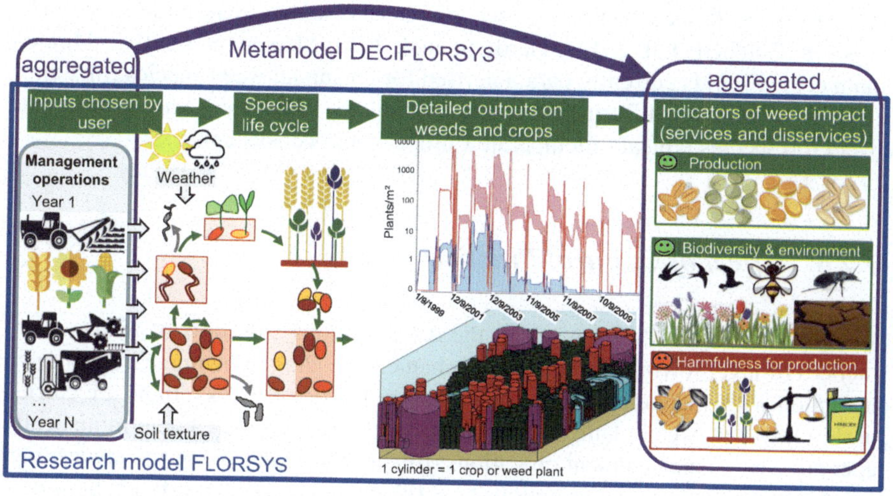


**Fig. 11.1** General representation of the (1) research model FLORSYS (Colbach et al. 2014c; Gardarin et al. 2012; Mézière et al. 2015; Munier-Jolain et al. 2013) which simulates crop growth and weed dynamics from cropping system, weather and soil inputs based on a mechanistic representation of biophysical processes at a daily time step (3D representation), and the (2) metamodel DECIFLORSYS which directly estimates weed services and disservices from cropping system inputs (Colas et al. 2020) (Nathalie Colbach © 2018)

## 11.2 FLORSYS: The "Virtual Field" Model

### 11.2.1 Input Variables

The input variables of FLORSYS (Fig. 11.1) consist of (1) a description of the simulated field (i.e. daily weather, latitude and soil characteristics); (2) all the simulated crops and management operations in the field (including dates, tools and options); and (3) the initial weed seed bank size and composition which is either measured on soil samples or, more feasible, estimated from regional flora assessments (Colbach et al. 2016).

### 11.2.2 Weed and Crop Life Cycle

The input variables influence the annual life cycle which applies to both annual weeds and crops, with a daily time step (Fig. 11.1). Pre-emergent stages (e.g. surviving, dormant and germinating seeds, emerging seedlings) are driven by soil structure, temperature and water potential. The crop–weed canopy is represented in 3D with an individual though simplified representation of each crop and weed plant. Post-emergent processes (e.g. photosynthesis, respiration, growth, etiolation) are driven by light availability and air temperature. At plant maturity, weed seeds are added to the soil seed bank; crop seeds are harvested to determine crop yield. FLORSYS (Colbach et al. 2014c; Gardarin et al. 2012; Mézière et al. 2015; Munier-Jolain et al. 2013) is currently parameterised for 25 annual weed species (including different populations differing in terms of herbicide resistance) and 33 cash and cover crop species typical of temperate European agroecosystems.

### *11.2.3 Effect of Cultural Practices*

Life cycle processes also depend on the dates, options and tools of management practices (tillage, sowing, herbicides, mechanical weeding, mowing, harvesting), in interaction with weather and soil conditions on the day the operations are carried out. For instance, weed plant survival probabilities are calculated deterministically depending on (1) management operations (tillage, herbicides, mechanical weeding, mowing, harvesting) and their options (e.g. tillage depth, tool, speed), (2) biophysical environment (e.g. soil moisture, canopy density) as well as (3) plant morphology and stage. The actual survival of each plant is determined stochastically by comparing this probability to a random probability. Survival after herbicide spraying also depends on plant genotype.

### *11.2.4 Parameterising Many Contrasting Species*

A mechanistic approach is important to ensure that a model allows continuously synthesising knowledge (Colbach 2010) but it requires an enormous amount of parameters, which hinders the addition of new species to the model. This is the reason why Gardarin et al. (2012, 2016) developed a new methodology based on functional relationships to estimate difficult-to-measure model parameters from easily measured species traits, databases and/or expert knowledge. The validity of this approach was checked on weed species, for the critical emergence stage (Gardarin 2008) and at a multiannual scale (Colbach et al. 2016; Pointurier et al. submitted to Ecological Modelling).

### *11.2.5 Assessing Crop Production and Weed (Dis)Services*

To simplify the comparison of cropping systems and to make simulations more accessible to policy makers, crop advisors and farmers, the detailed outputs are translated into indicators assessing crop production, as well as weed-borne agroecological services and disservices. FLORSYS production indicators comprise crop yield in terms of weight and energy (Fig. 11.2). Indicators of weed disservices describe weed harmfulness for crop production and were developed in cooperation with

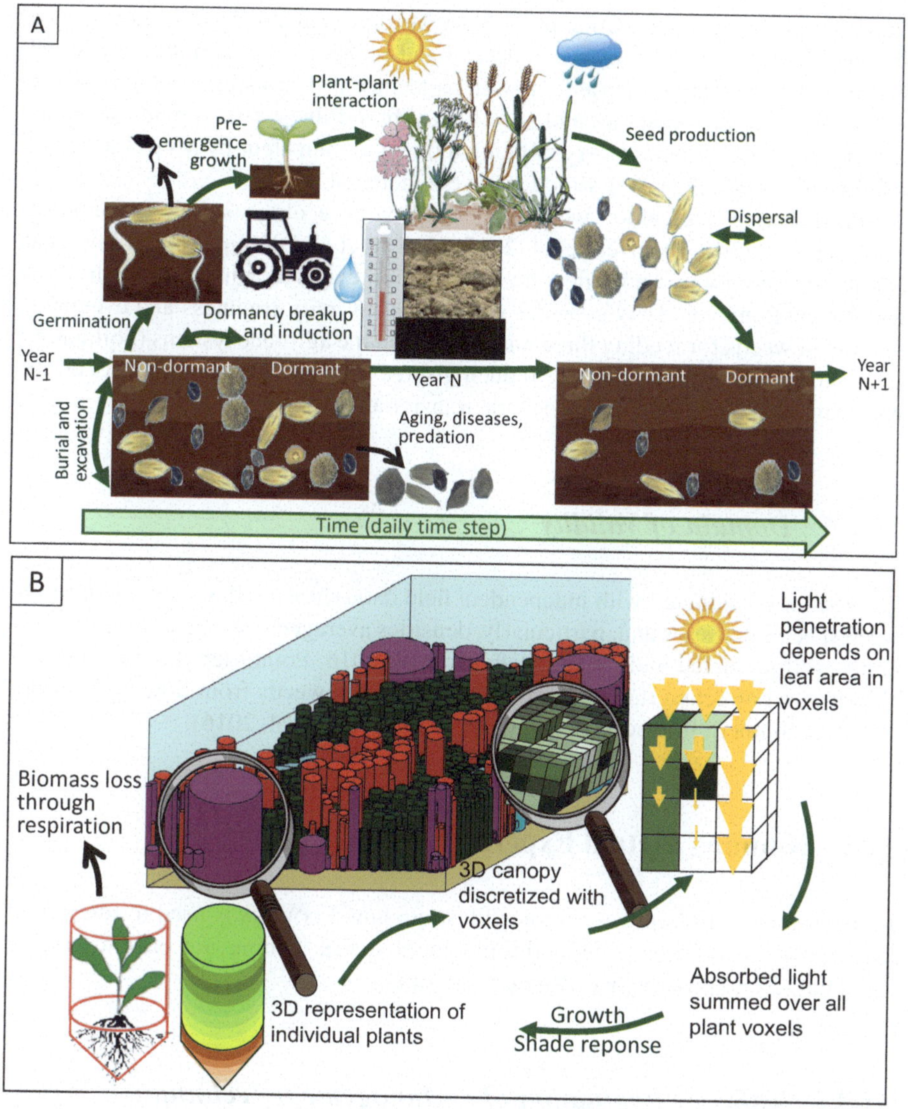


**Fig. 11.2** Simplified representation of spatio-temporal stages and processes in the FLORSYS model (Colbach et al. 2014c; Gardarin et al. 2012; Munier-Jolain et al. 2013). (**a**) Temporal representation of annual life cycle of crops and weeds, showing the 1D representation of the soil seed bank. (**b**) 3D individual-based representation of the crop–weed canopy, focussing on plant–plant competition for light (Nathalie Colbach © 2019)

farmers and crop advisors (Colas et al. 2020; Mézière et al. 2015). Direct (crop yield loss and harvest pollution by weed debris) and indirect weed harmfulness (weed-borne pests) affecting crop yield, as well as technical (harvesting problems due to weeds blocking the combine) and sociological harmfulness (weed field infestation as a proxy of the farmer's worry of being thought incompetent by his peers, even if there is no actual effect over yield loss) were included.

Weed-service indicators were developed with ecologists and agronomists (Colbach et al. 2020; Mézière et al. 2015; Moreau et al. 2020 in press at European Journal of Agronomy) and reflect the contribution that weeds make to biodiversity and the environment. They consider weed plant diversity (richness and evenness), the role of weeds for feeding three major guilds in the agro-ecosystems (pollinators, farm birds and carabids) and for reducing three physical farming impacts on the environment (nitrate leaching, pesticide transfer and soil erosion).

### *11.2.6 Domain of Validity*

FLORSYS was evaluated with independent field data, showing that crop yields, daily weed species densities and, particularly, densities averaged over the years were generally well predicted and ranked Colbach et al. 2016, Pointurier et al. submitted). However, a corrective function was required to keep weeds from flowering during winter in southern France (e.g. below 46°N) (Colbach et al. 2016).

## 11.3 Running Virtual Experiments

In this section, different case studies illustrate how FLORSYS is used to run virtual experiments at different temporal and spatial scales, aiming not only to control weed but also to promote weed-based services.

### *11.3.1 Efficacy Evaluation of a Management Technique*

Integrated crop production methods often delay seeding to enhance weed control. For example, delayed sowing in winter crops allows more time for false (or stale) seed bed technique favouring autumnal weeds to emerge during the summer fallow, thus resulting in a reduced weed seed bank at crop sowing and, hopefully, in a lower weed emergence inside the crop (Moss and Clarke 1994). However, this strategy is only efficient if the targeted weed seeds are not dormant before crop sowing. Moreover, its efficiency varies considerably with environmental conditions, mainly with soil moisture. Indeed, false seed bed techniques work best when the targeted weed seeds are moist. Conversely, if the sowing operation is combined with a superficial tillage and carried out when the soil is moist, tillage triggers additional weed germination, resulting in increased weed emergence inside the crop (Fig. 11.3).

To evaluate the success rate of delayed sowing and the risk of unexpected side effects, a virtual experiment was carried out with FLORSYS (Table 11.1). Five wheat sowing dates were tested in two French regions, and each was repeated with ten different weather series. The initial weed seed bank consisted mostly of *Alopecurus myosuroides*, an autumnal grass weed typical of winter-crop rotations in Eastern France. The frequency analysis of the simulation output showed that delayed sowing indeed decreased weed emergence in crops in both regions, in average. But, in Northern France, sowing had to be delayed until 7 Nov. to avoid all risk of increasing weed emergence. In Burgundy, where the soil often is too dry for germination in early October, a residual risk of increased weed emergence persisted until mid-November (Table 11.1).

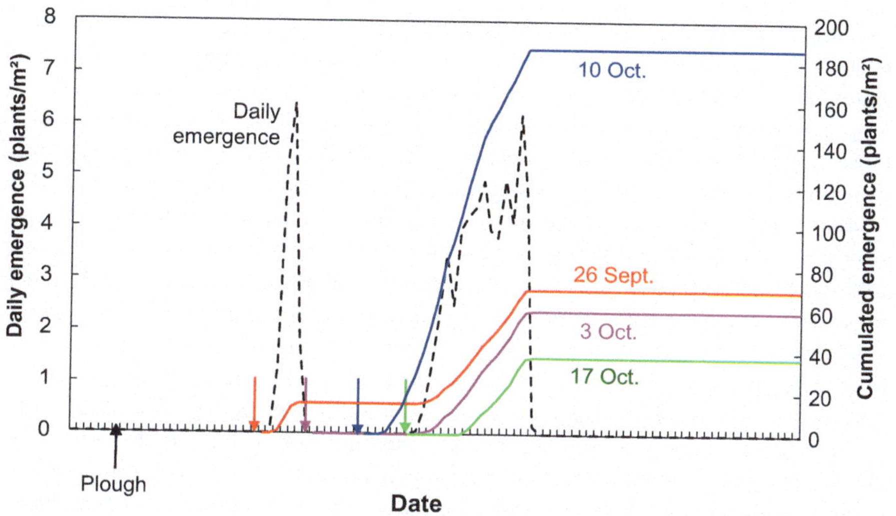


**Fig. 11.3** Effect of the last tillage date associated with the sowing operation (carried out on 26 Sept.; 3 Oct.; 10 Oct.; or 17 Oct.) on the cumulated autumnal emergence of grass weeds (e.g. *Alopecurus myosuroides* Huds.) in winter wheat in Burgundy simulated with the monospecific prototype of FLORSYS. The arrows indicate the sowing date relative to the daily weed emergence (dashed line) in case of the earliest sowing (26 Sept.). Grey areas indicate days where the soil was too dry for germination. Delayed crop sowing allows to avoid the earliest weed emergence flush and reduces in-crop weed emergence (sowing on 3 Oct. and 17 Oct. vs 26 Sept.). If the crop is sown shortly after the soil was remoistened by rain (10 Oct.), the associated tillage triggers a germination flush resulting in a huge increase in weed emergence after sowing. No additional triggering occurs if the soil is tilled in dry conditions (3 Oct.) or in continuously moist soil (17 Oct.) (based on Colbach et al. 2005) (Nathalie Colbach ©)

**Table 11.1** Effect of delayed winter-wheat sowing (combined with a power harrow) on autumnal grass weed emergence (e.g. *Alopecurus myosuroides* Huds.) in the crop simulated with the monospecific prototype of FLORSYS. Probability of occurrence (% years) that weed emergence increases or decreases relatively to the initial sowing date on 3 Oct. (based on Colbach et al. 2014a)

| Sowing date | Northern France | | Burgundy (Eastern France) | |
|---|---|---|---|---|
| | Probabilityof occurrence(%) | | Probabilityof occurrence(%) | |
| | Weed emergence | | Weed emergence | |
| | decreased e10% | increased e 10% | decreased bye 10% | increased bye 10% |
| 3 Oct. | Initial sowing date | | | |
| 10 Oct. | 7 | 7 | 14 | 14 |
| 17 Oct. | 14 | 0 | 14 | 14 |
| 24 Oct. | 7 | 14 | 0 | 36 |
| 31 Oct. | 29 | 7 | 14 | 14 |
| 7 Nov. | 43 | 0 | 79 | 14 |
| 14 Nov | 86 | 0 | 79 | 14 |

### *11.3.2 Multicriteria Long-Term Evaluation of Cropping Systems*

The main interest in using a model such as FLORSYS lies in the long-term and multicriteria assessment of comprehensive cropping systems. Figure 11.4 shows an example of model use in interaction with crop advisors, to assess the advantages of crop diversification, introducing spring crops into the usual 3-year winter rotation consisting of oilseed rape, wheat and barley. The analysis of the yield-loss dynamics over time demonstrated the necessity to evaluate innovations in the long term. For example, spring pea presented the highest yield loss of all tested crops (Fig. 11.3.A), but the yield loss in the following wheat crop was consistently lower than in wheat following oilseed rape or sunflower, even though both these crops presented a much lower yield loss than spring pea. Consequently, in average for the long-term evaluated time horizon, the rotation including spring pea performed much better in terms of crop production and weed harmfulness than the 3-year reference rotation and as good as the 5-year rotation including sunflower and spring barley (Fig. 11.5b).

In addition to conventional biodiversity and harmfulness criteria, FLORSYS also allowed to assess performance indicators that are almost impossible to evaluate under field conditions, such as weed-based food offer for pollinators or farm birds. In the present example, crop diversification allowed to improve all analysed performance indicators (i.e. increased biodiversity and crop production while reducing weed harmfulness and herbicide use).

This approach is invaluable to assess innovations before they are actually authorised and introduced into cropping systems, for instance to evaluate the impact of genetically modified herbicide-tolerant crops and the accompanying changes in cropping practices on biodiversity (Bürger et al. 2015; Colbach et al. 2017b), herbicide resistance in weeds (Colbach et al. 2017c; Sester et al. 2006) or harvest quality, for instance in terms of fatty acid content (Baux et al. 2011) or genetic impurities (Sausse et al. 2013).

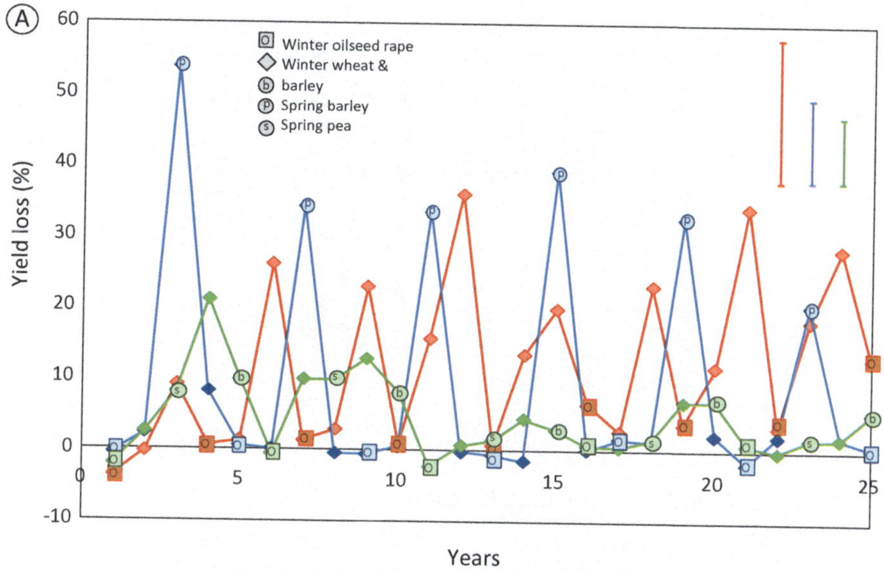


(B)

| Cropping system | Weed contribution to biodiversity: Wild plant species richness (number of species) | Weed contribution to biodiversity: Species evenness | Weed contribution to biodiversity: Bee food | Crop production: Yield (MJ/ha) | Damage to crop production: Yield loss (%) | Damage to crop production: Harvest pollution | Damage to crop production: Field infestation (T/ha) | Herbicide use (TFI[§]) |
|---|---|---|---|---|---|---|---|---|
| OWB Plough | 11.35 C | 0.19 C | 1.06 BA | 69391 B | 11.87 A | 2.70 A | 0.24 BA | 1.72 B |
| OWB noPlough | 10.19 D | 0.23 B | 1.04 BA | 68695 B | 12.55 A | 2.83 A | 0.29 A | 1.79 A |
| OWpW Plough | 11.99 B | 0.23 B | 1.00 B | 95980 A | 10.54 A | 1.81 B | 0.14 C | 1.39 D |
| OWpW noPlough | 12.22 BA | 0.24 B | 1.03 BA | 96804 A | 9.50 A | 1.82 B | 0.14 C | 1.57 C |
| OWsWb Plough | 12.59 A | 0.25 BA | 1.10 A | 98957 A | 4.36 B | 1.97 B | 0.18 BC | 1.12 F |
| OWsWb noPlough | 12.43 BA | 0.27 A | 1.06 BA | 98955 A | 4.19 B | 1.94 B | 0.18 BC | 1.38 E |

[§] TFI is treatment frequency index (a herbicide at full dosage over whole field = 1)

**Fig. 11.4** Effect of crop diversification on crop yield loss due to weeds simulated with FLORSYS. Annual means averaged over 10 weather repetitions with a Burgundy pedoclimate as a function of time (vertical bars show intra-annual standard-deviation averaged over time) (**a**) and multicriteria evaluation of weed (dis)services averaged over rotation (**b**) for winter oilseed rape/winter wheat/winter barley (OWB, red line); winter oilseed rape/winter wheat/spring pea/winter wheat (OWpW, blue line); winter oilseed rape/winter wheat/sunflower/winter wheat/spring barley (OWsWb, green line). Means followed by the same letter are not significantly different at $p = 0.05$, using a least-significant difference test. (Colbach and Cordeau 2018b; Colbach et al. 2010) (Nathalie Colbach ©). [§] TFI is treatment frequency index (a herbicide at full dosage over whole field = 1)

### *11.3.3 Upscale to Landscapes*

Model-based evaluation is also helpful when upscaling from the field to the landscape level. Switching scales could be necessary when weeds disperse to neighbour fields (via seeds or pollen) as the management of a given field will influence what happens in neighbour fields. Pollen dispersal is an issue if the immigrant genes change the fitness of the native population, which is particularly the case for herbicide resistance. Even without propagule exchange, working at the field cluster or landscape scale can be pertinent when there are trade-offs between crop production

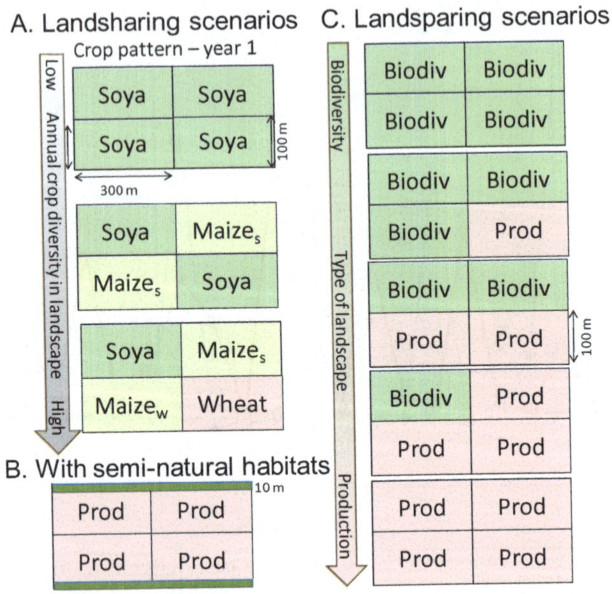


**Fig. 11.5** Landscape patterns that were virtually experimented with FLORSYS. Small landscapes consisted of four 3-ha fields and a typical pedoclimate from Aquitaine (south-western France). (**a**) Landsharing scenarios based on a single diverse rotation (soybean/maize/wheat/maize), differing in the number of crops present each year in the landscape; (**b**) Landsparing scenarios with fields aiming to maximise crop production ("Prod") but converting part of the field into permanent grass strips (green 10 m wide strips); and (**c**) Landsparing scenarios with varying proportions of contrasting cropping systems in the landscape, either aiming to maximise biodiversity ("Biodiv") or crop production ("Prod") (based on Colbach et al. 2018) (Nathalie Colbach ©)

and biodiversity conservation. In such a case scenario, models can contribute to decide whether landsharing or landsparing is more adequate (Colbach et al. 2018). Indeed, semi-natural habitats and landscape crop patterns contribute to weed dynamics by locating favourable habitats, both in time and in space (Petit et al. 2013).

FLORSYS allows tackling some of these questions by simulating several fields and/or semi-natural habitats in parallel (Colbach et al. 2018). At seed shed, weed seeds as well as shattered crop seeds are dispersed from a source plot to neighbouring habitats. Seed dispersal distance increases with weed plant height and decreases with seed mass; and it is higher for seeds dispersed by animals and wind than for those dispersed by gravity (Colbach et al. 2018; Thomson et al. 2011). The dispersed seeds then colonise new fields and habitats or integrate existing populations, contributing to wild plant biodiversity but also negatively affecting crops.

The spatially explicit model proposed by Colbach et al. (2018) allows to virtually experiment different landscape management scenarios, aiming to reconcile crop production with biodiversity conservation, either at field (landsharing) or landscape (landsparing) levels. Three series of scenarios were simulated over 28 years and 10 weather repetitions, using maize-based cropping systems (Fig. 11.5). The

**Table 11.2** Effect of landsharing and landsparing scenarios (Fig. 11.5) on weed (dis)service indicators at the landscape scale (Colbach et al. 2018)

| Landscape pattern | Weed-related biodiversity in landscape | | | Crop production (MJ/ha) | Weed harmfulness in crops | |
|---|---|---|---|---|---|---|
| | Species richness (number of species) | Bird food | Bee food | | Yield loss (%) | Harvest pollution |
| **A. Landsharing scenarios: Annual crop pattern in landscape grown with soybean/maize/wheat/maize** | | | | | | |
| One crop per year | 11.22 g | 3.55 i | 0.66 f | 68344 d | 22.68 e | 1.20 f |
| Two crops per year | 12.04 e | 4.16 gh | 0.91 e | 60184 e | 33.92 c | 1.60 e |
| All crops per year | 12.94 d | 4.31 f | 1.11 d | 51920 f | 44.31 a | 2.04 c |
| **B. With semi-natural habitats in landscape grown with high-production cropping systems** | | | | | | |
| 10% grass strips | 10.65 h | 7.13 d | 0.55 g | 90161 b | 0.14 h | 0.98 g |
| **C. Landsparing scenarios: % fields with high-production vs. with high-biodiversity cropping systems (production - biodiversity %)** | | | | | | |
| 0 – 100 % | 15.72 a | 9.58 a | 2.78 a | 59257 e | 40.83 b | 2.78 a |
| 25 – 75 % | 14.88 b | 8.90 b | 2.23 b | 70045 d | 30.30 d | 2.38 b |
| 50 – 50 % | 13.26 c | 8.02 c | 1.57 c | 80603 c | 19.69 f | 1.88 d |
| 75 – 25 % | 11.72 f | 7.11 d | 0.94 e | 90257 b | 9.93 g | 1.21 f |
| 100 – 0 % | 8.44 i | 5.36 e | 0.18 h | 100452 a | -0.10 h | 0.00 h |

Comparison of means after analyses of variance of weed (dis)service indicators simulated with FLORSYS and averaged over the field cluster as a function of landscape scenario, weather repetition, time, and the interaction between scenario and time. Means of a given column followed by the same letter are not significantly different at $p = 0.05$ (least significant difference test)

simulations showed that landsparing scenarios were better than landsharing, resulting in high crop production and medium biodiversity at the landscape scale (Table 11.2). Landsharing scenarios always produced less biodiversity and less crop production. The more crops were grown each year in the landscape, the more the weed impact on production and biodiversity increased.

## *11.3.4 Disentangling Effects*

Many of the evaluations of the previous sections would also be possible with empirical models. One of the major advantages of process-based models such as FLORSYS is their ability to disentangle complex interactions in the agroecosystem, often better than in situ field experiments could do. For instance, yield loss is notoriously difficult to assess in field conditions because it is next to impossible to produce a weed-free control identical to the weed-infested treatments except for the presence of weeds. Similarly, farm-field networks (i.e. a large number of farm fields that are monitored in terms of management practices and, for example, crop yield or weed infestation) are usually inadequate to assess the effect of individual management techniques, even influential ones such as herbicides, because farmers reason each technique depending on other cropping-system components (e.g. mechanical weeding, tillage and rotation) (Colbach and Cordeau 2018b).

FLORSYS was used to unravel some of these interactions by monitoring a virtual farm-field network covering contrasting production situations with several hundred cropping systems recorded in seven regions ranging from northern France to northern Spain (Colbach and Cordeau 2018a). The effect of herbicides was discriminated from that of other management practices by comparing the simulated weed floras and yields of the recorded cropping systems to those of these same systems minus herbicides (and without any other changes in practices). Moreover, the authors differentiated the relative effects of weeds and management practices on crop production by comparing the yields of simulations run with and without weeds. Also, management practices effects on weeds were separated from their reciprocals (i.e. effects of weeds on management practices) by simulating the recorded cropping systems without adapting the practices to the simulated weed floras. Long-term weed harmfulness was also assessed by looking at weed dynamics and weed-caused yield loss over succeeding years instead of considering only annual data. As a result, Colbach and Cordeau (2018a) were able to show that (1) weed–crop biomass ratio at crop flowering was the best indicator of the year's yield loss (Fig. 11.6); (2) herbicide use intensity was not correlated to either weed variables or yield loss, because farmers compensated reduced herbicide use by other preventive (e.g. false seed bed techniques) and curative measures (e.g. mechanical weeding); (3) average weed biomass during crop growth and yield loss increased by +116 and + 62% (averaged over rotation) respectively

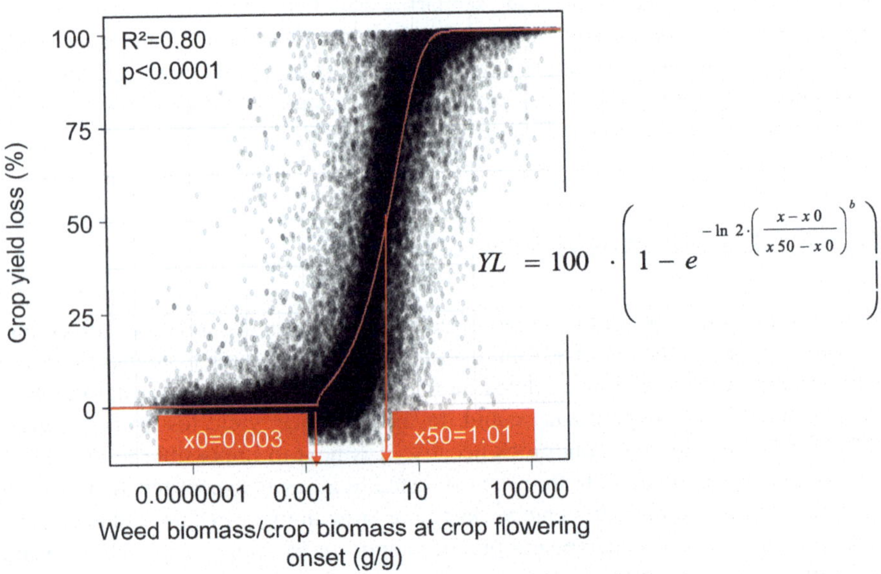


**Fig. 11.6** Generic function predicting grain yield loss (%, i.e. 100 t t$^{-1}$) in annual crops from the ratio of weed biomass vs crop biomass at the onset of crop flowering established on a virtual farm-field network simulated with FLORSYS. Each data point is 1 year of one cropping system and one weather repetition out of a total of 272 cropping systems × 30 years × 10 weather repetitions. Red line fitted to data with non-linear regression (Nathalie Colbach © 2018) (based on Colbach and Cordeau 2018a)

when herbicides were eliminated without any other change in management practices; and (4) effects were more visible at multiannual (rotation) than the annual scales.

This kind of virtual farm-field network can be used for other purposes, for instance to track innovations among farming practices. FlorSys simulations demonstrated the importance of crop diversity in rotations to control weed harmfulness with few or no herbicides (Colbach and Cordeau 2018a). Three types of strategies could be identified among the investigated farmers' cropping systems, which differed in terms of rotation, tillage strategy and so on. The strategy based on a summer-crop monoculture relied heavily on herbicides as well as mechanical weeding to limit weed-caused yield loss. Conversely, two other strategies diversified crops, with longer rotations, crop mixtures, cover crops and temporary grassland. Combined with well-reasoned tillage, crop diversification allowed for reducing herbicide use while limiting yield loss.

More generally, these farm-field networks allow to identify which management techniques drive weed (dis)services. The previous studies simulated actual cropping systems practiced by farmers or proposed by crop advisors, but the network can also be extended with randomly constructed systems to run sensitivity analyses. This was actually the approach used when metamodelling FlorSys into a decision-support system (see section 11.4.1) where various statistical methods were used to identify the most influential management techniques. Table 11.3 shows an example of a ranking of management techniques in terms of their effect on weed contribution for protecting the soil from erosion and nitrate leaching. This analysis shows that tillage, particularly deep and/or inverting operations, was the major determinant of weed-based soil protection while rotation and herbicides had much less impact.

The same simulation approach, combined with statistical methods usual in ecology, such as RLQ or fourth corner, can identify crop and weed traits that drive crop production and weed (dis)services (Colbach et al. 2014d; Colbach et al. 2017a; Colbach et al. 2017b; Colbach et al. 2019). Table 11.4 shows an example where the aim was to identify which weed-morphology parameters that drive weed harmfulness for crop production in average over many contrasting cropping systems and pedoclimates. The most damaging weeds in terms of crop yield loss and harvest pollution were the ones that occupied space earlier and faster, starting with a large leaf area at emergence and with larger and/or thinner leaves (larger SLA). Later in the season, shading neighbours with taller plants per unit biomass (larger SPH) also becomes important, but lateral space occupation is still an issue, as wider heavier plants (larger HPW) with a uniform leaf area distribution (lower MLH) are more damaging. When shaded, the damaging weeds react by shifting their leaves towards (increase in MLH).

**Table 11.3** Identification of key management techniques influencing weed (dis)services in a virtual farm-field network consisting of several hundred cropping systems from six French and Spanish pedoclimates simulated with FLORSYS over 27 years and 10 weather repetitions (based on Moreau et al. 2020). Example of weed-based protection from soil erosion and nitrate leaching averaged over 27 simulated years, identifying key techniques with LASSO regressions

| | Regression parameter value | |
|---|---|---|
| Cropping system component | Nitrate leaching | Soil erosion |
| Years between successive direct sowings | −0.273 | −0.356 |
| Tillage depth (cm) | −0.717 | |
| Frequency of mouldboard ploughing in winter (operations/year) | −1.03 | |
| *Superficial tillage (operations/year)* | | |
| Total | −1.24 | |
| In winter (operations/year) | −1.12 | −12.3 |
| With disks | −9.91 | −28.0 |
| With a chisel | −4.49 | −15.7 |
| With a power harrow | −0.46 | −23.9 |
| With a rotavator | | −30.7 |
| Days from harvest to first tillage | | 0.0109 |
| Residue shredding height (cm) | −0.104 | 0.558 |
| Days from last rolling operation to cash-crop sowing | −0.0040 | −0.0630 |
| *Rotation: proportion of* | | |
| Flax | | 1285 |
| Triticale | | 35.8 |
| Oilseed rape | | −30.4 |
| Barley | 6.56 | |
| Spring crops | −2.43 | |
| Pea | −10.1 | |
| Cropping-season diversity[a] | | 16.1 |
| Frequency of cover crops (years/years) | −4.57 | 58.3 |
| Duration of cover crops (months/12 months) | −21.2 | . |
| Sowing date of spring crops | . | 0.0784 |
| Harvest date of spring crops | −0.0340 | |
| *Herbicides[b]: Number of treatments per year with* | | |
| Multi-entry herbicides | −0.507 | |
| Pseudo-root-only herbicides | 8.24 | |
| Root only herbicides | −0.0853 | |

Only techniques with a significant effect are shown ($P < 0.05$). For the nitrate-leaching indicator, $n = 2306$ and $R^2 = 0.69$. For the soil-erosion indicator, $n = 2590$ and $R^2 = 0.61$

[a]Proportion of crop years where previous and current cash crops differ in terms of winter, summer and multiannual crops

[b]Herbicides can enter plants via leaves ("foliar"), shoot tips during emergence ("pseudo-root") or roots ("root"). Multiple entry modes are possible ("multi-mode")

**Table 11.4** Which weed plant-morphology parameters drive weed harmfulness for crop production?

| Weed parameter | Stage (BBCH scale) | Yield loss | Harvest pollution | Weed harmfulness | |
|---|---|---|---|---|---|
| | | | | Low | High |
| A. Early growth parameters | | | | | |
| Leaf area at emergence ($cm^2$) | | 0.21 | 0.17 |  |  |
| B. Potential plant morphology parameters in unshaded conditions | | | | |  |
| Specific leaf area SLA ($cm^2$/g) | 0 | 0.19 | 0.16 |  |  |
| | 1 | 0.19 | 0.14 | | |
| | 8 | -0.21 | | | |
| | 9 | -0.22 | | | |
| | 10 | -0.20 | | | |
| Specific plant height SPH (cm/g) | 7 | 0.19 | 0.17 |  |  |
| | 8 | 0.20 | 0.18 | | |
| | 9 | 0.20 | 0.17 | | |
| | 10 | 0.19 | 0.16 | | |
| Increase in plant width with plant biomass HPW (no unit) | 9 | | -0.15 |  |  |
| | 10 | -0.20 | -0.17 | | |
| Median relative leaf height MLH (cm/cm) | 7 | -0.19 | -0.16 |  |  |
| C. Parameters driving weed shading response | | | | |  |
| Increase in MLH | 6 | | 0.15 |  |  |
| | 7 | | 0.15 | | |
| | 10 | | 0.15 | | |

Pearson correlation coefficients $r$ (and $p$-values) between weed parameters and annual weed harmfulness indicators estimated with RLQ and fourth corner analyses via weed plant density. Only correlations exceeding 0.10 and significant at $p = 0.05$ were kept in the table. Cells with correlations were coloured from green (−1) to 1 (red), depending on coefficient values. The pictures in the last two columns illustrate the morphological characteristics of weeds resulting in respectively low or high weed harmfulness for crop production (based on Colbach et al. 2019) (Nathalie Colbach ©)

## 11.4 Decision Support with Stakeholders

### *11.4.1 DeciFlorSys: A Decision Support Tool*

The previous sections show examples of how FlorSys can be used to track innovations in existing farming practices, test, diagnose and fine-tune prospective cropping systems proposed by farmers and crop advisors, produce expert knowledge for policy makers and so on. However, FlorSys still remains a research model inadequate for direct use in participatory workshops as (1) it requires numerous input variables to be assigned and parameters to be tuned, (2) its mechanistic and individual-based approach induces a high algorithmic complexity and very slow simulations and (3) it evaluates cropping-system candidates rather than actually designing these candidates (Colbach 2010). In order to address these limitations, we transformed FLORSYS into a Decision Support System (DECIFLORSYS). DECIFLORSYS gathers three operational tools, each addressing one of the above issues. The three tools all derive from the metamodelling of FLORSYS using sensitivity analyses and machine learning, and they were co-designed with future users (Colas et al. 2020). Instead of using detailed inputs, DECIFLORSYS uses aggregated inputs corresponding to meta-decision rules at the rotation scale (e.g. proportion of spring crops in rotation, frequency of mouldboard ploughing) (Fig. 11.7). It directly predicts weed (dis)service indicators, without calculating detailed crop and weed variables (Fig. 11.1). The three DECIFLORSYS tools are (1) a table showing the cropping system components to be changed as a priority, (2) decision trees showing how to combine management practices to reach a given goal in terms of weed (dis)services and (3) a predictor based on random forests (AI technique) that calculate the performance of the cropping-system prototypes, with a much faster response time than FLORSYS and easier to handle than the parent model.

While the DECIFLORSYS predictor is as good as FLORSYS to rank cropping systems, it cannot adequately evaluate effects that strongly interact with pedoclimatic conditions, such as the effect of tillage timing with respect to soil moisture (Colas 2018).

### *11.4.2 Use of Models to Promote Integrated Weed Management*

Both FLORSYS and its derivate, DECIFLORSYS, have been used by our research team and crop advisors in participatory workshops with farmers. Implicating farmers in cropping-system design is essential, as innovations proposed by scientists are often disregarded by farmers because they are incompatible with farming constraints (Meynard et al. 2018) or with farmers' risk perception and management (Wilson et al. 2008). Crop advisors can also be reticent to promote the necessary changes (Pasquier and Angevin 2017). In this context, models are invaluable teaching tools to propagate knowledge and promote innovations via training sessions, participatory workshops and role-playing games (Hossard et al. 2013; Martin et al. 2011; Meylan et al. 2013; Sausse et al. 2013). This is particularly true for easy-to-use models such as DECIFLORSYS, which allow stakeholders to directly and immediately see the consequences of changes in their practices in their particular production situation.

Using models with farmers and crop advisors is somewhat different than when using them for research purposes (Fig. 11.8c vs. b). During the workshops, farmers start from their own experience, they are implicated in all steps and get an immediate feedback, all of which makes the resulting solutions more acceptable to them. Conversely, the risk of missing highly performant solutions and remaining inside current conventions is much higher. The best approach for investigating a larger

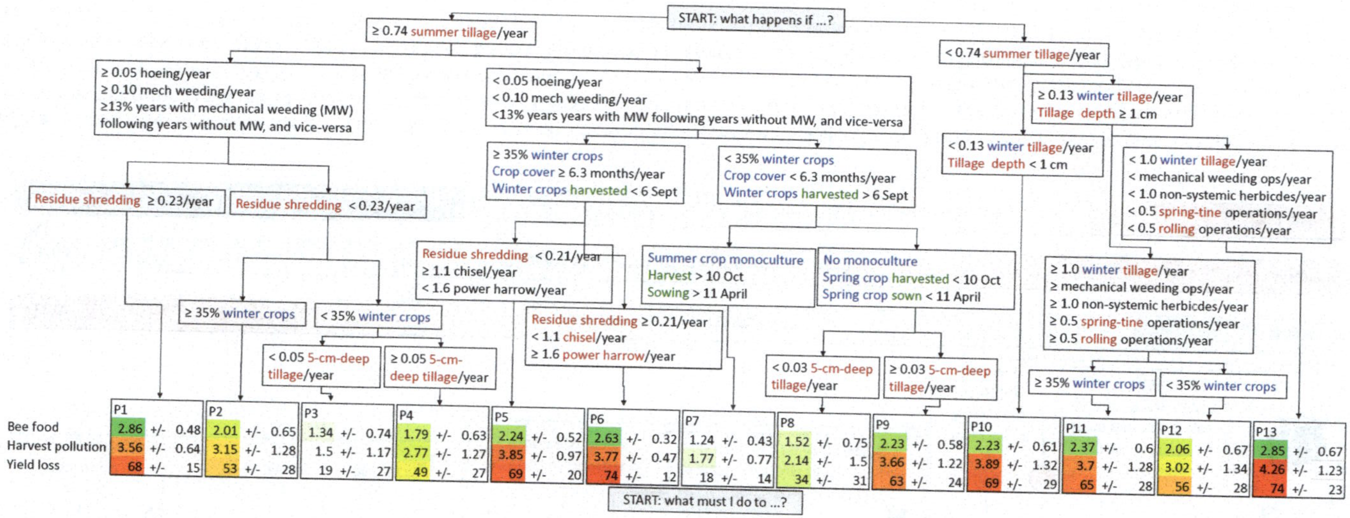


**Fig. 11.7** Decision tree identifying cropping system rules for reaching a series of 13 weed impact profiles based on a multivariate regression tree linking weed impact indicators simulated by FlorSys to combinations of cropping system variables (blue: rotation, green: sowing/harvest, brown: tillage, black: weed control), based on 4350 cropping systems from 20 French regions. Indicator values averaged over years and weather repetitions were coloured from white (minimum) to green (maximum) for biodiversity, from white to red (maximum) for harmfulness to crop production. Uncoloured cells show standard-error including weather effects and variability among systems in a branch. Cross-validation error = 0.106, for indicator range of variation rescaled to [0,1]. The tree can be read top-down to get an idea of the performance of a proposed combination of management practices, or bottom-up to identify the practices corresponding to a target performance (Floriane Colas © 2019) (based on Colas 2018)

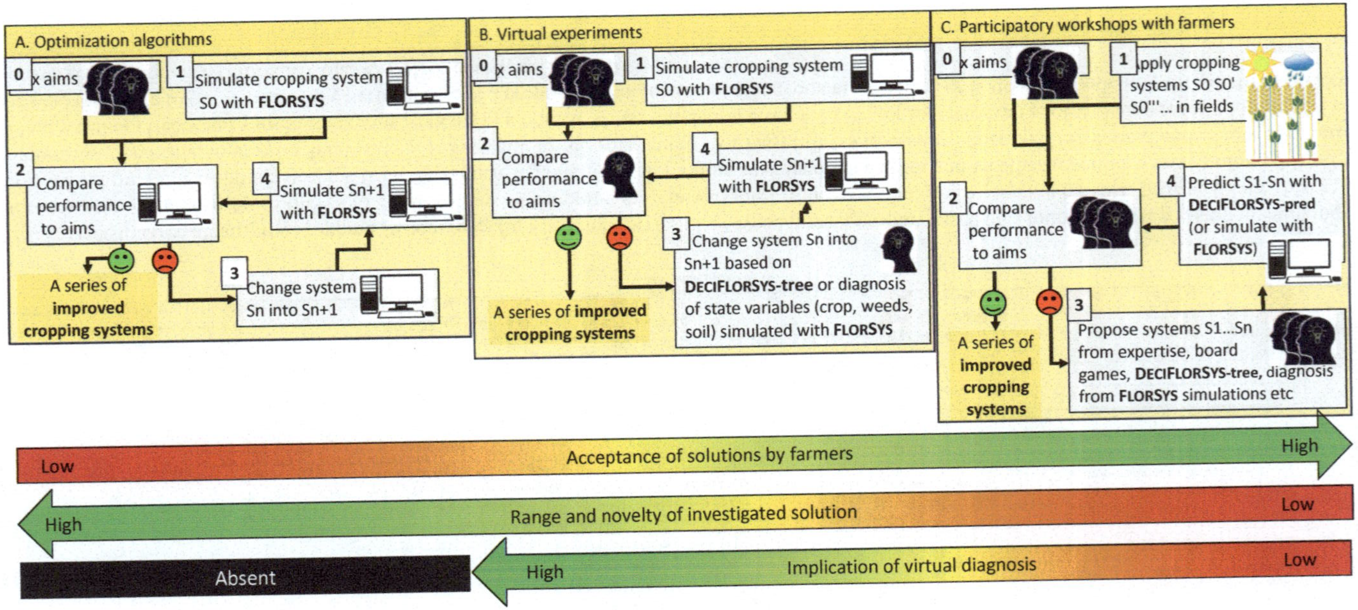


**Fig. 11.8** Three ways to use the FLORSYS tools to design innovative cropping systems with a step-by-step improvement on an initial cropping system S0. (**a**) Optimisation algorithms manage all steps in interaction with FLORSYS, except fixing the aims and constraints for the novel cropping systems, which are determined by a group of experts. (**b**) When running virtual experiments, experts fix aims and constraints, compare the simulated performance of the systems to these aims and propose innovative systems, from expert knowledge and the decision tree of DECIFLORSYS. (**c**) Participatory workshops including farmers often start with a system that performed badly in the field; innovative systems are proposed by a group of interacting farmers and other experts using a variety of tools and these systems are evaluated by the predictor component of DECIFLORSYS to benefit from an immediate feedback that sets off another round of system design (Nathalie Colbach © 2019)

range of possible solutions is automatic optimisation algorithms (Fig. 11.8a) which have already been used with simpler and faster models than FLORSYS (Bergez et al. 2010) and are now being adapted to FLORSYS (Maillot et al. 2019).

## 11.5 Discussion and Conclusion

FLORSYS is one of the very few process-based models that include all the key mechanisms that are relevant for cropping system and does this at a sufficiently precise scale to produce realistic results. To overcome the trade-off between process analysis and decision aid (Colbach 2010), the detailed simulated outputs were aggregated into indicators to support decisions (Bockstaller et al. 2008), and the knowledge synthesised in the mechanistic research model was further extracted and summarised as the empirical (meta)model DECIFLORSYS which is easier to use. This dual approach allowed us to synthesise knowledge on the functioning and effects of crop diversification at different levels of detail and make it available to different stakeholders, consisting of scientists, crop advisors, farmers and policy makers. It is also essential to continue including new knowledge, by adding new crop and weed species as well as management techniques.

All these advantages are subject to the model's prediction quality, which must be confirmed by comparing model simulations to independent observations or expertise (model evaluation). This step is even more crucial for a mechanistic model aggregating data and models from different teams and disciplines, to make sure that the new entity produces consistent results. Though this step has been carried out for FLORSYS, it also pointed to a major drawback of complex mechanistic models (i.e. the difficulty to find adequate data for evaluating the model and its many submodels).

The possibility of continuous model evolution is crucial as, despite its complexity, FLORSYS (and its derivate DECIFLORSYS) neglect several processes that are essential for the more innovative cropping systems, particularly in a context of crop diversification, input reduction and climate change. For instance, including competition for soil nitrogen would improve the assessment of legumes or drought-resistant crops in rotations and mixtures.

The synthesis of the various case studies demonstrated the usefulness of FLORSYS not only for synthesizing knowledge on biophysical processes implicated in cropping system effects but above all for producing emerging knowledge on the functioning of the agroecosystem, and for promoting this knowledge among farmers. In terms of integrated weed management, the many studies carried out with FLORSYS to date demonstrated that, generally, (1) weed damage can be controlled with few or no herbicides if the cropping system is consistently redesigned, (2) many conclusions in terms of crop diversification only have a local validity, (3) which proves the need for flexible rules and (4) the usefulness of models such as FLORSYS and optimisation algorithms to establish these rules.

## References

Baux A, Colbach N, Pellet D (2011) Crop management for optimal low-linolenic rapeseed oil production—field experiments and modelling. Eur J Agron 35:144–153

Bergez JE, Colbach N, Crespo O, Garcia F, Jeuffroy MH, Justes E, Loyce C, Munier-Jolain N, Sadok W (2010) Designing crop management systems by simulation. Eur J Agron 32:3–9

Bockstaller C, Guichard L, Makowski D, Aveline A, Girardin P, Plantureux S (2008) Agri-environmental indicators to assess cropping and farming systems. A review. Agron Sustain Dev 28:139–149

Bürger J, Granger S, Guyot SHM, Messéan A, Colbach N (2015) Simulation study of the impact of changed cropping practices in conventional and GM maize on weeds and associated biodiversity. Agr Syst 137:51–63. https://doi.org/10.1016/j.agsy.2015.03.009

Colas F (2018) Co-développement d'un modèle d'aide à la décision pour la gestion intégrée de la flore adventice. Métamodélisation et analyse de sensibilité d'un modèle mécaniste complexe (FLORSYS) des effets des systèmes de culture sur les services et disservices écosystémiques de la flore adventice. PhD Thesis, Univ. Bourgogne Franche-Comté, Dijon, France

Colas F, Cordeau S, Granger S, Jeuffroy M-H, Pointurier O, Queyrel W, Rodriguez A, Villerd J, Colbach N (2020) Co-development of a decision support system fo integrated weed managment: contribution from future users. Eur J Agron 114:in press. https://doi.org/10.1016/j.eja.2020.126010

Colbach N (2010) Modelling cropping system effects on crop pest dynamics: how to compromise between process analysis and decision aid. Plant Sci 179:1–13

Colbach N, Cordeau S (2018a) Reduced herbicide use does not increase crop yield loss if it is compensated by alternative preventive and curative measures. Eur J Agron 94:67–78. https://doi.org/10.1016/j.eja.2017.12.008

Colbach N, Cordeau S (2018b) Réduire les herbicides sans perte de rendement. Phytoma 717:8–12

Colbach N, Dürr C, Roger-Estrade J, Caneill J (2005) How to model the effects of farming practices on weed emergence. Weed Res 45:2–17

Colbach N, Schneider A, Ballot R, Vivier C (2010) Diversifying cereal-based rotations to improve weed control. Evaluation with the ALOMYSYS model quantifying the effect of cropping systems on a grass weed. OCL 17:292–300

Colbach N, Biju-Duval L, Gardarin A, Granger S, Guyot SHM, Mézière D, Munier-Jolain NM, Petit S (2014a) The role of models for multicriteria evaluation and multiobjective design of cropping systems for managing weeds. Weed Res 54:541–555. https://doi.org/10.1111/wre.12112

Colbach N, Busset H, Roger-Estrade J, Caneill J (2014b) Predictive modelling of weed seed movement in response to superficial tillage tools. Soil Tillage Res 138:1–8

Colbach N, Collard A, Guyot SHM, Mézière D, Munier-Jolain NM (2014c) Assessing innovative sowing patterns for integrated weed management with a 3D crop: weed competition model. Eur J Agron 53:74–89. https://doi.org/10.1016/j.eja.2013.09.019

Colbach N, Granger S, Guyot SHM, Mézière D (2014d) A trait-based approach to explain weed species response to agricultural practices in a simulation study with a cropping system model. Agric Ecosyst Environ 183:197–204. https://doi.org/10.1016/j.agee.2013.11.013

Colbach N, Bertrand M, Busset H, Colas F, Dugué F, Farcy P, Fried G, Granger S, Meunier D, Munier-Jolain NM, Noilhan C, Strbik F, Gardarin A (2016) Uncertainty analysis and evaluation of a complex, multi-specific weed dynamics model with diverse and incomplete data sets. Environ Model Software 86:184–203. https://doi.org/10.1016/j.envsoft.2016.09.020

Colbach N, Bockstaller C, Colas F, Gibot-Leclerc S, Moreau D, Pointurier O, Villerd J (2017a) Assessing broomrape risk due to weeds in cropping systems with an indicator based on a simulation model. Ecol Indic 82:280–292. https://doi.org/10.1016/j.ecolind.2017.05.070

Colbach N, Darmency H, Fernier A, Granger S, Le Corre V, Messéan A (2017b) Simulating changes in cropping practices in conventional and glyphosate-resistant maize. II. Effect on

weed impacts on crop production and biodiversity. Environ Sci Pollut Res 24:13121–13135. https://doi.org/10.1007/s11356-017-8591-7
Colbach N, Fernier A, Le Corre V, Messéan A, Darmency H (2017c) Simulating changes in cropping practices in conventional and glyphosate-resistant maize. I. Effects on weeds. Environ Sci Pollut Res 24:11582–11600. https://doi.org/10.1007/s11356-017-8591-7
Colbach N, Cordeau S, Garrido A, Granger S, Laughlin D, Ricci B, Thomson F, Messéan A (2018) Landsharing vs landsparing: how to reconcile crop production and biodiversity? A simulation study focusing on weed impacts. Agric Ecosyst Environ 251:203–217. https://doi.org/10.1016/j.agee.2017.09.005
Colbach N, Gardarin A, Moreau D (2019) The response of weed and crop species to shading: which parameters explain weed impacts on crop production? Field Crop Res 238:45–55
Colbach N, Chauvel B, Messéan A, Villerd J, Bockstaller C (2020) Assessing services and dysservices related to weed pollen. A simulation study. Ecol Indicators. https://doi.org/10.1016/j.ecolind.2020.106635
Fried G, Norton LR, Reboud X (2008) Environmental and management factors determining weed species composition and diversity in France. Agric Ecosyst Environ 128(1–2):68–76. https://doi.org/10.1016/j.agee.2008.05.003
Gardarin A (2008) Modélisation des effets des systèmes de culture sur la levée des adventices à partir de relations fonctionnelles utilisant les traits des espèces. PhD Thesis, Université de Bourgogne, Dijon, France
Gardarin A, Dürr C, Colbach N (2012) Modeling the dynamics and emergence of a multispecies weed seed bank with species traits. Ecol Model 240:123–138. https://doi.org/10.1016/j.ecolmodel.2012.05.004
Gardarin A, Coste F, Wagner M-H, Dürr C (2016) How do seed and seedling traits influence germination and emergence parameters in crop species? A comparative analysis. Seed Sci Res 26(4):317–331. https://doi.org/10.1017/s0960258516000210
Hossard L, Jeuffroy MH, Pelzer E, Pinochet X, Souchere V (2013) A participatory approach to design spatial scenarios of cropping systems and assess their effects on phoma stem canker management at a regional scale. Environ Model Softw 48:17–26. https://doi.org/10.1016/j.envsoft.2013.05.014
Lewis J (1973) Longevity of crop and weed seeds: survival after 20 years in soil. Weed Res 13(2):179–191. https://doi.org/10.1111/j.1365-3180.1973.tb01262.x
Liebman M, Dyck E (1993) Crop rotation and intercropping strategies for weed management. Ecol Appl 3:92–122
Liebman M, Gallandt ER (1997) Many little hammers: ecological management of crop-weed interactions. In: Jackson LE (ed) Ecology in agriculture. Academic Press, New York, pp 291–343. https://doi.org/10.1016/B978-012378260-1/50010-5
Maillot T, Mion M, Vioix J-B, Colbach N (2019) Conception de systèmes de cultures par algorithmes d'optimisation. In: Colbach N, Angevin F, Bockstaller C, Chauvel B, Denieul C, Moreau D, Omon B, Pellet D, Rodriguez A, Trannoy L, Volan S, Vuillemin F (eds.). Gestion des adventices dans un contexte de changement - Séminaire CoSAC Paris, France, 32–35
Martin G, Felten B, Duru M (2011) Forage rummy: a game to support the participatory design of adapted livestock systems. Environ Model Softw 26:1442–1453
Meylan L, Merot A, Gary C, Rapidel B (2013) Combining a typology and a conceptual model of cropping system to explore the diversity of relationships between ecosystem services: the case of erosion control in coffee-based agroforestry systems in Costa Rica. Agr Syst 118:52–64
Meynard JM, Charrier F, Fares M, Le Bail M, Magrini MB, Charlier A, Messean A (2018) Sociotechnical lock-in hinders crop diversification in France. Agron Sustain Dev 38(5):13. https://doi.org/10.1007/s13593-018-0535-1
Mézière D, Petit S, Granger S, Biju-Duval L, Colbach N (2015) Developing a set of simulation-based indicators to assess harmfulness and contribution to biodiversity of weed communities in cropping systems. Ecol Indic 48:157–170. https://doi.org/10.1016/j.ecolind.2014.07.028

Moreau D, Pointurier O, Nicolardot B, Villerd J, Colbach N (2020) In which cropping systems can residual weeds reduce nitrate leaching and soil erosion? Eur J Agron, in press

Moss SR, Clarke JH (1994) Guidelines for the prevention and control of herbicide-resistant black-grass (*Alopecurus myosuroides* Huds.). Crop Prot 13:230–234

Munier-Jolain NM, Guyot SHM, Colbach N (2013) A 3D model for light interception in heterogeneous crop:weed canopies. Model structure and evaluation. Ecol Model 250:101–110. https://doi.org/10.1016/j.ecolmodel.2012.10.023

Munier-Jolain NM, Collard A, Busset H, Guyot SHM, Colbach N (2014) Modelling the morphological plasticity of weeds in multi-specific canopies. Field Crop Res 155:90–98. https://doi.org/10.1016/j.fcr.2013.09.018

Pasquier C, Angevin F (2017) Freins et leviers à la réduction de l'usage d'herbicides en grande culture. Paper presented at the Gestion des adventices dans un contexte de changement: Connaissances, méthodes et outils pour l'élaboration de stratégies innovantes, Séminaire de restitution à mi-parcours du projet de recherche ANR CoSAC, Paris, France

Petit S, Alignier A, Colbach N, Joannon A, Le Coeur D, Thenail C (2013) Weed dispersal by farming at various spatial scales. A review. Agron Sustain Dev 33:205–217

Pointurier O, Moreau D, Colbach N (submitted) Reconciling a 3D individual-based root architecture model with large-scale and multi-annual simulations. Case study with a weed dynamics model. Ecol Modelling

Sausse C, Le Bail M, Lecroart B, Remy B, Messean A (2013) How to manage the coexistence between genetically modified and conventional crops in grain and oilseed collection areas? Elaboration of scenarios using role playing games. Land Use Policy 30(1):719–729. https://doi.org/10.1016/j.landusepol.2012.05.018

Sester M, Angevin F, Choimet C, Messéan A (2006) Coexistence of genetically-modified and non-genetically-modified beet crops in France and Germany: the case of tolerance to a non-selective herbicide. In: Messean A, Angevin F, Gómez-Barbero M, Menrad K, Rodríguez-Cerezo E (eds) New case studies on the coexistence of GM and non-GM crops in European agriculture. Joint Research Center of the European Commission, Technical Report Series. EUR 22102 EN, Sevilla, Spain

Thomson FJ, Moles AT, Auld TD, Kingsford RT (2011) Seed dispersal distance is more strongly correlated with plant height than with seed mass. J Ecol 99(6):1299–1307. https://doi.org/10.1111/j.1365-2745.2011.01867.x

Wilson RS, Tucker MA, Hooker NH, LeJeune JT, Doohan D (2008) Perceptions and beliefs about weed management: perspectives of Ohio grain and produce farmers. Weed Technol 22(2):339–350. https://doi.org/10.1614/WT-07-143.1

# Chapter 12
# Ryegrass Integrated Management (RIM)–Based Decision Support System

**Joel Torra and Marta Monjardino**

**Abstract** RIM, or "Ryegrass Integrated Management", is a model-based decision support system (DSS) for weed management that aims to deliver key recommendations to manage herbicide resistance (HR). Its success as a DSS for HR management is proven by the various adaptations of RIM to a range of weed species and/or cropping systems: Multispecies RIM for *Lolium rigidum* and *Raphanus raphanistrum* in Australia, Wild Radish RIM for *Raphanus raphanistrum* in Australia, PIM for *Papaver rhoeas* in Spain, RIMPhil for *Echinochloa crus-galli* in the Philippines, BYGUM for *Echinochloa colona* in Australia, PAM for *Amaranthus palmeri* in the USA, Brome RIM for *Bromus* spp. in Australia, Barley Grass RIM for *Hordeum glaucum* in Australia, SA-RIM for *Lolium rigidum* in South Africa and DK-RIM for *Lolium multiflorum* in Denmark. This chapter will describe the rationale, structure and strengths of these RIM-based DSS to manage HR.

**Keywords** Adoption · Bioeconomic model · Herbicide resistance · IWM · *Lolium rigidum* · Annual ryegrass · Simulation modeling

## 12.1 Introduction

Cropping enterprises in Australia, as in many parts of the world, are heavily dependent on herbicides for weed control. However, during the 1990s, the phenomenon of herbicide resistance in prominent crop weeds increased dramatically, especially in

J. Torra (✉)
Department d'Hortofructicultura, Botànica i Jardineria, Agrotecnio, Universitat de Lleida, Lleida, Spain
e-mail: joel.torra@udl.cat

M. Monjardino
CSIRO Agriculture and Food, Waite Campus, Adelaide, SA, Australia
e-mail: marta.monjardino@csiro.au


G. R. Chantre, J. L. González-Andújar (eds.), *Decision Support Systems for Weed Management*, https://doi.org/10.1007/978-3-030-44402-0_12

the dominant weed species of Australian cropping, *Lolium rigidum* Gaud. (annual ryegrass) (Owen et al. 2014).

*Lolium rigidum* has been able to develop resistance to a broad spectrum of sites of action (SoA) in Australia, including almost all herbicide chemistries currently available. Therefore, it can evolve resistance to any existing SoA. Moreover, multiple and cross-resistant populations are common by means of enhanced metabolism, giving unpredictable resistance patterns (Busi and Powles 2016).

The evolution of herbicide-resistant populations of *L. rigidum* led Australian weed researchers to promote an integrated weed management (IWM) approach that combines chemical, physical and biological practices with the aim to kill existing weeds, prevent seed set and deplete the seed bank (e.g. Matthews et al. 1996; Powles et al. 1997).

A consequence of herbicide resistance is that integration with non-herbicide methods for weed control is required. Alternatives to herbicides include a return to integrated strategies that had been minimised as a result of herbicide efficacy, as well as the potential for innovative new practices. IWM strategies rely on ecological processes, physical and cultural control methods, and a reduced range of herbicides of types that, in combination, are sustainable (Powles and Bowran 2000). Thus, by necessity, many Australian farmers started adopting diverse combinations of weed control measures but faced a number of difficulties in their decision-making about the implementation of IWM programmes. These include (1) they may be unfamiliar and inexperienced with a number of the control options; (2) strategies must be evaluated over the longer term; (3) the long-term impacts of multiple control options are difficult to predict; (4) the impacts of individual methods within an integrated strategy are difficult to interpret from field observations; (5) some strategies have indirect and direct costs; and (6) there are many possible combinations of methods to be considered. Given these difficulties, IWM within farming systems benefited from a decision support system (DSS) approach for farm advisors and farmers.

DSS development has focussed on the tactical/strategic planning problems, where alternative weed control scenarios can be tested and compared according to their economic output over a long-term horizon (Pannell et al. 2004; Torra et al. 2010). Some DSS, such as the RIM-based tools, do not include automated optimisation features nor provide management recommendations. Instead, users can easily experiment with options and visually compare the consequences of their management choices (Lacoste and Powles 2015).

## 12.2 The Original RIM Model

### *12.2.1 Motivation for Model Development*

RIM—Ryegrass Integrated Management—is a model-based DSS originally developed for testing the biological and economic performance of integrated *L. rigidum* management strategies for dryland broadacre systems of the southern Australian grain belt. The model simulates a comprehensive set of weed management methods, including both herbicide and non-herbicide options.

Simpler models that pre-dated and led to the development of RIM are described by Bennett and Pannell (1998), Gorddard et al. (1995, 1996), Schmidt and Pannell (1996) and Stewart (1993), all from Western Australia. Apart from these publications, there is little in the literature on economic aspects of herbicide resistance management, an exception being Orson (1999).

RIM's first phase of development spanned several years starting in the late 1990s which culminated in 1999 with RIM's first official launch and in 2004 with RIM final version (version RIM 2004), described in Pannell et al. (2004).

The core of RIM, including baseline data, results from the collective effort of many scientists within various institutions, including the University of Western Australia (UWA), the Australian Herbicide Resistance Initiative (AHRI), the Department of Agriculture and Food of Western Australia (DAFWA), the University of Adelaide, the Cooperative Research Centre for Weed Management (Weeds CRC) and the CSIRO. More recently, significant model improvements led to the 2013 launch of redeveloped RIM (version RIM 2013) as an open-access product (Lacoste and Powles 2015, 2016).

### *12.2.2 Model Description*

RIM is a dynamic simulation model that combines weed, crop and pasture biology, agronomy and economics in more than 500 parameters, many of which are adjustable by the user (Fig. 12.1). Specification of values for each of these parameters was a major task in its development. Sources of data and information were numerous and diverse. Economic parameters were obtained from an existing whole-farm economic model (Kingwell and Pannell 1987), and updated from budget guides published for farmers. Parameters for control effectiveness of weed control options were estimated based on long-term field experiments designed to evaluate their effects and from other field trials conducted by DAFWA. Parameters for weed competition functions were calibrated in cooperation with specialists from DAFWA to provide relationships consistent with field trial evidence.

RIM users select a crop–pasture sequence over a 10-year period and specify a feasible agronomic strategy that combines a range of chemical and non-chemical practices, based on pre-defined control efficacy values. The main outputs of the model are financial gross margins ($/ha) and weed plant and seed numbers per hectare, which change with the strategy selected. Other outputs include weed control expenses and income for each enterprise. Figure 12.2 provides a snapshot of the core page of the RIM model. A more detailed description of RIM can be found in Pannell et al. (2004) and Lacoste and Powles (2014, 2015).

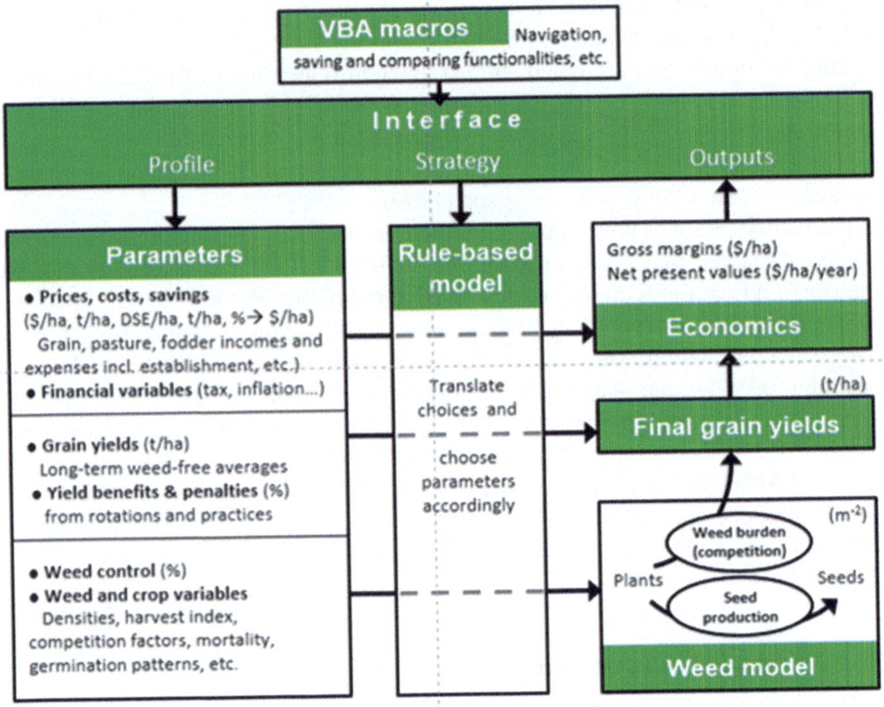


**Fig. 12.1** Key relationships between the main components of the RIM model (Source: Lacoste and Powles 2015)

### 12.2.3 RIM Adaptations to Other Weed Species and Cropping Systems

Since the development of the original RIM model in 2002, several versions have been created for a range of herbicide-resistant weeds and cropping systems around the world (Table 12.1). Weed species modelled so far include *L. rigidum* in Australia and South Africa; *Raphanus raphanistrum*, *Echinochloa colona*, *Bromus* sp. and *Hordeum glaucum* in Australia; *Papaver rhoeas* in Spain; *Echinochloa crus-galli* in the Philippines; *Amaranthus palmeri* in the USA; and *Lolium multiflorum* in Denmark. Development of RIM adaptations to several other weed species is currently underway in Australia and Denmark, and there are plans for a Laos version as well (based on RIMPhil).

All RIM versions are deterministic and do not represent annual variations in weather, yield, prices, costs and herbicide performance. They represent only a single field and a single weed species, with the exception of Multispecies RIM that represent two weeds: a grass species and a broadleaf species. With the

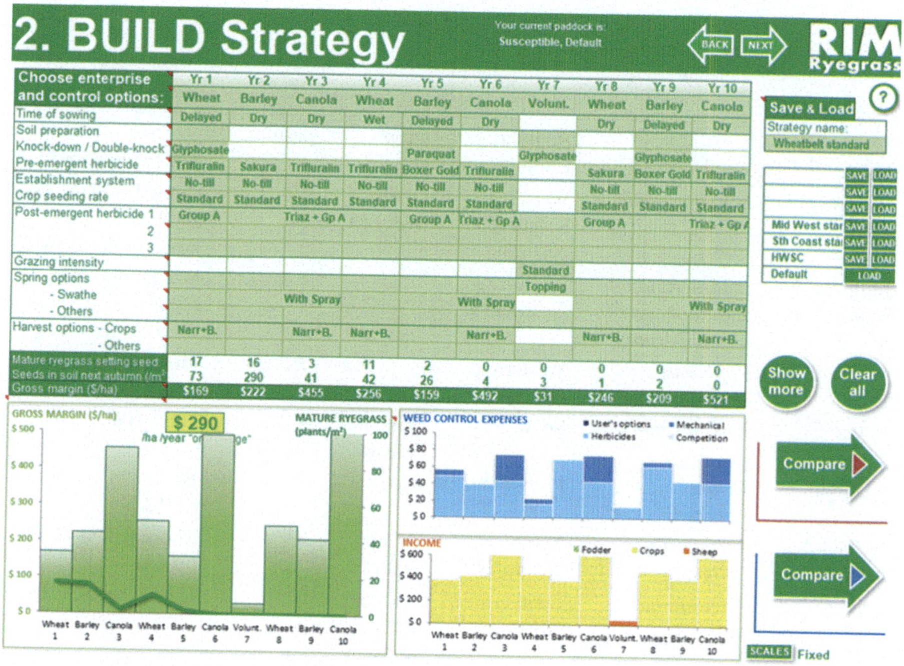


**Fig. 12.2** Snapshot of the core page of the Ryegrass RIM model: example of strategy building and associated key bioeconomic outputs

expansion of RIM and the desire to retain the RIM brand, the original model became known as Ryegrass RIM.

### 12.2.4 Model Validation

The biological output of the Western Australian version of Ryegrass RIM (Pannell et al. 2004) has been validated against field data, as reported in Draper and Roy (2002). In addition, sensitivity analysis is a useful approach for evaluating the significance of data deficiencies and facilitating a more focussed research effort (Pannell 1997), and users can readily alter the biological and economic parameter values to suit their particular situation.

### 12.2.5 Model Applications: Examples of Scenario Analyses

The RIM model has been used in a range of bioeconomic analyses of *L. rigidum* management, including in a specific research project testing a novel metaheuristic optimisation technique, compressed annealing (CA), implemented in RIM as an

**Table 12.1** Existing versions of the RIM model adapted to different herbicide-resistant weed species and cropping systems worldwide

| Name | Weed species | Region/ Country | Crop system | No. of crop/ pasture options | Key publications |
|---|---|---|---|---|---|
| Ryegrass RIM | *Lolium rigidum* | Western/ Southern Australia | Grain–livestock | 7 | Pannell et al. (2004) and Lacoste and Powles (2015, 2016) |
| Multispecies RIM | *Lolium rigidum* and *Raphanus raphanistrum* | Western Australia | | 7 | Monjardino et al. (2003) |
| Wild radish RIM | *Raphanus raphanistrum* | Western Australia | | 7 | Monjardino et al. (unpublished) |
| PIM | *Papaver rhoeas* | Spain | Winter cereals | 3 | Torra et al. (2010) |
| RIMPhil | *Echinochloa crus-galli* | Philippines | Rice | 1 | Beltran et al. (2012a) |
| BYGUM | *Echinochloa colona* | Northern Australia | Cotton–grain | 8 | Thornby and Werth (2015) |
| PAM | *Amaranthus palmeri* | Southern USA | Cotton, corn, soybean | 6 | Lindsay et al. (2017) |
| SA-RIM | *Lolium rigidum* | Western South Africa | Winter cereals | 4 | Spammer (2018) |
| Brome RIM | *Bromus* sp. | Southern Australia | Grain–livestock | 7 | Monjardino and Llewellyn (2018) |
| Barley grass RIM | *Hordeum glaucum* | Southern Australia | | 7 | |
| DK-RIM | *Lolium multiflorum* | Denmark | Grain–pasture | 8 | |

optimisation algorithm (Doole and Pannell 2008a). While CA is currently unavailable to all RIM users, it was effectively used to evaluate the potential economic value of a range of strategies to control *L. rigidum* in the Western Australian dryland agriculture, such as the following:

- Crop rotations including *Medicago sativa* L. (lucerne or alfalfa) phases (Doole and Pannell 2008b);
- Crop rotations including a popular pasture species, *Ornithopus sativus* Brot. cv. Cadiz (French serradella) (Doole et al. 2009);
- Ungrazed pasture fallows grown tactically between crop phases (Doole and Weetman 2009);
- Crop rotations including the newly developed *Trifolium dasyurum* C. Presl (Eastern star clover), a legume pasture species with delayed germination suited to short pasture phases (Doole and Revell 2010).

### 12.2.6 Adoption and Current Status

The RIM model can be used either as a management, training or research tool (i.e. hypothesis generator). It is therefore aimed at a wide audience in the agricultural field, including the following:

- Farmers attempting to make weed management decisions;
- Private consultants, extension agents or agribusiness agronomists wishing to provide advice to their clients;
- Facilitators running RIM workshops with groups of farmers;
- Researchers, students and others wishing to understand the management of a particular weed species within farming systems.

The tool is most often used in workshops with a combination of farmers, advisers and agronomists with the goal to provide answers to a range of questions, such as the following:

- Which combination of control options and crop rotations provides the best overall weed management system in the long-term?
- How fast can a specific weed problem develop?
- How can income be maintained if herbicides become ineffective?
- If a pasture phase is included, how long should it be for?
- Is it worth investing in specific machinery? Is a particular treatment (e.g. green manuring) a profitable practice? If so, under what circumstances?

Between 2003 and 2005 several RIM workshops were conducted across the states of Western Australia—WA, South Australia—SA, Victoria—Vic and New South Wales—NSW. These workshops included the release of an updated RIM package with a pea crop and the option for top crop treatment. Some RIM workshops were delivered as part of the national adviser IWM training programme (118 advisers, 6 workshops NSW, Vic, WA). Another series of RIM workshops on the topic "Breaking the bank—managing weeds for the future" was delivered to 200 farmers and 82 advisors as part of the Victorian DPI State Focus 2004.

Most of these workshops were concluded with questionnaires for evaluating RIM among its target audience, that is, farmers and consultants of the southern Australian grain belt. Key findings following the evaluation of 10 herbicide resistance workshops using RIM were as follows:

- Using RIM was stated as a highlight of the workshops, which were highly valued overall. Almost 90% of participants thought RIM was useful and a good learning experience;
- Participants saw RIM as an engaging, accessible "hands-on" tool;
- Particularly praised was the possibility of exploring scenarios through simulation, and the group interaction and discussions;
- 80% of the participants said attending the workshop changed their perception about herbicide resistance and as many specified they may change their crop–

weed management as a result, particularly specific techniques and/or increasing the overall system diversity;
- Suggested changes to RIM revolved around adding or developing more options (management and enterprises).

A broader evaluation of extension on management of herbicide-resistant weeds by means of RIM workshops confirmed that "changes in the perceived short-term economic value of some weed management practices did occur where the broader value of practices to the farming system, not necessarily relating to weed control, could be demonstrated. This also led to more growers deciding to adopt those practices" (Llewellyn et al. 2005).

Since the 2003–2005 workshops, uptake of the RIM model and manual package roughly doubled, reaching a record of 500 copies in 2005 (Lacoste et al. 2013). Between 2005 and 2012, records show that, while less than 10 copies of RIM 2004 were requested in total, 10–15 IWM workshops were delivered annually by consultants in WA, NSW and SA with RIM as a key component. In addition, a few RIM sessions a year were run with students in universities in Australia (e.g. UWA, Charles Sturt University, Cleve Area School), as well as the UK and Canada. In 2006, RIM featured prominently in the programme of an EWRS-funded International Workshop on Bio-economic Modelling for Weed Management held in Denmark. During 2017–2018, eight workshops have been conducted with farmers, advisers and or/researchers using the new Brome and Barley Grass RIM tools.

Despite a lack of knowledge of the actual number of people who have adopted the RIM models suite in Australia and beyond, recent records of the number of RIM workshops delivered, count of model downloads from the Australian Herbicide Resistance Initiative website (https://ahri.uwa.edu.au/), count of podcast listeners and number of RIM-related publications and citations provide an indication of at least the interest generated in the tools (Table 12.2).

## 12.3 Multispecies RIM: *Lolium rigidum* and *Raphanus raphanistrum* in Australia

### *12.3.1 Motivation for Model Development*

*Lolium rigidum* and *Raphanus raphanistrum* L. (wild radish) dominate and co-exist throughout southern Australian dryland cropping regions (Owen et al. 2014; Walsh et al. 2001). *R. raphanistrum* and *L. rigidum* are economically important weeds of crops in many parts of the world, especially Australia, and herbicide-resistant populations are now widespread in the cropping regions of Western Australia (Lu et al. 2019).

Widespread herbicide resistance in both these species has led to the need to adopt even more diverse and complex IWM practices, which require evaluation of their impact on the farming system. Therefore, a multispecies version of the RIM model

**Table 12.2** Key indicators of dissemination, use and adoption of the Australian adaptations of RIM

| Name | Total number of workshops delivered | Count of model download since online release | Number of webinar/ podcast[a] listeners | RIM-related publications | Number of citations[b] |
|---|---|---|---|---|---|
| Ryegrass RIM | 70–100 | 328[c] | – | Pannell et al. (2004), Doole and Pannell (2008b), Doole et al. (2009), Doole and Weetman (2009); Doole and Revell (2010) and Lacoste and Powles (2014, 2015, 2016) | 272 |
| Multispecies RIM | – | – | – | Monjardino et al. (2003, 2004a, b, 2005) | 109 |
| Wild radish RIM | – | – | – | Monjardino et al. (unpublished) | – |
| BYGUM | – | 255 | 90 | Thornby and Werth (2015) | – |
| PAM | – | 182 | – | Lindsay et al. (2017) | 3 |
| Brome RIM | 6 | 264 | 143 | Monjardino and Llewellyn (2018) and Llewellyn et al. (2018) | 2 |
| Barley grass RIM | 2 | 250 | 119 | Monjardino and Llewellyn (2018) | 2 |

[a]https://weedsmart.org.au/webinars/rick-llewellyn-describes-new-brome-rim/ (launched 20/09/2017). https://ahri.uwa.edu.au/podcast/new-barley-grass-and-brome-rim/ (launched 06//09/2017); https://ahri.uwa.edu.au/podcast/barley-grass-rim-is-now-available/ (launched 01/08/2018)

[b]Google Scholar (accessed 27 Feb 2019)

[c]Since the counter feature was included in 2017—model uptake expected to be much higher since first model release in the early 2000s

was developed to compare long-term economic and weed population outcomes of different integrated management scenarios.

### 12.3.2 Key Model Updates and Changes

The original single-species Ryegrass RIM model was extended to include *R. raphanistrum* biology and additional weed management practices used to control this weed species. Changes were made to key biological processes such as seed germination, production and mortality, plant growth, as well as inter- and intra-species competition to account for the multiple weed effects in the crop field.

Careful selection of a multispecies competition approach for use in the Multispecies RIM model resulted in a single crop yield function approach, capable of representing different crop plant densities and the realistic features of crop–weed and weed–weed competition, including the fact that the effects of a weed on crop yield depend on the density of another species, and that high densities of different weeds result in different minimum crop yields (Monjardino et al. 2003).

The Multispecies RIM accounts for a broader range of weed control options, including herbicides for both grass and broadleaf control. Each control treatment has its own impact on weed mortality and seed set, based on the control efficacy parameters represented in the model for *L. rigidum* and *R. raphanistrum*. The Multispecies RIM model is not a resistance model as it excludes the genetics of resistance. However, it evaluates the effects of resistance by allowing the user to specify the herbicide resistance status of both weed species with respect to each of nine herbicide mode-of-action groups.

### 12.3.3 Model Applications: Examples of Scenario Analysis

Multispecies RIM has been used to evaluate weed management scenarios for co-existing herbicide-resistant species by investigating the implications of different crop–pasture rotational sequences and varying herbicide availability. Examples of scenario analyses include the economic value of including non-cropping phases in the rotation, such as haying/green manuring (Monjardino et al. 2004a) and pasture phases (Monjardino et al. 2004b), and the value of glyphosate-resistant canola (Monjardino et al. 2005) in the management of *L. rigidum* and *R. raphanistrum* in a Western Australian farming system.

Results indicated that, while the inclusion of hay/green manuring did not generally increase returns (except in cases of extreme weed numbers and/or high levels of herbicide resistance), involving occasional 3-year phases of pasture in the sequence was competitive with the best continuous cropping rotation, particularly where herbicide resistance was at high levels. In contrast, the clear benefits of glyphosate-resistant canola over the commonly grown triazine-resistant canola would need to be weighed up against potential risks to marketability (due to consumer resistance) and risks of increased weed resistance to glyphosate (due to increased selection pressure). Overall, the Multispecies RIM analyses revealed that economic difference between the scenarios is less due to differences in weed densities than to differences in total weed control costs.

### 12.3.4 Adoption and Current Status

Widespread use of Multispecies RIM has been limited due to lack of effective calibration. The *L. rigidum* section of the model has undergone a fairly extensive validation process, particularly in regard to input data, individual functions and testing of population dynamics (Draper and Roy 2002). However, the *R. raphanistrum* section was less thoroughly validated given that only limited relevant research data was available at the time of development. Specifically, more data are needed in the areas of *R. raphanistrum* population dynamics, weed–crop competition and weed control by non-herbicide methods. Multispecies competition also requires new research attention. In addition, Multispecies RIM would greatly benefit from an upgrade to the current, more user-friendly format (Lacoste and Powles 2015).

## 12.4 Wild Radish RIM: *Raphanus raphanistrum* in Australia

### 12.4.1 Motivation for Model Development

Wild Radish RIM was developed in the sequence of Multispecies RIM, when the team identified the opportunity to easily create a similar tool designed specifically for the evaluation of long-term management strategies for the control of *R. raphanistrum* in western Australian broadacre agriculture.

### 12.4.2 Key Model Updates and Changes

Wild Radish RIM was essentially created by removal of all parameters and relationships relating to *L. rigidum* and weed–weed interactions present in the Multispecies RIM. Likewise, considerable effort was expended on data collection for *R. raphanistrum*, but there are still areas where the available biological information is relatively weak, or has not been updated since model development in the early 2000s. This seems inevitable in such a comprehensive model.

### 12.4.3 Model Applications: Examples of Scenario Analysis

Wild Radish RIM has been used to investigate the impacts of changing farm enterprise sequences for *R. raphanistrum* control. That analysis concluded that the choice of crop–pasture sequence had major implications for the management of this weed species. In particular, inclusion of pasture phases had the potential to reduce reliance on selective herbicides and other practices, despite poor economic returns given the assumed market conditions at the time. Overall, the unpublished results indicated the potential for more diverse rotations to allow for more flexible management of *R. raphanistrum*.

### 12.4.4 Adoption and Current Status

Wild Radish RIM has never been released to potential users outside the research team, partly because *R. raphanistrum* often coexists with *L. rigidum* in cropping fields, thus giving Multispecies RIM a relative advantage over the single-weed model. In addition, Wild Radish RIM would also benefit from stronger validation of the *R. raphanistrum* component, as well as an upgrade to the newest format (Lacoste and Powles 2015).

## 12.5 PIM: *Papaver rhoeas* in Spain

### 12.5.1 Motivation for Model Development

*Papaver rhoeas* L. (common corn poppy) is the most important dicotyledonous weed species infesting winter cereals in southern Europe. Because of high fecundity, highly persistent seeds and an extended period of germination, *P. rhoeas* is difficult to control and can substantially reduce grain yields (Torra and Recasens 2008). Control has become worse with the appearance of herbicide resistance to ALS inhibitors and/or synthetic auxins in several European countries (Heap 2019). Resistance mechanisms to group B are quite well established (Délye et al. 2011; Rey-Caballero et al. 2017a), and significant advances have occurred for group O (Rey-Caballero et al. 2016; Torra et al. 2017). In Spain, it is estimated that 40% of the dry-land cropping fields infested with *P. rhoeas* harbour some type of herbicide-resistant biotype (CPRH 2017). This scenario was the key motivation for developing a RIM-based DSS, the Poppy Integrated Management (PIM), to help Spanish farmers better manage this weed species (Torra et al. 2010).

### 12.5.2 Key Model Updates and Changes

PIM simulates the population dynamics of *P. rhoeas* over a 20-year period within a single cereal field. The model operates biologically at the level of nine periods in which the agronomic year is divided based on timing of control treatments, tillage operations and sowing dates. In this version, some structural modifications were made compared with the original RIM model (Pannell et al. 2004) in order to better accommodate the biology of *P. rhoeas* and the Spanish agronomic context.

The most significant change in PIM was the inclusion of four seed bank layers of 5 cm down to 20 cm because it was considered necessary to simulate seed movement in the soil profile associated with the different soil tillage systems that can be present in Spain (Torra et al. 2010). In brief, the model draws from matrices of seed movement from one layer to another as a result of tillage operations. Different rates of seed bank decline for each soil layer depending on soil cultivation were also incorporated into PIM, along with different emergence rates in cultivated versus uncultivated soil (Cirujeda et al. 2006, 2008).

Users of PIM might specify the crop sequence (barley, wheat and fallow) and any feasible combination of 38 different weed management practices, including selective herbicides (14), non-selective herbicides (1), biological and cultural treatments (11) and user-defined treatments (1).

### *12.5.3 Model Applications: Examples of Scenario Analysis*

This DSS may be used to evaluate weed management scenarios by investigating the implications of different tillage, fallow and cereal rotational sequences, as well as constraints on herbicide availability. Model validation showed that PIM was sufficiently accurate for predicting *P. rhoeas* population dynamics within a single season (Torra et al. 2010). The simulation of three different scenarios (mouldboard ploughing, minimum tillage and zero tillage) showed that profit increased as tillage operations were reduced, with the best income for the zero-tillage scenario. Conversely, seed bank depletion improved in the two scenarios with tillage. A combined scenario (7 years with zero tillage, 2 with minimum tillage and 1 with ploughing) was a good compromise between profitability and *P. rhoeas* management, representing an 83% seed bank reduction over 10 years (Torra et al. 2010).

### *12.5.4 Adoption and Current Status*

This version is an experimental tool that has never been released to potential users. Due to lack of resources, its update and maintenance has ceased, making impossible the delivery to farmers and other stakeholders. A decade after its creation, PIM has remained mostly unchanged with its use restricted to a limited number of universities and private educators (J. Torra, personal communication). An upgrade should be undertaken considering that new herbicides have become available and others banned in Europe, or crops that have since become part of the rotations, such as winter oilseed rape and field pea. Likewise, a new validation process should be undertaken, particularly in regard to weed population dynamics into the mid- to long-term, as several relevant field studies have since been conducted (Torra et al. 2011, 2018; Rey-Caballero et al. 2017b).

## 12.6 RIMPhil: *Echinochloa crus-galli* Complex in the Philippines

### *12.6.1 Motivation for Model Development*

*Echinochloa crus-galli* L. (annual barnyard grass) is the most harmful weed species in rice crops due to its rapid development to seed shed, high phenotypic plasticity, high reproduction capacity, germination flexibility, and strong competitive ability (Beltran et al. 2012a). Moreover, *E. crus-galli* is the second weed species in the world resistant to at least 10 different SoA (Heap 2019). In addition, *E. crus-galli* is one of the most serious weeds of rice crops in the Philippines, where butachlor (group K3)- and propanil (group C2)-resistant populations have been reported in some important rice-growing areas (Juliano et al. 2010). Consequently, rice farmers throughout the Philippines were encouraged to use IWM strategies, which led to the development of RIM Philippines (RIMPhil), a bioeconomic model to inform weed management decisions and to analyse the implications of IWM programmes for rice farmers in the Philippines (Beltran et al. 2012a).

### *12.6.2 Key Model Updates and Changes*

RIMPhil simulates predicted effects on the *E. crus-galli* population, grain yield and profit over 5, 10, 15 and 20-year periods a rice field. The RIMPhil model incorporates around 300 parameters, typical for a lowland irrigated rice farm, but could readily be adapted to similar rice production systems in other countries (Beltran et al. 2012a). The model user defines the maximum number of applications of each group of herbicides that can be used prior to the onset of herbicide resistance, as in the original RIM (Pannell et al. 2004). RIMPhil includes herbicide (selective, non-selective, pre-emergence, early or late post-emergence) and non-herbicide weed controls. Non-chemical tactics include different types of stale-seedbed preparation (also with chemical control), tillage, seeding (transplanted or direct wet-seeded rice), seed quality, seeding rates and manual or mechanical weeding at different timings. Even though some crop rotation options are available in these farming systems, RIMPhil focusses on a single crop, rice (Beltran et al. 2012a).

The model assumes that the weed has different emergence flushes during the growing season. Potential weed-free yield is higher in the dry than in the wet cropping season. Cohorts that emerge and survive in direct-seeded rice are more competitive than in transplanted crops. However, the biological output of RIMPhil has not been validated given the absence of reliable information, particularly over a series of years (Beltran et al. 2012a). On the other hand, RIMPhil demonstrates that the framework in which RIM was developed is flexible enough to adapt it to very different farming/weed scenarios.

### *12.6.3 Model Applications: Examples of Scenario Analysis*

Bioeconomic analyses indicated that a mixture of chemical and non-chemical treatments can provide good *E. crus-galli* control in Philippine rice crops, maximising long-term profit while lowering weed plant densities and seed banks (Beltran et al. 2012a). The model is particularly useful for testing scenarios of labour intensification and higher resistance to herbicides (Beltran et al. 2012b). Outputs of such analyses indicated that herbicides would become increasingly attractive relative to manual weeding as labour cost increases and also that the onset of herbicide resistance would result in substantial losses in farm profit.

### *12.6.4 Adoption and Current Status*

The authors are unaware of the adoption and current status of RIMPhil or of future updates and improvements, apart from those detailed in the published literature. A potential project is under consideration for adapting RIMPhil to the Laos context.

## 12.7 BYGUM: *Echinochloa colona* in Australia

### *12.7.1 Motivation for Model Development*

Glyphosate is the cornerstone of chemical weed management for cotton growers in Australia in the last decade. However, the appearance of several glyphosate-resistant weeds, such as *Echinochloa colona* L. (awnless barnyard grass), is challenging weed management in these cropping systems. The referred species is one of the most common in cotton fields, and glyphosate resistance is now widespread. Moreover, populations resistant to photosystem II inhibitors (atrazine) are also reported in the country (Heap 2019). This motivated using the basis of RIM to create a new tool, the BarnYard Grass Understanding and Management (BYGUM) to help growers' engagement in IWM thinking and strategy development for subtropical Australian cotton–grain farming systems (Thornby and Werth 2015).

### *12.7.2 Key Model Updates and Changes*

BYGUM extends the framework of RIM to northern subtropical Australian farming systems, where winter and summer crops (including irrigated and rainfed cotton, sorghum, corn and mung beans) and fallows, as well as new cover crops, are all important components of the system. In BYGUM 1 year from the original RIM is divided into summer and winter periods, with all choices for each period as for a usual RIM year; therefore, 5 years is the time frame evaluated. BYGUM also allows for more weed control applications than RIM, including different herbicides and cultivation. BYGUM evaluates 5-year rotations including testing the weed management and economic value of fallows, winter and summer cropping, cover crops, tillage, different herbicide options and herbicide resistance management. Some northern region-specific enhancements are included such as subtropical crop choices, barnyard grass seed bank, competition, ecology parameters and more freedom in weed control applications.

### *12.7.3 Model Applications: Examples of Scenario Analysis*

This RIM-based DSS has been used to assess the remaining value of glyphosate in rotations with glyphosate-resistant *E. colona*, both with strong resistance (5% efficacy) and moderate resistance (40% efficacy). The result was that after 5 years the system was still profitable if using high levels of crop competition to keep seed production (per escaping weed) low, while adding other tactics to effectively reduce the number of surviving plants to moderate/low levels (i.e. less than 14 plants per square meter at the end of the fifth season). BYGUM has been also used to test the value of a cover crop, showing that the benefit came, as expected, in the following crop, where the seed bank was driven down and the final weed density was kept low (www.cottoninfo.com.au).

### *12.7.4 Adoption and Current Status*

To date, BYGUM has been used for creating extension materials around specific weed management issues. It has been successfully delivered to industry in a series of workshops over a 3-year period (2015–2018). Approximately 90 industry agronomists attended the workshops and received a copy of BYGUM subsequently. Also, there have been 255 individual user downloads so far. Despite no active plans to create a new version of BYGUM, new funding to develop a multi-species version is highly sought after.

## 12.8 PAM: *Amaranthus palmeri* in the USA

### *12.8.1 Motivation for Model Development*

Several attributes confer *Amaranthus palmeri* S. Wats. (Palmer amaranth) the capacity to become the most troublesome weed in row crops, especially in cotton and soybean, on much of the American continent. Such attributes include dioecy, C4 photosynthetic pathway, high growth rate and reproduction capacity, genetic variability, multiple cohorts and stress tolerance (Palma-Bautista et al. 2019). This species has become even more concerning and widespread due to the evolution of multiple herbicide-resistant biotypes to glyphosate, ALS inhibitors, protoporphyrinogen oxidase inhibitors and/or triazines (Heap 2019). As a result, *A. palmeri* is very hard to control, and a zero-threshold strategy has already been suggested for its management (Norsworthy et al. 2014). However, before the development of thc Palmar Amaranth Management (PAM) model, no DSS was available to help demonstrate farmers the long-term biological and economic viability of IWM strategies of this species (Lindsay et al. 2017).

### 12.8.2 Key Model Updates and Changes

Like RIM, PAM plans in a 10-year horizon to allow the user to simulate consecutive 3-year crop rotations including mouldboard ploughing in Autumn (Fig. 12.3). *A. palmeri* management tactics include chemical and non-chemical approaches, such as crop rotation, row spacing, cover crops, seedbed preparation tillage practices, mouldboard ploughing and harvest-time weed seed control, among others. The economic component of PAM was designed to replicate southern US crop production practices, and like RIM, uses crop budgeting and discounting techniques to determine the overall profitability of weed management strategies. However, a new key aspect of PAM is its ability to demonstrate the magnitude of long-term benefits (net present value, NPV) vs. potential short-term losses (Lindsay et al. 2017).

Another new feature is that the user can monitor the degree of diversity of weed control options employed as an indirect assessment of risk of resistance evolution, as well as the timing of escapes (Fig. 12.3). The resistance risk assessment uses

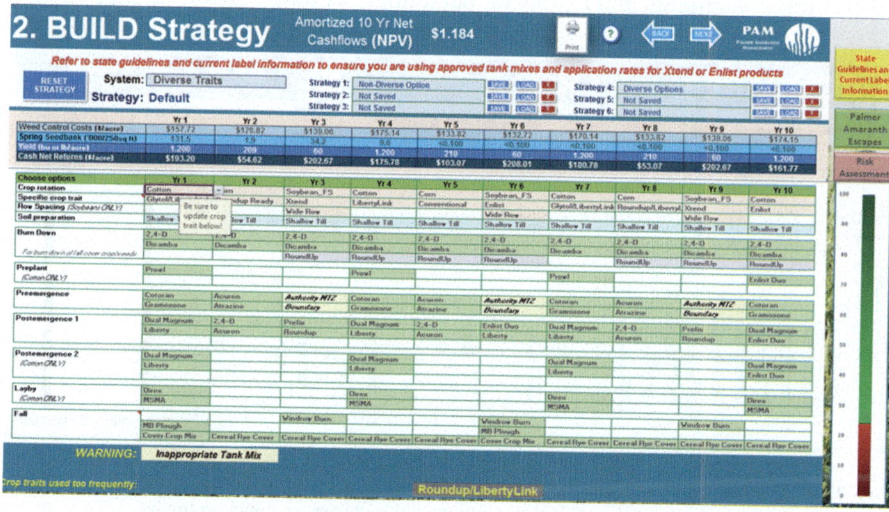


**Fig. 12.3** Snapshot of the core page of the PAM model: example of strategy building and associated risk assessment and weed escapes on left side

23-parameter model with higher percentage scores indicating a lack of diversity in weed control tactics and more risk of resistance evolution. The PAM model is currently available for download at http://agribusiness.uark.edu/decision-support-software.php#PAM.

### 12.8.3 Model Applications: Examples of Scenario Analysis

So far, two contrasting strategies have been compared to show the usefulness of PAM: a "Non-Diverse Options" strategy based on no-tillage and relying heavily on herbicides and planting only a few crop traits; and a "Diverse Options" strategy to reflect a diversified strategy that uses both chemical and nonchemical management options, such as fall cultural practices to drive seedbank near to zero and limit *A. palmeri* escapes (Fig. 12.4). Results showed that yields for the "Non-Diverse Options" strategy were volatile (and net returns became negative with years) compared with the "Diverse Options" strategy, which maintained relatively constant yield potential near 100%, with net returns in the positive range of 415 to 880 USD per ha. Furthermore, the "Non-Diverse Options" showed 53% risk of resistance, whereas the "Diverse Options" strategy showed only 38% risk, as shown in Fig. 12.4 (Lindsay et al. 2017).

### 12.8.4 Adoption and Current Status

The PAM model has been widely circulated through various outreach outlets and is currently used in extension activities. The major target audience for this tool is crop consultants and extension personnel, who run various scenarios and

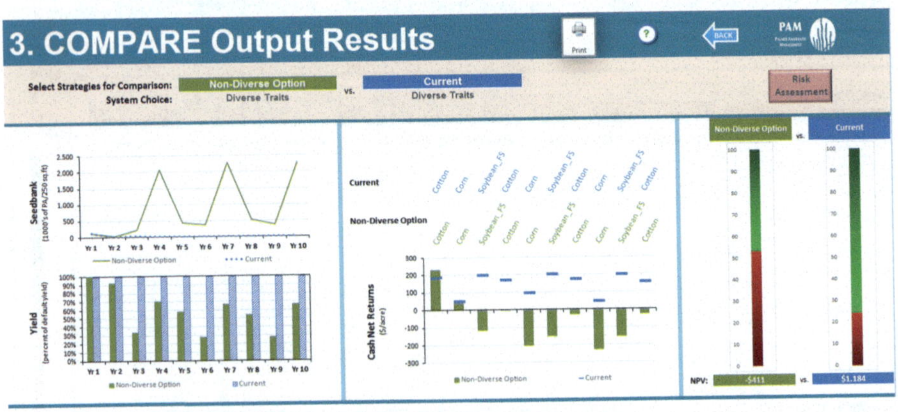


**Fig. 12.4** Snapshot of two contrasting strategies ("Non-Diverse Options" versus "Diverse Options") and associated key bioeconomic outputs and risks assessments on the left

use the information in their weed management planning and outreach activities, while some progressive farmers also use this by themselves. A number of weed management and crop production scenarios can be simulated by the software and the outputs are used to quantify and demonstrate the long-term benefits of diversified weed management. The analytics data and downloads reports indicate that the PAM software was downloaded 182 times, with the majority of users located across the southern USA. A recent user-survey has indicated that the most common users of the software were crop consultants, extension personnel and sales representatives, who collectively impact several hundreds of thousands of acres across the South since each of them serve several farmers. Thus, the tool has been widely impactful in management decision making in the southern region.

## 12.9 SA-RIM: *Lolium rigidum* in South Africa

### *12.9.1 Motivation for Model Development*

This RIM adaption is focussed on winter cereal farming in the Central Swartland area of the Western Cape Province in South Africa. *L. rigidum* is the most important weed affecting crop production in the area, and farmers are heavily reliant on chemical control. This heavy reliance on herbicides together with mal-practice regarding their application has caused *L. rigidum* to develop herbicide resistance. To date, resistance to glyphosate, paraquat, and ACCase and ALS inhibitors have been reported in South Africa (Heap 2019). That was why researchers from this area in South Africa were motivated to adapt RIM to this winter cereal farming system and develop the South Africa RIM (SA-RIM) (Spammer 2018).

### *12.9.2 Key Model Updates and Changes*

This RIM adaptation targets the same key weed species in a similar farming system with a Mediterranean-type climate as found in Western Australia; therefore, not many changes were required compared to the original RIM. A group of experts validated the parameters and assumptions of the Australian version when no data from South Africa were available or reliable. Some of the main differences in SA-RIM include calculations not being based on one-hectare units, nor farm size being an option, like it is done in the Ryegrass RIM. Also, yield benefits of 20–30% are assumed in canola after legumes in this version but not in the original RIM. Moreover, maximum *L. rigidum* density was increased in SA-RIM because South African farmers are able to crop with a higher weed burden in their fields than their Australian counterparts. Finally, annual self-regenerating pasture legumes are very common in the crop rotation in this South

African region with more alkaline soils, so a decision was made to change subterranean clover to *Medicago* spp. as a rotational pasture option in SA-RIM.

### 12.9.3 Model Applications: Examples of Scenario Analysis

So far, SA-RIM has only been used to assess the negative effect of *L. rigidum* herbicide resistance on the profitability of winter cereals. A non-resistance scenario was compared to two different resistance scenarios. While resistance always caused the highest reductions of gross margin under wheat monoculture, when any crop rotation was chosen these detrimental effects were alleviated, or even eliminated in a 10-year horizon in all simulations. Finally, the 4-year wheat *Medicago* spp. rotation achieved the highest level of *L. rigidum* control in all three scenarios.

### 12.9.4 Adoption and Current Status

Since SA-RIM was developed in 2018, no more updates are expected, and its current status has not changed.

## 12.10 Brome RIM: *Bromus* Sp. in Australia

### 12.10.1 Motivation for Model Development

Australian grain farms have the highest level of adoption of conservation cropping systems worldwide. These systems are built on three principles of minimum soil disturbance (i.e. minimum/zero tillage), permanent soil cover (retained stubble, crop/pasture cover) and diversity in rotations. Currently most broadacre grain farms routinely retain a majority of crop residues, and their ambition is to maximise stubble retention provided that any impacts on crop performance and profitability can be managed. As a result, a new research initiative was funded to take a farming systems approach to maintain profitable farming business with retained stubble, including grass weed management in retained stubble CSIRO, part of a Grains Research and Development Corporation (GRDC) initiative. In particular, herbicide-resistant grass weeds, such as *Bromus* sp. L. (brome grass), *Hordeum glaucum* and *L. rigidum,* were considered a significant threat in most southern farming systems of Australia. Therefore, the Brome RIM model was developed (along with Barley Grass RIM and a southern version of Ryegrass RIM) to improve understanding of the impact of contemporary stubble management systems such as cutting heights, windrowing and inter-row sowing, on the seed bank dynamics of these weeds so that their threat can be better managed.

### 12.10.2 Key Model Updates and Changes

Brome RIM represents the population dynamics of *Bromus* spp. over a period of up to 10 years. Key biological factors that drive the pattern of *Bromus* sp. population change over time include weed and seedbank population dynamics (i.e. germination rates during the growing season, and natural mortality rates of seedlings, dormant seeds during season and of seeds over summer), as well as weed–crop competition effects for each crop type. Overall, *Bromus* sp. has high seed dormancy and longevity in the soil, is an aggressive germinator and is a very competitive grass weed against all crops, especially cereals (Kleeman and Gill 2009).

As part of the CSIRO/GRDC initiative, additional trials were conducted in 2017 to collect biological data to inform Brome RIM. These complement the trials being run through grower groups including EPARF, UNFS and MSF on *Bromus* sp. Additional *Bromus* sp. focussed trials run by CSIRO include (I) interactive effect of pre-emergent herbicide and crop row placement on *Bromus* sp. control and (II) efficacy of current and potential new pre-emergent herbicide options for *Bromus* sp. control (in collaboration with the University of Adelaide) (e.g. Kleeman and Gill 2008; Boutsalis et al. 2014).

Based on feedback by workshop users, Brome RIM was further adjusted to achieve the following:

- Replace "Cadiz serradella" pasture with a grazing vetch option (both legumes) in order to get grazing value from vetch before it is brown manured. Inputs and costs were left unchanged.
- Replace the original "clover" pasture with a "clover/*Medicago* spp." option to allow for different soil types—inputs and costs left unchanged.
- Include "higher crop density" as a more generic option, for example to show the impact of a better crop establishment on sandy soils.

### 12.10.3 Model Applications: Examples of Scenario Analysis

The new Brome RIM model has been used to illustrate the potential for improving crop establishment and thus competition against weeds that are becoming increasingly difficult to manage in the southeastern dryland cropping regions of Australia (Monjardino and Llewellyn 2018). The results of this analysis indicate the potential for greater crop competition to reduce weed seedbanks and improve profit, at least for the default rotation and management strategy used in the models. This is particularly relevant in the case of *Bromus* sp., which is often found on sandy soils where achieving strong crop establishment can be difficult.

Brome RIM has also been used in a brief analysis of crop sequencing for the Mallee (SA), which showed that weed management is a major driver of the overall profitability of crop sequences, in particular that broadleaf, Clearfield and hay crops are effective tools to control *Bromus* sp. and improve the profitability of Mallee farming systems (Moodie 2017).

Finally, Brome RIM has been used to assess the cost-effectiveness of *Bromus* sp. management by using fenceless spatial grazing technology in sheep. Results demonstrate the ability to use virtual fencing within a field to focus grazing pressure on a weedy area while successfully excluding livestock from areas where high stocking density would cause environmental damage and usually prevent the use of high grazing pressure as a weed control tool (Llewellyn et al. 2018).

### *12.10.4 Adoption and Current Status*

Since online release in mid-2018, Brome RIM has been downloaded 264 times. In addition, it has been used in a series of five workshops with farmers and consultant agronomists in South Australia and Victoria.

## 12.11 Barley Grass RIM: *Hordeum glaucum* in Australia

### *12.11.1 Motivation for Model Development*

*Hordeum glaucum* Steud. (barley grass), like *Bromus* sp. and *L. rigidum*, is an important weed in the conservation farming systems of southern Australia with focus on no-till, stubble retention and crop rotations (Llewellyn et al. 2015). Managing these grass weeds is an increasingly challenging task requiring integration of several practices, often in low-rainfall, low-cost cropping environments with relatively limited herbicide options within cereal crops, and emerging herbicide-resistance risk (Owen et al. 2015). *Hordeum* spp. populations with delayed germination are now demanding a more strategic IWM approach be applied, hence the development of Barley Grass RIM.

### *12.11.2 Key Model Updates and Changes*

Barley Grass RIM represents the population dynamics of *Hordeum* spp. over a period of up to 10 years. As for all RIM models, key biological factors that drive the population dynamics of the focus weed species change over time, including germination, growth, mortality and weed–crop competition effects for each crop type.

*Hordeum* spp. is considered a difficult weed to control due to its high seed germination potential, high seed dormancy and longevity in the soil, high dispersion potential, and a particularly short growing season, which allows it to set seed even in the driest of seasons. In addition, it is an alternate host for a number of cereal diseases, such as take-all, harbour scald, net blotch and stripe rust (e.g. barley grass); its seed causes stock health problems, such as damage to skin, gums and eyes of sheep, fatal injuries in lambs, decreased bodyweight and reduced wool quality (e.g. barley grass); and it has only limited control with post-emergence herbicides.

As part of the CSIRO/GRDC initiative, regional trials that were conducted in 2017 to collect weed data also informed the Barley Grass RIM. These complement the trials being run through grower groups including EPARF, UNFS and MSF on *Hordeum* spp. (e.g. Mudge 2016; Fleet et al. 2017).

Barley Grass RIM has undergone similar adjustments to pasture and crop density representation as described for Brome RIM.

### *12.11.3 Model Applications: Examples of Scenario Analysis*

Like Brome RIM, the new Barley Grass RIM model has been used to assess the effect of increased crop competition on reducing reliance on selective herbicides, lowering weed seed banks and increasing profitability over time in dryland cropping regions of Australia increasingly threatened by *Hordeum* spp. (Monjardino and Llewellyn 2018). This brief analysis highlights the long-term biological and economic benefits of potential innovations that could improve crop competition, demonstrating potential for analyses around specific technologies, such as high vigour cultivars (e.g. Mudge 2016), soil wetter agents and on-row sowing in non-wetting soils (McBeath et al. 2017) that may increase the relative advantage of the crop against weeds.

### *12.11.4 Adoption and Current Status*

Since online release in late 2018, Barley Grass RIM has been downloaded 250 times.

## 12.12 DK-RIM: *Lolium multiflorum* in Denmark

### *12.12.1 Motivation for Model Development*

*Lolium multiflorum* Lam. (Italian ryegrass) is an increasing problem in Danish crop rotations with large proportions of winter cereals. Moreover, this species has developed resistance to ALS and ACCase inhibiting herbicides in Denmark (Heap 2019). With this background the Australian DSS for herbicide resistance management, RIM was adjusted to Danish conditions by Aarhus University to develop DK-RIM (Sønderskov 2018).

### 12.12.2 Key Model Updates and Changes

One of the major changes in DK-RIM relative to the original RIM is offering the possibility of two sowing seasons; autumn and spring sown crops. Furthermore, DK-RIM does not include animal production and has a limited economic component around purchases of equipment. A broader variety of crops has been implemented compared to the original RIM, including winter wheat, winter barley, winter rye, spring barley, spring wheat, spring oat, winter oilseed rape and legumes. In addition, grass and grass clover can be included in the crop rotation, but are not cash-crops in the system. The main tools for the management of *L. multiflorum* include crop rotation, soil tillage, and the option to alternate herbicide modes of action. There are no harvest control options in Denmark, hence also not in DK-RIM. Options for desiccation and crop sacrifice have been built into this version of the model.

### 12.12.3 Model Applications: Examples of Scenario Analysis

So far, DK-RIM has only provided an overview about which combination of control measures and crop rotations, such as spring or grass options, is the optimal strategy in the long term to manage *L. multiflorum* in Denmark. Also, DK-RIM allows users to choose between ACCase, ALS or metabolic resistant populations to understand which the best management practices are when different resistance profiles can be present.

### 12.12.4 Adoption and Current Status

DK-RIM has only had limited adoption due to its recent release but it has much potential for use in similar Scandinavian cropping systems. However, these authors are unaware of future updates and/or improvements. This DSS is in Danish, but an English translation of DK-RIM is available by contacting the developer. English documentation for DK-RIM was released in early 2019, and a scientific publication is planned. Users can download DK-RIM, along with a Danish users guide, from the web page https://www.landbrugsinfo.dk/Planteavl/Sider/pl_19_AU_DK_RIM_bekaempelse_italiensk_rajgraes.aspx.

**Table 12.3** Ongoing adaptations and future releases of the RIM model

| Name | Weed species | Country | Crop system |
|---|---|---|---|
| DK-RIM 2 | *Alopecurus myosuroides* | Denmark | Winter cereals |
| RIMLao | *Echinochloa crus-galli* | Southern Laos | Rice |

## 12.13 Future Versions and Adaptations

The future release of RIM updates, as well as versions and adaptations for other weeds/cropping systems/countries is summarised in Table 12.3 and described in the sections below.

### *12.13.1 Blackgrass DK-RIM, Denmark*

*Alopecurus myosuroides* Huds. (Blackgrass) is one of the most important grass-weeds in North-western Europe and is also the most important herbicide-resistant weed species in European agricultural systems (Keshtkar et al. 2015). In Denmark, populations with multiple resistance to the three SoA ACCase, ALS and Microtubule assembly inhibitors are prevalent. The widespread occurrence of particularly NTSR is a severe challenge to the effective management of *A. myosuroides*. In this country this challenge is even more prominent due to few SoA being available for its control mainly due to national regulation on groundwater protection (Keshtkar et al. 2015). This scenario is motivating the development of Blackgrass DK-RIM, a version of the *L. multiflorum* DK-RIM, by the same developers of the former Danish RIM-based DSS.

Information on the life of the weed seed bank (persistence) is critical for the growers to develop management practices for herbicide-resistant populations or for difficult-to-control weeds. At this stage, information on the rate of decline of seed banks for the nominated locally important weeds under field conditions is not available. Therefore, this project will undertake studies to quantify the rate of decline of weed seed banks under field conditions. In addition, there will also be investigation into seed production, dormancy, establishment pattern and phenology of the target weed species.

Weed biology information and practical management options will be used to update the RIM model, which will undergo further adaptation to accommodate the summer, weeds. The new Summer Weeds RIM will then be used by consultants and farmers to further understand the impact of their management decisions on these weeds.

### *12.13.2 RIMLao, Laos PDR*

A proposed research project would focus on adapting RIMPhil to the conditions of southern Laos in order to increase the capacity of smallholder farmers to reduce the impacts of weeds on rice systems in that region. Rice weeds, such as *E. crus-galli*, reduce yield and quality of rice, and limit adoption of new and improved crop establishment techniques, such as direct-seeding. Most smallholder farmers still use hand-weeding for weed control in Laos, but labour is scarce and costly, and alternative control methods are urgently required. To manage weeds effectively there is a need to introduce farmers and advisors to a range of cultural, physical and chemical methods for integrated weed management. Questions around, for example, whether the adoption of sustainable direct-seed rice production without or with minimal use of herbicides would be economically viable in the long term could be best answered using a RIM-type DSS.

## 12.14 Conclusions

The RIM-based models are tools aiming to aid complex farm managerial decisions and communication thereof. The software fulfils this objective by allowing users to assess the impacts of various management strategies in a relative manner. However, according to Lacoste and Powles (2015), "RIM only deals with a partial aspect of the herbicide resistance problem and does not replace expert judgment and direct observations. As such, it should be remembered that RIM is not a forecast model aiming to provide exact predictions. RIM is built with compromises, with accessibility taking precedence over representativeness, simplicity over accuracy, and modelling efficiency over complexity."

The key strengths of RIM can be summarised as follows:

- Its user-friendly platform allows farmers and industry professionals to conveniently test and compare the long-term performance and profitability of numerous weed control options.
- Its decision support system can aid the delivery of key recommendations among the agricultural community for cropping systems threatened by herbicide resistance.
- Its research and training potential are particularly useful in workshops and student fora with the goal to provide answers to a range of weed management questions involving complex trade-offs.

But to achieve the end use of RIM as a practical and effective DSS, a number of simplifications, exclusions and/or compromises in the assumptions behind RIM were deemed necessary. These include the following:

- Deterministic approach (vs. stochastic), where there is no representation of annual variations in weather, yield, prices, costs, herbicide performance, control efficacies, soil, germination, biomass growth, animal behaviour, etc.

- Crops and pastures are generic, i.e. enterprises do not represent a given cultivar or species but rather a type of enterprise, defined by adjustable characteristics and management specifics.
- Single field and single weed species are used in all model adaptations, except the Multispecies RIM that represents a grass species and a broadleaf species.
- Weed genetics and the evolution mechanisms of herbicide resistance are not modelled. Simulation of a resistant weed can be represented by altering the control efficacy of herbicides.
- The environmental impact of weeds and/or management practices is only partially represented, as is the case of increased risk of soil erosion by specific physical practices e.g. full-cut seeding and mouldboard plough.

Overall, RIM is a unique decision support tool with potential to help with the extension effort to advocate sustainable practices when the rise of herbicide resistance poses a serious challenge to the agricultural industry, in Australia and worldwide.

## References

Beltran JC, Pannell DJ, Doole GJ, White B (2012a) A bioeconomic model for analysis of integrated weed management strategies for annual barnyardgrass (*Echinochloa crus-galli* complex) in Philippine rice farming systems. Agr Syst 112:1–10

Beltran JC, Pannell DJ, Doole GJ (2012b) Economic implications of herbicide resistance and high labour costs for management of annual barnyardgrass (*Echinochloa crus-galli*) in Philippine rice farming systems. Crop Prot 31:31–39

Bennett AL, Pannell DJ (1998) Economic evaluation of a weed-activated sprayer for herbicide application to patchy weed populations. Aust J Agric Res Econ 42:389–408

Boutsalis P, Kleeman S, Gill G, Preston P (2014) A hidden threat: widespread Group B herbicide resistance in brome across South-Eastern Australia. In: Baker M (ed) Proceedings of the 19th Australasian Weeds Conference. Tasmanian Weeds Society, Hobart, pp 202–205

Busi R, Powles SB (2016) Cross-resistance to prosulfocarb + S-metolachlor and pyroxasulfone selected by either herbicide in Lolium rigidum. Pest Manag Sci 72:1664–1672

Comité para la Prevención de Resistencias a Herbicidas, CPRH (2017) Meeting of the working group of the Spanish Weed Science Society, 2nd February, Lleida, Spain. http://semh.net/grupos-de-trabajo/cprh/

Cirujeda A, Recasens J, Taberner (2006) Dormancy cycle and viability of buried seeds of *Papaver rhoeas*. Weed Res 46:327–334

Cirujeda A, Recasens J, Torra J, Taberner A (2008) A germination study of herbicide-resistant field poppies in Spain. Agron Sustain Dev 28:207–220

Délye C, Pernin F, Scarabel F (2011) Evolution and diversity of the mechanisms endowing resistance to herbicides inhibiting acetolactate-synthase (ALS) in corn poppy (*Papaver rhoeas* L.). Plant Sci 180(2011):333–342

Doole GJ, Pannell DJ (2008a) Optimisation of a large, constrained simulation model using compressed annealing. J Agric Econ 59(1):188–206

Doole GJ, Pannell DJ (2008b) Role and value of including lucerne (*Medicago sativa* L.) phases in crop rotations for the management of herbicide-resistant *Lolium rigidum* in Western Australia. Crop Prot 27:497–504

Doole GJ, Revell CK (2010) Delayed pasture germination allows improved rigid ryegrass (Lolium rigidum) control through grazing and broad-spectrum herbicide application. Crop Prot 29(2):153–162

Doole GJ, Weetman E (2009) Tactical management of pasture fallows in Western Australian cropping systems. Agr Syst 102:24–32
Doole GJ, Pannell DJ, Revell CK (2009) Economic contribution of French serradella (*Ornithopus sativus* Brot.) pasture to integrated weed management in Western Australian mixed farming systems: an application of compressed annealing. Aust J Agric Resour Econ 53:193–212
Draper A, Roy B (2002) Ryegrass RIM model stands the test of IWM field trial data. AgriBusiness Crop Updates 2002:49–50
Fleet B, Cook A, Richter I, Shepperd W, Preston C, Gill G (2017) Management of group A herbicide resistant barley grass in pasture phase. GRDC Eyre peninsula farming systems 2016. https://eparf.com.au/wp-content/uploads/2017/04/39.-Managment-of-group-Aherbicide-resistant-barley-grass-in-pasture-phase.pdf. Accessed 15 Feb 2019
Gorddard RJ, Pannell DJ, Hertzler GL (1995) An optimal control model of integrated weed management under herbicide resistance. Aust J Agric Econ 39:71–87
Gorddard RJ, Pannell DJ, Hertzler GL (1996) Economic evaluation of strategies for management of herbicide resistance. Agr Syst 51:281–298
Heap I (2019) International survey of herbicide resistant weeds [online]. http://www.weedscience.org/In.asp. Accessed 10 Jan 2019
Juliano LM, Casimero MC, Llewellyn R (2010) Multiple herbicide resistance in barnyardgrass (*Echinochloa crus-galli*) in direct-seeded rice in the Philippines. Int J Pest Manage 56:299–307
Keshtkar E, Mathiassen SK, Moss SR, Kudsk P (2015) Resistance profile of herbicide-resistant Alopecurus myosuroides (black-grass) populations in Denmark. Crop Prot 69:83–89
Kingwell RS, Pannell DJ (eds) (1987) MIDAS, a bioeconomic model of a dryland farm system. Pudoc, Wageningen, 207 pp
Kleeman S, Gill G (2008) Applications of metribuzin for the control of rigid brome (*Bromus rigidus*) in no-till crops of southern Australia. Weed Technol 22:34–37
Kleeman S, Gill G (2009) Population ecology and management of rigid brome (*Bromus rigidus*) in Australian cropping systems. Weed Sci 57:202–207
Lacoste M, Powles S (2014) Upgrading the RIM model for improved support of integrated weed management extension efforts in cropping systems. Weed Technol 28(4):703–720
Lacoste M, Powles S (2015) RIM: anatomy of a weed management decision support system for adaptation and wider application. Weed Sci 63(3):676–689
Lacoste M, Powles S (2016) Beyond modelling: considering user-centred and post-development aspects to ensure the success of a decision support system. Comput Electron Agric 121:260–268
Lacoste M, Llewellyn R, Powles S, Pannell D (2013) RIM 2004 workshops: evaluation—farmers and consultants surveys. Australian Herbicide Resistance Initiative & School of Agricultural and Resource Economics, The University of Western Australia, Perth, 8 pp
Lindsay K, Popp M, Norsworthy J, Bagavathiannan M, Powles S, Lacoste M (2017) PAM: decision support for long-term palmer Amaranth (*Amaranthus palmeri*) control. Weed Technol 31(6):915–927
Llewellyn R, Pannell D, Lindner R, Powles S (2005) Targeting key perceptions when planning and evaluating extension. Aust J Exp Agric 45:1627–1633
Llewellyn RS, Ronning D, Ouzman J, Walker SR, Mayfield A, Clarke M (2015) Impact of weeds on Australian grain production and adoption of tillage practices. GRDC and CSIRO, Canberra, ACT, 160 pp
Llewellyn R, Lee C, Marini D, Belson S, Ouzman J, Monjardino M (2018) Cost effective weed management using targeted sheep grazing technology. A report for the Australian Government, Department of Agriculture and Water Resources. CSIRO, Canberra, ACT
Lu H, Yu Q, Han H, Owen M, Powles SB (2019) Metribuzin resistance in a wild radish (*Raphanus raphanistrum*) population via both psbA gene mutation and enhanced metabolism. J Agric Food Chem 67:1353–1359
Matthews JM, Llewellyn R, Powles S, Reeves T (1996) Integrated Weed Management for the control of herbicide resistant annual ryegrass. Report no. UA201. University of Adelaide, Adelaide

McBeath T, Llewellyn R, Gupta V, Davoren B, Shoobridge W, Kroker S (2017) On-row sowing for brome competition on non-wetting sands. Mallee Sustainable Farming. http://www.msfp.org.au/row-sowing-brome-competition-nonwetting-sands. Accessed 15 Feb 2019

Monjardino M, Llewellyn R (2018) Newly-developed RIM models for the integrated management of brome and barley grasses. In: Proceedings of the 21st Australasian Weeds Conference, 9–12 Sept 2018, Sydney, Australia

Monjardino M, Pannell DJ, Powles SB (2003) A multi-species bio-economic model for integrated management of *Lolium rigidum* and *Raphanus raphanistrum* in Australian farming systems. Weed Sci 51:798–809

Monjardino M, Pannell DJ, Powles SB (2004a) The economic value of haying and green manuring in the integrated management of annual ryegrass and wild radish in a Western Australian farming system. Aust J Exp Agric 44(8):1195–1203

Monjardino M, Pannell DJ, Powles SB (2004b) The economic value of pasture phases in the integrated management of annual ryegrass and wild radish in a Western Australian farming system. Aust J Exp Agric 43(12):1–7

Monjardino M, Pannell DJ, Powles SB (2005) The economic value of glyphosate-resistant canola in the management of two widespread crop weeds in a Western Australian farming system. Agr Syst 84:297–315

Monjardino M, Pannell DJ, Powles SB (unpublished). Wild Radish RIM: a bio-economic model for integrated management of *Raphanus raphanistrum* in Australian cropping.

Moodie M (2017) Crop sequencing considerations for Mallee. MSF—Mallee Sustainable Farming. https://www.msfp.org.au/wp-content/uploads/Crop-Sequencing-Considerations-for-the-Mallee-Michael-Moodie-Full.pdf

Mudge B (2016) Over dependence on agrichemicals—barley grass trial. Unpublished report. Barry Mudge Consulting for Upper North Farming Systems

Norsworthy JK, Griffith G, Griffin T, Bagavathiannan M, Gbur EE (2014) In-field movement of glyphosate-resistant palmer amaranth (*Amaranthus palmeri*) and its impact on cotton lint yield: evidence supporting a zero-threshold strategy. Weed Sci 62:237–249

Orson J (1999) The cost to the farmer of herbicide resistance. Weed Technol 13:607–611

Owen MJ, Martinez NJ, Powles SB (2014) Multiple herbicide-resistant *Lolium rigidum* (annual ryegrass) now dominates across the Western Australian grain belt. Weed Res 54:314–324

Owen MJ, Martinez NJ, Powles SB (2015) Herbicide resistance in *Bromus* and *Hordeum* spp. in the western Australian grain belt. Crop Pasture Sci 66:466–473

Palma-Bautista C, Torra J, García MJ, Bracamonte ER, Rojano-Delgado AM, Alcántara-de la Cruz R, De Prado R (2019) First study of resistance mechanisms to glyphosate detected in *Amaranthus palmeri* harvested in GR soybean in Argentine. J Agric Food Chem 67:1052–1060

Pannell DJ (1997) Sensitivity analysis of normative economic models: theorethical framework and practical strategies. Agric Econ 16:139–152

Pannell DJ, Stewart V, Bennett A, Monjardino M, Schmidt C, Powles SB (2004) RIM: a bio-economic model for integrated weed management. Agr Syst 79:305–325

Powles SB, Bowran DG (2000) Crop weed management systems. In: Sindel BM (ed) Australian weed management systems. R. G. and F. J. Richardson, Melbourne, pp 287–306

Powles SB, Preston C, Bryan IB, Jutsum AR (1997) Herbicide resistance: impact and management. Adv Agron 58:57–93

Rey-Caballero J, Menéndez J, Giné-Bordonaba J, Salas M, Alcántara R, Torra J (2016) Unravelling the resistance mechanisms to 2,4-D (2,4-dichlorophenoxyacetic acid) in corn poppy (*Papaver rhoeas*). Pestic Biochem Physiol 133:67–72

Rey-Caballero J, Menéndez J, Osuna MD, Salas M, Torra J (2017a) Target-site and non-target-site resistance mechanisms to ALS inhibiting herbicides in *Papaver rhoeas*. Pestic Biochem Physiol 138:57–65

Rey-Caballero J, Royo-Esnal A, Recasens J, González I, Torra J (2017b) Management options for multiple herbicide–resistant corn poppy (*Papaver rhoeas*) in Spain. Weed Sci 65:295–304

Schmidt C, Pannell DJ (1996) Economic issues in management of herbicide-resistant weeds. Rev Market Agric Econ 64(3):301–308

Sønderskov M (2018) DK-RIM, integreret bekæempelse af Italiensk rajgræs–Brugervjledning Flakkebjerg. Aarhus Universitet, Agroøkologi

Spammer Z (2018) The financial and managerial implications of herbicide resistance of annual ryegrass in the Central Swartland. Master of Commerce, Stellenbosch University, South Africa

Stewart VA (1993) Economics of integrated strategies for managing herbicide resistant ryegrass under continuous cropping. B.Sc. Thesis, University of Western Australia

Thornby D, Werth J (2015) BYGUM—a new tool for BarnYard Grass Understanding and Management. In: 2nd Australian Cotton Research Conference, Toowoomba

Torra J, Recasens J (2008) Demography of corn poppy (*Papaver rhoeas*) in relation to emergence time and crop competition. Weed Sci 56:826–833

Torra J, Cirujeda A, Recasens J, Taberner A, Powles SB (2010) PIM (PoppyIntegrated management): a bio-economic model for *Papaver rhoeas* in rainfed cropping systems. Weed Res 50:127–139

Torra J, Royo-Esnal A, Recasens J (2011) Management of herbicide-resistant *Papaver rhoeas* in dry land cereal fields. Agron Sustain Dev 31:483–490

Torra J, Rey-Caballero J, Royo-Esnal A, Salas M, De Prado R (2017) Enhanced 2,4-D metabolism in two resistant *Papaver rhoeas* populations from Spain. Front Plant Sci 8:1584

Torra J, Royo-Esnal A, Rey-Caballero J, Recasens J, Salas M (2018) Management of herbicide resistant Corn Poppy (*Papaver rhoeas*) under different tillage systems does not change the frequency of resistant plants. Weed Sci 66:764–772

Walsh MJ, Duane RD, Powles SB (2001) High frequency of chlorsulfuron resistant wild radish (*Raphanus raphanistrum* L.) populations across the Western Australian wheatbelt. Weed Technol 15(2):199–203

# Chapter 13
# IPMwise: A Decision Support System for Multispecies Weed Control in Cereal Crops

José M. Montull, Andreu Taberner, Ole Bøjer, and Per Rydahl

**Abstract** Integrated weed control is mandatory in the current legislative framework for sustainable plant protection programmes. The advent of synthetic pesticides in the 1950s allowed farmers to simplify cropping systems and forego more complicated crop protection strategies, especially in cereal production. Moreover, the awareness of the necessity to decrease pesticide use has been raised considerably since the mid-1980s in Europe. In this work, a Danish Decision Support System (DSS) for Field-Specific Crop Management is presented. This DSS, known as Crop Protection Online (CPO) and later IPMwise, optimizes herbicide weed control by providing recommendations of specific herbicide solutions to achieve a required control level. It has been developed since the 1980s, and the actual version (IPMwise) has recently been adapted to the edaphic and climatic conditions of Spain.

The adaptation process required (1) generation of dose–response curves for Spanish-relevant weed species and (2) calculation and adjustment of the shift of the dose–response curves according to phenological stages of the weed species. IPMwise was validated in winter cereal field trials from 2010 to 2018 and in maize from 2016 to 2018. IPMwise recommendations were compared to the efficacies obtained with standard herbicide treatments decided by local advisors. In 84% of the evaluated cases, efficacies were equal to or higher than those predicted by advisors. Thus, IPMwise is a robust DSS tool showing the potential to decrease amounts of applied herbicides by at least 30% in Spanish cereal agricultural systems.

**Keywords** Field-Specific Crop Management · Integrated weed control · Legislative framework · Crop Protection Online · Dose-response curves · Winter cereal field trials · DSS tool

J. M. Montull (✉) · A. Taberner
University of Lleida-Agrotecnio, Lleida, Spain
e-mail: josemaria.montull@udl.cat

O. Bøjer · P. Rydahl
IPM Consult ApS, Stenlille, Denmark
e-mail: ole.bojer@ipmconsult.dk; per.rydahl@ipmconsult.dk


G. R. Chantre, J. L. González-Andújar (eds.), *Decision Support Systems for Weed Management*, https://doi.org/10.1007/978-3-030-44402-0_13

## 13.1 Introduction

Agriculture and livestock farming require continuous evolution towards more efficient production processes. This is applicable for all inputs, but due to various reasons, such as social acceptance and economics, it is especially applicable for pesticide use.

Pesticide use has increased exponentially due to its excellent cost–benefit relationship compared to other agricultural practices since its appearance in the market in the 1950s. For example, herbicide consumption in Spain was 6,326 metric tonnes in 1995 reaching 14,179 metric tonnes by 2013 (FAOSTAT 2016), being the majority of this consumption related to winter cereal crops which in Spain totalizes about six million hectares (MAPAMA 2019).

Their excellent cost–benefit relationship in economic terms has led to a misuse of pesticides with a negative impact on non-target organisms and contamination of groundwater and surface water. However, the use of herbicides for nearly 60 years has created a deep knowledge of their advantages and disadvantages. For example, it is known that the effect of herbicides can be modulated more easily than other weed control methods. Thus, optimizing their use should consider both improving the economic efficiency of farms and also decreasing the negative environmental impacts of their widespread use. In addition, the evolution of society entails greater awareness of environmental issues, which has resulted in various directives and regulations in Europe. Within this legal framework various initiatives such as the Network of Excellence ENDURE and the PURE-IPM Project have been developed in order to promote tools, such as DSS for European farmers with the objective of a more sustainable use of pesticides.

Following this philosophy, CPOWeeds has been developed since the 1980s and the actual version (IPMwise) has currently been adapted to edaphic and climatic conditions in Spain. Conversely to other Spanish DSS for weed management (see Gonzalez-Andujar et al. 2010a, b; Torra et al. 2010), IPMwise does not focus only on specific weed species. Instead, the aim of IPMwise is to generate several alternatives for tactical control, based on specific aspects that the user (advisors/farmers) can choose. The IPMwise response is based specifically on the weed community present in the plot, helping to assess the best spraying options from a list including most of the authorized herbicides for a given crop. In this way, IPMwise supports decision-making to improve current weed management practices.

In Spain, climatic conditions vary widely within the different parts of the country. It has been defined nine agroclimatic conditions taking into account the rainfall and temperature values of each locality (GENVCE 2015). Regarding temperature, the following areas have been established:

- Cold areas, with an average temperature below 11 °C in April.
- Temperate zones, with an average temperature between 11 and 13 °C in April.
- Warm areas, with an average temperature above 13 °C in April.

Regarding rainfall:

- Semi-arid zones: Areas where annual rainfall is <500 mm.
- Subhumid zones: Areas with an annual rainfall >500 mm and <700 mm.
- Humid zones: Areas with an annual rainfall >700 mm.

Compared to Spanish agroclimatic conditions, Denmark presents colder environments, with an average temperature in April below 7 °C with subhumid to humid zones depending on the distance to the sea. For that reason, the main issue to use this technology in Spain was to verify whether the assumptions calculated for northern Europe remained valid for Spanish conditions considering different herbicide options and different weed communities.

## 13.2 Towards a Sustainable Use of Herbicides Using Precision Agriculture Techniques

In Western Europe, there has been a tradition to recommend herbicide treatment programmes on a regional level, as supported by regional advisors. This allowed for some differentiation, for example, in terms of adjustments of treatment programmes according to dominating weed species in each region.

A next step has been to encourage and equip farmers and advisors to promote decision-making and treatments at both farm and field levels. In this case, farmers may need support to (1) identify weed infestations, (2) to quantify needs for control and (3) identify suitable alternative accompanying options for a sustainable weed control.

Precision Agriculture (PA) techniques aiming to increase spatial resolution have been widely examined in terms of design and field validation of CPOWeeds and more recently IPMwise.

Results from field trials, comparing traditional weed management methods at a regional scale (used as reference) and recommendations from IPMwise have demonstrated the following reductions of the input of herbicides:

- In Denmark, at least 40% in cereal crops and 20% in crops sown in a wide row distance like sugar beet and maize (Rydahl 2003; Sønderskov et al. 2014).
- In Norway, about 30% mainly in spring cereal crops (Netland 2005).
- In Spain, about 30% in winter cereal crops and maize (Montull et al. 2014).

Currently, only in Denmark, 1500 farmers/advisors use CPOWeeds (Rydahl, personal communication). In addition, it is implemented to varying degrees in Norway, Estonia, Poland and Germany in one or more crops. In these countries, the validation tests have showed that the recommendations were robust (Sønderskov et al. 2015). However, the potential of herbicide reductions varies between countries and depends on the weed species present in the fields and also on management decisions (Been et al. 2010). Furthermore, an ongoing project develops CPOWeeds for control in maize in Germany, Italy and Slovenia with a module for mechanical measures included (Rydahl et al. 2015).

In this case, the DSS is used to make a Field-Specific Crop Management, which is a form of precision agriculture approach whereby decisions on resource application and agronomic practices are improved to better match soil and crop requirements as they vary between the fields. In this case, the differences in weed infestation for each field require different doses of herbicides or even different herbicides. Actual PA implementation in the 1980s started when farmers integrated newly developed fertilizers capable of deploying variable rate application technology with maps that showed the spatial variability of soil chemical properties.

In Denmark, the PA challenge for IWM is being addressed in a project, which have the acronym name “RoboWeedMaPS”, and which runs in 2017–2020. This project aims to establish a chain of commercial products, which includes the following steps:

- Collecting digital pictures to determine weed infestations.
- Production of weed-maps on the species level.
- Automatic calculation by IPMwise of needs for control and accompanying herbicide products and dose rates, to produce digital spray-maps, which are readable to controller units, which may be integrated with sprayers.
- Site-specific spraying (PA spraying), by injections sprayers, which may regulate mixing on the fly, and which exist in various technical versions.

This project fits completely with the basic definition of PA: to apply the right treatment in the right place at the right time (Gebbers and Adamchuk 2010), or as mentioned in the area of crop protection: application should be as much as necessary but as little as possible (Been et al. 2010). A next step, which is more difficult to achieve with current technology, would be also to apply the required amount of herbicide plant by plant, to centimetre accuracy. However, although it seems relatively simple, optimizing the application of plant protection products is a complex decision and it is affected by many variables, such as the crop type, growth stage of the weeds and crops, weather and soil conditions, treatment cost, expected return, expected sales price and also long-term profitability.

That is why in the past 30 years various DSS for crop protection have been developed, in particular trying to optimize the use of herbicides as it is shown in the next section.

## 13.3 The Danish Crop Protection Online (CPOWeeds)

A general challenge, in order to achieve a more rational use of herbicides, is the estimation in a simple and quick way of both the need for weed control and the expected efficacy of control measures. To achieve this goal, since the 1980s, scientists from Aarhus University began to develop what is now the most widely used DSS for IWM in Europe, the Crop Protection Online-Weeds (Rydahl 2003). It was initially designed only for spring cereals, such as spring barley (Rydahl and Pedersen

2003) and it was first released in 1989 and further commercialised since 1991 (Rydahl 2003; Kudsk 2008a, b).

Up to date, four generations have been produced, which reflect ongoing developments in Information Technology (IT) and IWM. The third generation, which was released in 2002, and which will be operational outside Denmark until March 2020, is called Crop Protection Online (CPOWeeds). Basically, the goal of CPOWeeds is to optimise herbicide use by combining type and dosages according to the weed infestation levels in a given crop considering either by lowest TFI or lowest price depending on farmer/advisor decision. This system is owned jointly by Aarhus University and SEGES/Danish Farmers Unions. The fourth generation tool is called IPMwise, which differ from CPOWeeds in terms of IT standards, new agronomical features including non-chemical control, resistance management and PA-facilities like the availability to create spraying maps. As CPOWeeds and IPMwise contain several common features, both systems will be referred according to the periods of time, where these were used. The neutral term DSS is used to refer indistinctly to any of the systems.

In CPOWeeds, target efficacies are estimated based upon both crop and weed species densities and growth stages. The general principle is that high competitiveness and weed densities induce high target efficacies, while the less competitive weed species and low densities calls for lower target efficacies. The aim is to set a target efficacy level, which ensures yield and prevent excessive build-up of the soil weed seed banks but still enables reduced doses.

In the user-interface eleven criteria are integrated to define the weed scenario in a particular field. These criteria are season, crop type and density, potential yield, weed species densities, phenological stages of both crop and weeds, temperature and water stress. When the user has provided this information, the programme calculates the level of control required for every weed species and accompanying recommendations for control.

The latest update of IPMwise in Spain has been the integration of herbicide resistance management. This is a major concern issue in weed control. In this case, resistant weed biotypes are incorporated by creating separate weed biotypes, which are supplied with dose-response calculations, which show very low efficacy of herbicides with the mode of action, for which they are resistant. In IPMwise, additional measures have been included to reduce the risk of new herbicide resistance. This has been done by integration of measures to avoid unilateral use of active substances of herbicides (mode-of-actions, MoAs), which are considered risky in terms of generation of herbicide resistance on the national level (HRAC 2019).

A major advantage of these DSS is that technical control recommendations are based on models and parameters, which are easy to understand, and with a straightforward association to weed biology and herbicides' behaviour (Rydahl 2003). The optimization process is based on mathematical models as shown in Streibig et al. (1998) and Jensen and Kudsk (1988). Despite being designed in Denmark, the cost optimization models and TFI have succeeded in adapting to Spanish conditions.

## 13.4 High Doses vs. Low Doses in the Development of Resistance Cases: An Open Debate

Technically, one of the greatest controversies on the development and large-scale use of some DSS's has been related to those cases where recommendations point out lower doses than the maximum recommended rates by manufacturers. In these cases, farmers may fear yield losses and/or weed seedbank density increase in the long term (Kudsk 2014).

Bulk data of herbicide efficacy available to farmers merely classifies weed species as "controlled", "partly controlled" or "not controlled" using the maximum registered dose rates. For most herbicides, the group of weeds classified as "controlled" consist of species easily managed with doses considerably lower than the maximum label rate and species satisfactorily controlled at full recommended doses (Kudsk 1989, 2008a, b). Despite the fact that dose-response data for most commonly used herbicides is a necessary input for improvement of the decision-making process, such information is rarely available.

As no direct and linear relationship between herbicide dose and efficacy exists, it must be wondered, whether the dose rate debate should focus on dose rather on herbicide efficacy. In fact, depending on the application time and the weed species, a given herbicide dose can produce very different efficacies. Therefore, idealistically, debates on the importance of herbicide dose rates should conveniently be exchanged by a debate on the importance of achieved efficacy. In fact, when scientific articles mention low doses use and the generation of resistant weed biotypes, the results show that biotypes were selected among individuals which survived doses that caused about 30% mortality (Busi et al. 2012; Yu et al. 2012). Hence, these works are about low dose selection but also low efficacies that in no case could be acceptable from an agronomic point of view.

According to herbicide product labels, even the registered dose rate will probably never provide 100% efficacy. According to herbicide products and dose-response calculations included in IPMwise in Denmark, susceptible species could be controlled with reduction of 10–20% of the registered label dose rate. These results support the authors' viewpoint that a set-off in levels efficacy should be made rather than concentrating on application doses. Also, when a scientific paper discusses about resistance management in relation to high or low doses, it is merely in relation to the maximum authorized dose in each country. However, the maximum amount authorized for diclofop-methyl in Australia is 375 g a.i./ha, while in Spain and France is 630 and 900 g a.i./ha, respectively. In Australia, prosulfocarb maximum authorised rate is 2000 g a.i./ha while in Spain is 4800 g a.i./ha. This fact also depends on the registration criteria and the active dose for each country. For that reason, there are fewer differences for the new authorized herbicides. However, herbicide efficacy varies between environments due to many factors, such as crop density and weed species characteristics, so it is unclear whether and to what extent these differences in "regulated rates" can be translated into real differences in an "effective application rate" at the level of individual weeds and thus into real differences in control levels (Renton et al. 2011). Hence, globally, there are countries with greater potential (than others) to optimize the pesticide use, especially when having a high maximum authorized rate.

Here again, the DSS plays an important role because they allow to adjust the applied dose to obtain a particular efficacy without increasing the selection pressure that can favour resistance evolution. Finally, DSS can offer the option to select or avoid a given MOA according to information risk on weed resistant biotypes.

## 13.5 Developing the IPMwise in Spain

Taking into account these pros and cons, and the potential to prevent and manage resistant biotypes, the general objective was to develop an IPMwise version for Spanish agroclimatic conditions, common herbicides and weed species. The existing and newly developed algorithms were used to tune up a version for northern Spain agricultural systems.

The first issue was to determine the parameters for dose–response curves of key weed species. The second issue was to validate the CPOWeeds concept under Spanish conditions.

### *13.5.1 Obtaining Dose-Response Parameters for Key Weeds*

In many cases, the maximum authorized rate is based on environmental parameters or safety restrictions for farmer's use. Chemical companies need to demonstrate that their products are safe for workers and the environment at the suggested maximum rates. A secondary aspect is related to efficacy. For these reasons, if there are specific cases in which the amount of applied product can be diminished, the evaluation is not necessarily done. This occurs since it is difficult to have enough data to know the behaviour of plant protection products at doses different than the maximum. This is an important issue for optimizing herbicide use given that efficacy can vary considerably depending on weed species and environmental conditions (Minkey and Moore 1996).

In the article 14 of the Directive 2009/128/CE it is said that *Member States shall establish or support the establishment of necessary conditions for the implementation of Integrated Pest Management. In particular, they shall ensure that professional users have at their disposal information and tools for pest monitoring and decision-making, as well as, advisory services on IPM.*

In this context, DSS allow advisors to make an adjusted recommendation, considering both environmental as well as economic goals. To build a proper DSS it is necessary to gather enough data regarding herbicide efficacy for every weed species under all possible field conditions (climate, soil, growth stage, etc.).

One way to model all this data is using dose–response curves, which are the basis of CPOWeeds (Sønderskov et al. 2014). The use of these dose–response curves allows applying the optimal solution in a given agronomic context.

The first step to adapt IPMwise in Spain consisted of the determination of dose–response curves for relevant herbicide–weed species combinations in standard conditions. The second step consists in determining the magnitude of the parallel displacement which is specific for each herbicide and is based on experimental data (Rydahl 2003). The efficacy response is assumed to be well explained by the logistic model proposed by Seefeldt et al. (1995). Although, it is assumed that the maximum response available for an infinite dose is 100%, and the minimum response is equal to 0 at dose = 0.

An alternative version of this equation can be obtained as follows:

$$Y = \frac{1}{1 + \exp(-2(a_{\text{w}} + b_{\text{h}}(\log\left(\frac{x}{r_{\text{s}} * r_{\text{t}} * r_{\text{w}}}\right)} \tag{13.1}$$

where.

$Y$: relative efficacy (%),

$b_{\text{h}}$: "slope" in inflection point, herbicide specific,

$a_{\text{w}}$: horizontal displacement specific for each weed species,

$x$: dose (g/ha of 1 or more a.i.'s),

$r_{\text{s}}$, $r_{\text{t}}$, $r_{\text{w}}$: relative potencies or "dose factors", quantifies the relative effect of growth stage, temperature and drought stress, respectively, as compared to a reference point.

A graphical illustration of this function is provided in Fig. 13.1.

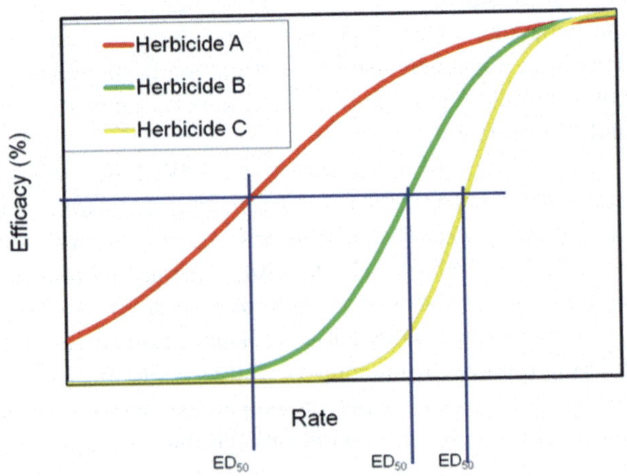


**Fig. 13.1** Herbicide dose-response functions for herbicides A, B and C

Theoretically, if herbicides have the same site of action, then all other parameters should be equal; that is, their response curves should be similar (parallel) with a relative horizontal displacement. On the other hand, the assumption of similar curves is a necessary but not sufficient condition for assuming similar MOAs of compounds (Streibig et al. 1998). For these reasons, $b_h$ only depends on the mode of action of the herbicide; thus, such parameter is available for all herbicides (Kudsk 1989). Therefore, for a given weed species and herbicide, the unique unknown parameter would be the horizontal displacement of the curve, this value is the $a_w$-parameter.

Performing dose–response curves is the best way to manage big data generated after running field tests for new herbicide registration. Moreover, using too many tests under different climatic conditions allow for better estimation of herbicide behaviour.

It is needed efficacy data for each weed species and herbicide combination at different doses. In the first version of IPMwise Spain, 35 herbicides were introduced. Chemical companies interested in developing the DSS provided an important part of this data. Data was obtained from official field trials conducted by the companies to develop herbicides, and this raw efficacy data was used as the basis to perform statistical studies. Such field tests have proved to be very useful because generating this information in such a short time would have been virtually impossible. Usually, these trials are performed in several places, versus different, natural weed infestations under different climatic conditions. All this contribute to estimate, how various factors (as in Eq. 13.1) affect herbicide efficacy against different species. For the same reason, and by using proper statistical tools, it is possible to predict herbicide efficacy in a wide range of situations, which is one of the objectives of a DSS.

Such efficacy datasets are often available for 2–4 herbicide dose rates, which often include the registered dose rate, the highest dose and some systematically reduced dose rate. For example, ¼*X*, ½*X* and the registered dose rate (*X*). However, datasets with so few dose rates are not sufficient for using statistical methods to estimate the dose-response parameters ($b_h$ and $a_w$). For such limited datasets, a different method executed by the local experts is used to estimate $a_w$. Using an estimate of $b_h$, which is based on information on MOA, the achieved efficacy of the registered dose rates is used to produce the estimate of $a_w$. Using these estimates of $b_h$ and $a_w$, efficacy is subsequently simulated for 6–8 dose rates (including the dose rates, on which efficacy data is supported) and these are compared to the real data obtained in the field. Estimates of $a_w$ are changed until the estimated efficacy fits with the real efficacy values taking into account also requirements for ensuring the agronomical robustness (i.e. some safety margins need to be included).

As IPMwise calculates the herbicide dosage for the desired control level based on dose–response curves, all trials should be conducted with a minimum of four replicates. At the moment of spraying, certain weed phenological stage should be the standard application moment for each herbicide.

The $a_w$ parameter is the basis for starting the calculation of herbicide dose rates. If there is a minimum of one weed species which need to be controlled (by a certain level of target efficacy); the actual herbicide dose can be lowered accordingly by use of Eq. (13.1). and the actual estimates of $a_w$ and $b_h$, which are stored in the DSS database. In these cases, the DSS will recommend the exact dose to achieve the expected efficacy in the actual field. Conversely, if the estimated efficacy at the registered (maximum) dose rate cannot meet the efficacy targets, the DSS will recommend necessary tank-mixtures of two or more herbicides with the exact dose for each herbicide.

The second aspect for optimizing herbicide use is to know the effect of weed growth stage on herbicide performance. This is important, since it is known that annual weed species are generally more susceptible to herbicides at early growth stages, although some exceptions might be mentioned, for example *Galium aparine* L. vs mecoprop and fluroxypyr or *Avena* spp. vs some graminicides (Kudsk 2008b). Moreover, these differences in efficacy due to weed growth stages, differ also between herbicides. For example, one conclusion extracted from the previously cited study carried out by the EWRS Herbicide Optimization Working Group, stated that growth stage affects the performance of clodinafop but not mesosulfuron + iodosulfuron (EWRS Herbicide Dose Optimisation WG 2013).

Parallel dose–response curves could also be used to compare the effect of the growth stage on herbicide performance. With this aim, the relative herbicide potency (*R*-parameter) could be used (Streibig 2003):

$$R_i = \frac{\text{ED50}_1}{\text{ED50}_i} \tag{13.2}$$

being $\text{ED50}_1$ the parameter value calculated at standard weed growth stage and $\text{ED50}_i$ the parameter value for the other weed growth stages. The *R* in Eq. (13.2) represents $r_s$, $r_t$ and $r_w$ in Eq. (13.1).

In conclusion, IPMwise development in Spain required quantifying the efficacy variation of herbicides ($R_i$-parameter or $r_s$-parameter) depending on the five key-weed growth stages: 0–1 leaf, 2–3 leaves, 4–5 leaves, 6–8 leaves and >8 leaves. For this the local *R*-parameters for the dose-response function described by Seefeldt et al. (1995) were parameterized. As an example, *Bromus diandrus* behaviour vs [iodosulfuron methyl sodium 0.6% + mesosulfuron methyl 3%] can be seen in Fig. 13.2. As expected, brome grass is less susceptible to herbicides when sprayed at bigger growth stages, showing about 10% less herbicide efficacy at standard dose for both herbicides.

In cases, where efficacy data on different weed growth stages were sparse, default estimates from similar products or for other countries were used, ultimately worst-case estimates (i.e. the highest values of $R_i$ as observed in the countries yet involved in customization of the DSS for practical use).

When considering the numbers of countries, crops, weeds, herbicides and "agronomical conditions", yet included in the DSS, underlying datasets on efficacy will always be variable. Consequently, a general ambition of a DSS is to safely interpret different available datasets and supplementary information. If additional datasets

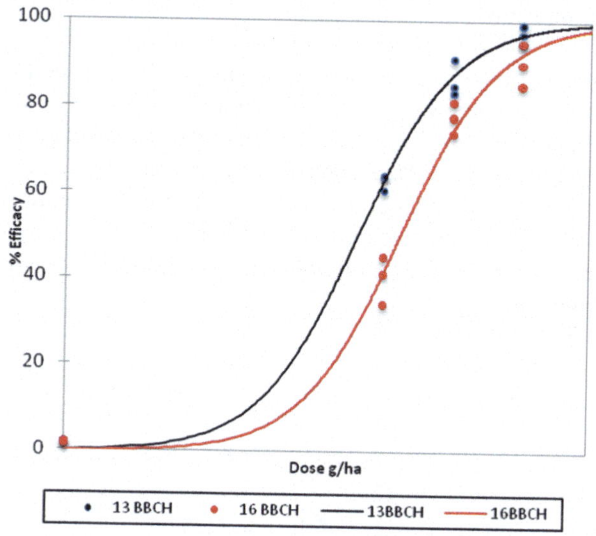


**Fig. 13.2** Estimated percentage of efficacy of iodosulfuron methyl sodium 0.6% + mesosulfuron methyl sodium 3% at 13BBCH in black and 16BBCH in red of brome grass

are provided later, updates will be made accordingly. In general, regular updating is required to follow the continuously changes in supply of herbicide products, and also changes in legal restrictions for use of previously registered products.

### *13.5.2 New Equations in IPMwise*

In order to tune up the IPMwise database to include more extreme weed infestations (i.e. weed density and growth stages) and to reduce time for calculation and thereby prepare for PA technology in real-time site-specific spraying, some database look-up functions have been replaced.

The new algorithms include calculations of the following:

- $Y$, which is target values of the relative efficacy (%).
- $r_s$, which is the relative potencies (dose factors) for classes of weed size.
- $r_t$, which is the relative potencies (dose factors) for classes of temperature on the day of herbicide application.

Estimates of $r_w$, which is the relative potencies (dose factors) for classes of water (drought) stress, are stored and looked up in the DSS database and depend on each active ingredient.

#### 13.5.2.1 *Y* Estimation

For combinations of crop–weed species, a threshold value is used to determine the class of weed density, where values of $Y > 0$, shall be initiated. For higher classes of weed density, values of *Y* will increase towards an upper asymptote, approaching total control (Fig. 13.3). As a 100% control can never be achieved or would require extremely high input of herbicide dose rates, a general maximum value of *Y* has been set to 97–98% for Spanish agricultural systems.

To calculate estimates of *Y*, a combination of a traditional thresholds and the following equation is used:

if $Density > DensityMin$

$$Y = EfficacyMin + \log\left(Density - DensityMin\right) * a \tag{13.3}$$

$$a = \left(EfficacyMax - EffiacyMin\right) / \log\left(DensityMax - DensityMin\right) \tag{13.4}$$

where local expert support estimates of the parameters:

*DensityMin* = minimum density that require control (threshold value, fixed for combinations of crop and weed species),
*DensityMax* = maximum density (fix on national level),
*EfficacyMin* = efficacy target at minimum density (fix on crop × weed level),
*EfficacyMax* = max efficacy target at maximum density (fixed national level).

The actual start and end points are determined by local experts and should include conditions which local farmers consider to be important (e.g. quantity and quality of yield, weed propagation, harvesting problems, cosmetic aspects [farmer's pride]).

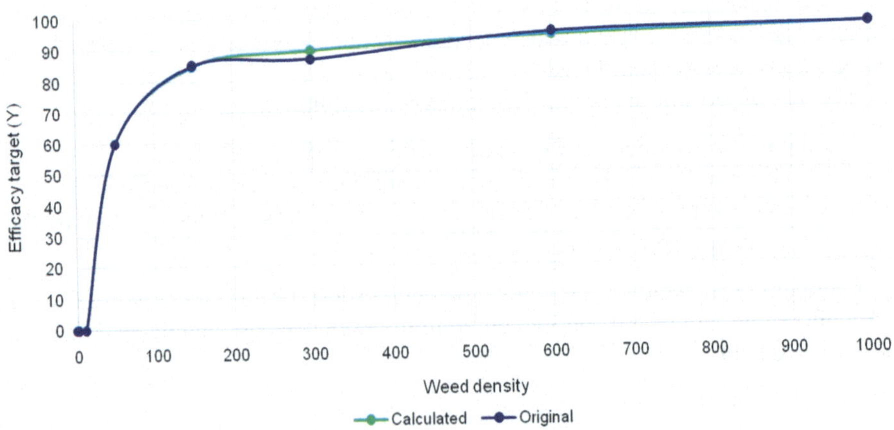


**Fig. 13.3** Estimated vs. observed herbicide efficacy as a function of weed density

In order to identify estimates of $Y$, which balances the above-mentioned aspects, actual values (i.e. start and end points) should idealistically be determined from field validation experiments, where values of $Y$ have been varied, and where yield and residual weed infestation, and additional conditions which farmers find important have been measured/evaluated. In Spain, initial values of $Y$ were used, which local experts considered to be safe and subsequently systematic reductions were used (e.g. 5%-point in maize and 20%-points in wheat). Field validation trials were used to identify versions suitable for official release (see Sect. 13.6).

#### 13.5.2.2 $r_s$ Estimation

In general, annual weeds will often be most susceptible at seedling stage, while perennials require downward translocation to reach an effective control (e.g. 3–6 leaves may be required at the time of spraying to ensure enough herbicide absorption and translocation).

To calculate estimates of the parameter $r_s$, which provided dose adjustment factors for the influence of classes of weed growth stages, Eq. (13.5) is used:

$$r_s = MinWeedSize\_par + e^C * \left(e^B * X\right) \tag{13.5}$$

where $X$ are original values/classes of weed growth stage; *MinWeedSize_par*, $e^C$ and $e^B$, are parameters stored in the model dataset running IPMwise.

Estimates have been calculated by Excel Solver for minimizing sum of squares of deviations between original and calculated values. The neutral level (=1.0) has been determined from the actual weed growth stage in efficacy data behind $b_h$ and $a_w$ (e.g. below the "3–4 leaves" weed growth stage).

The values of $r_s$ refer to weed growth stages, which were found in field experiments, where estimates of $a_w$ were produced. As an example, for early post-emergence herbicides, the standard is the 3–4 leaf stage. In cases where weeds were smaller or bigger (e.g. 0–2 leaves or 5–6 leaves), this point should be used as reference point instead (where $r_s$ has a neutral value of 1.0).

#### 13.5.2.3 $r_t$ Estimation

The efficacy of many herbicides depends on the general conditions for plant growth. Temperatures may affect efficacy, mainly at the time of application. According to results obtained from studies in climatic simulators in Denmark, temperature is relatively less important to efficacy compared to weed species and classes of weed size (estimates of $a_w$ and $r_s$). According to IPMwise, estimates of $r_t$ are typically 0.7–1.3 for different herbicides for extreme conditions. However, as temperatures (for a region) is often relatively normal, actual estimates will often be relatively smaller. Algorithms/equations used in Denmark are as follows:

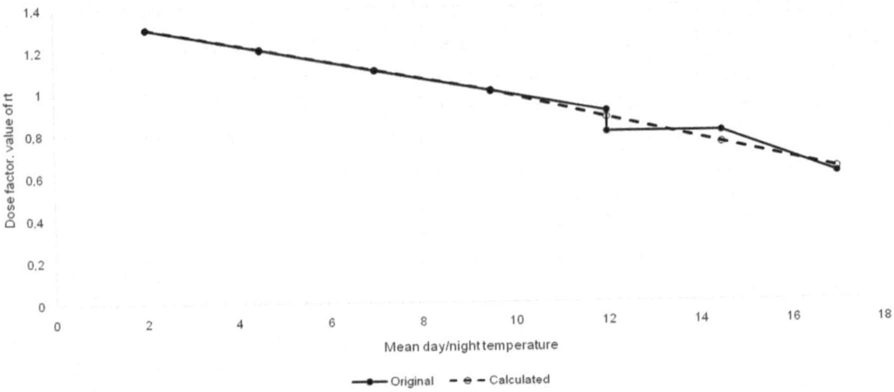


**Fig. 13.4** Original vs. estimated dose factor values ($r_t$) as a function of mean daily temperature

$$\text{for } x \leq 9.5: \quad r_t = TLowIntercept + TLowSlope * x \tag{13.6}$$

$$\text{for } x > 9.5: \quad r_t = THighIntercept + THighSlope * x \tag{13.7}$$

where $x$ is the mean day/night temperature (°C); *TLowIntercept*, *TLowSlope*, *THighIntercept* and *THighSlope* are calculated parameters.

The neutral level (=1.0) has been determined by actual temperatures, when the efficacy data behind "b-par/a-par" was created, in the example below 9.5 °C.

In Fig. 13.4, examples of original and calculated estimates of $r_t$ are provided. These have been calculated from efficacy data produced in small pots placed in climatic simulators. These differ for different herbicide products.

### 13.5.3 $r_w$ Estimation

In Spain and other countries of southern Europe, water is the most important environmental factor limiting plant growth. This affects also the growth of weeds, which may be simultaneously affected by both drought stress and herbicide application.

Based on the results obtained from efficacy trials in climatic simulators, it has been generally concluded that weed species which have no visual symptoms of withering will have normal susceptibility to herbicides. When visual symptoms of wilting can be detected, also a decreased susceptibility to herbicides may be expected (Rydahl, personal communication).

Striving generally for simplicity in DSS design, only three levels of water stress have been found suitable (i.e. "No water stress", "Slight water stress" and "Severe water stress"). Original estimates of $r_w$ are stored in the IPMwise database. In case of severe stress, estimates of $r_w$ may be in the range of 3–6, which are likely to have the implication that maximum doses may be exceeded, why IPMwise may be unable to provide options for treatment. Consequently, "Severe water stress" operates indirectly as a message to the user indicating that due to drought conditions, herbicides will not work.

### 13.5.4 *Common Features for Estimates of* $\mathbf{r}_s$, $\mathbf{r}_t$ *and* $\mathbf{r}_w$

Small pot trails behind estimation of $r_s$, $r_t$ and $r_w$, did not include all possible herbicide options, but instead selected different MOAs which according to literature were considered to differ in terms of affecting estimates of these three parameters. For example, estimates from studies on MCPA, were copied to other members of the HRAC group O (auxinic effect), or estimates from studies on tribenuron-methyl were copied to other members of the HRAC group B (ALS-enzyme inhibitors).

A total exclusion of these three parameters may be relevant in the so-called minor use crops, where efficacy data is really sparse. Eventually, only information from herbicide product labels may be available. In crops grown in bigger areas, more data may be available, and more refinements may be used, idealistically also with an increased potential for herbicide savings.

## 13.6 Field Validation

The last step was to validate the adjusted version of IPMwise under Spanish conditions.

Values of target efficacy were established by local experts' evaluation (Table 13.1), with the aim of being relatively safe, when considering all the aspects included in Sects. 13.3 and 13.5.2.1. In order to fine-tune these values, and to evaluate options for dose reduction (i.e. Herbicide input reduction), different prototypes were constructed and field tested. To that effect, the original targets values were systematically reduced. In wheat by up to 20%-points, and in maize by around 5%-points (i.e. only the estimates in Eqs. (13.3) and (13.4) of *DensityMin* and

**Table 13.1** Target efficacies (%) for different weed species and densities in IPMwise Spain

| Species | Efficacy required (%) for each density (plants/m$^2$) | | | | |
|---|---|---|---|---|---|
| | ½–1 | 2–10 | 11–40 | 41–150 | >150 |
| *Alopecurus myosuroides* | 0 | 85 | 85 | 90 | 95 |
| *Avena sterilis* | 0 | 75 | 85 | 90 | 95 |
| *Galium aparine* | 85 | 90 | 90 | 95 | 95 |
| *Lolium rigidum* | 0 | 85 | 85 | 90 | 95 |
| *Papaver rhoeas* | 0 | 80 | 85 | 90 | 95 |
| Abutilon theophrasti | 80 | 90 | 95 | 97 | 98 |
| Echinochloa crus-galli | 75 | 85 | 95 | 98 | 98 |

*EfficacyMin* were reduced, while the estimates of *DensityMax* and *EfficacyMax* were not reduced).

Based on actual weed infestations and the obtained values, CPOWeeds listed all possible solutions for a given weed composition in specific fields sorted by TFI.

Different trial setups were conducted with different winter cereals and maize crops from 2010 till 2018. Some field trials were performed to check for efficacy and others to check yields.

Different spraying times were considered depending on the standard moment for each herbicide, and in all cases, a standard treatment was chosen by local advisors for all fields to have a reference for the IPMwise solutions. The accuracy of the DSS predictions was estimated based upon weed counting 35 days after spraying in the efficacy trials.

For winter cereals, nine different species were used in the analyses and there were some differences in the accuracy among the species (Fig. 13.5). As shown in Fig. 13.5, observed values were equal to or higher than predicted for 84.2% of the sampling points. In addition, 88% of results are included in the range 5–10% error, with an average value of 2.35% difference between the observed and predicted values.

The average difference between predicted and observed efficacies for *A. sterilis*, *L. rigidum* and *P. rhoeas*, which are key species in this region, showed a difference just above 2%. This was similar to that obtained with *Anthemis arvensis* L., which is less commonly found in this region. For *Malcolmia africana* (L.) R. Br. and *A. myosuroides*, differences between observed and predicted values were as low as 0.7% and 0.15%. The largest differences between predicted and observed efficacies were found for *L. rigidum* and *P. rhoeas* in 2011, although these differences were

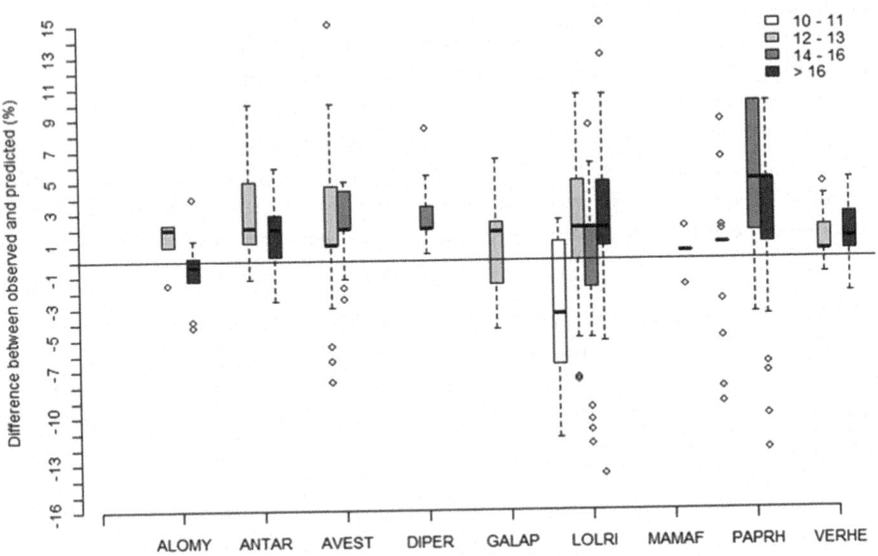


**Fig. 13.5** Difference between field observed values and predicted efficacies by the IPMwise for different weed species and growth stages (legend indicates growth stages)

not consistently positive or negative. Generally, the negative differences for *L. rigidum* were found for plants sprayed at the earliest stage (BBCH 10–13), whereas there was no tendency for *P. rhoeas* for dependence on growth stage.

Weed species growth stage at the time of application did not influence the robustness of the model recommendations, as the difference between predicted and observed efficacies were not significant ($P = 0.41$, $n = 4$). The model was designed to account for the developmental stage at the time of application and IPMwise was observed to adequately adjust the doses.

The observed efficacies were less consistent with the predicted efficacies in 2011 than in the other years. During 2011, the rainfall was only 33.7 mm compared to an average rainfall ≈110 mm in the period when the field tests were carried out (December 2010 till February 2011). In semi-arid climatic conditions, such as those typically found in the northern Spain, it is important to focus future studies on modelling the effect of water stress on herbicide effectiveness.

Water stress causes a shift of the dose–response curve to the right (i.e. greater dose required to achieve the same efficacy). This effect may vary according to the mode of absorption of the herbicide. In general, root absorbed herbicides, which need to be dissolved in the water phase are more affected by drought than foliar herbicides (Kudsk 2008a, b). This effect should be studied for each herbicide compound, as pre-emergent herbicides like pendimethalin which act also through contact could be more independent of soil moisture. In addition, performance could also vary depending on the formulation used (e.g. microencapsulated or slow release formulations). This kind of formulations may affect soil bioavailability as is the case with other herbicides such as mesotrione (Galán-Jiménez et al. 2015) or flufenacet (Gómez-Pantoja et al. 2015).

Yield trials support the results obtained in the efficacy trials, which show that IPMwise provide adequate weed control recommendations being agronomically robust. The yield of IPMwise treatment was equal to or even higher than the standard treatments as can be seen in Table 13.2.

**Table 13.2** Yield trials in winter cereals. Yields of IPMwise treatments are given as an interval as 4–6 different solutions were tested in each field

| Location | Treatment | Yield (kg ha$^{-1}$) |
|---|---|---|
| Termens | Standard | 10,450[a] |
| | IPMwise | 6403[b]–11,006[a] |
| Vimbodí 1 | Standard | 4082[a] |
| | IPMwise | 4168[a]–4793[a] |
| Vimbodí 2 | Standard | 4763[a] |
| | IPMwise | 6286[b]–7103[b] |

Lower-case letters indicate differences between standard treatment and IPMwise treatments

## 13.7 Future Challenges

Despite the fact that IPMwise is able to handle with herbicide-resistant biotypes, there were a few fields where weeds were controlled insufficiently, which was probably due to the lack of identification of resistant weed biotypes prior to spraying. Actually, a resistance prevention initiative has been developed in IPMwise, which aims at limiting the development of more resistant weed species by systematically altering herbicides' MoA among weed generations. These precautions are used according to reports from the Herbicide Resistance Action Committee which monitor and report developments in combinations of MOAs and weeds where incidents have been found (HRAC 2019). However, in case the number of combinations of MOAs and weed species where resistance cases increase drastically, the IPMwise will require to formulate more drastic measures, eventually without use of herbicides.

In the future, feedback from users will be important to adjust the target efficacies to levels that will provide sufficient control in all situations. The present target efficacies were estimated by experts, but experiences from Denmark has shown that adjustments are necessary through the initial implementation period as it is difficult to account for all influencing factors. The final conclusion is that the use of this tool allows for an optimization of herbicide applications, adjusting the applied dose rate with a very high robustness for the conditions of northern Spain and it has a potential to reduce the amount of applied herbicides ≥30%.

## References

Been TH, Consiglio A, Evans N, Gouache D, Gutsche V, Jensen JE, Kapsa J, Levay N, Munier-Jolain N, Nibouche S, Raynal M, Rydahl P (2010) Review of new technologies critical to effective implementation of Decision Support Systems (DSS's) and Farm Management Systems (FMS's), Critical policy studies. Aarhus University, Denmark

Busi R, Gaines TA, Walsh MJ, Powles SB (2012) Understanding the potential for resistance evolution to the new herbicide pyroxasulfone: field selection at high doses versus recurrent selection at low doses. Weed Res 52:489–499. https://doi.org/10.1111/j.1365-3180.2012.00948.x

EWRS Herbicide Dose Optimisation WG (2013) Joint experiment background and results. https://www.google.com/url?sa=t&rct=j&q=&esrc=s&source=web&cd=2&ved=2ahUKEwjhgfHFkcjkAhW9DGMBHQoACGcQFjABegQIABAB&url=http%3A%2F%2Fwww.ewrs.org%2Fdoc%2FHTV%2FWG_Herbicide_Dose_Opt.pdf&usg=AOvVaw0Edl3O-XSkob86c-s1hnZgZ. Accessed 11 Sept 2019

FAOSTAT (2016) Pesticides used in selected countries. Rome. http://faostat3.fao.org. Accessed

Galán-Jiménez M, Undabeytia T, Morillo E, Florido M (2015) Síntesis y eficacia de formulaciones de liberación lenta del herbicida mesotrione. In: Actas del XV Congreso de la SEMh, pp. 79–84

Gebbers R, Adamchuk VI (2010) Precision agriculture and food security. Science 327(5967):828–831. https://doi.org/10.1126/science.1183899

GENVCE (2015) Evaluación agronómica y de la calidad de las nuevas variedades de cebada, trigo blando, trigo duro, triticale, avena y centeno en España. Campaña 2014–2015. www.genvce.org/repositorio/ba36/informe/1107/2/cereal-de-invierno.pdf?d=1

Gómez-Pantoja M, López-Fernández E, Florido M, Morillo E, Undabeytia T (2015) Evaluación de formulaciones de liberación controlada del herbicida flufenacet. In: Actas del XV Congreso de la SEMh, pp 85–91

Gonzalez-Andujar JL, Fernandez-Quintanilla C, Bastida F, Calvo R, Izquierdo J, Lezaun JA (2010a) Assessment of a decision support system for chemical control of annual ryegrass (*Lolium rigidum*) in winter cereals. Weed Res 51:304–309. https://doi.org/10.1111/j.1365-3180.2011.00842.x

Gonzalez-Andujar JL, Fernandez-Quintanilla C, Bastida F, Calvo R, Gonzalez-Diaz L, Izquierdo J, Urbano J (2010b) Field evaluation of a decision support system for herbicidal control of *Avena sterilis* ssp. *ludoviciana* in winter wheat. Weed Res 50(1):83–88. https://doi.org/10.1111/j.1365-3180.2009.00744.x

Herbicide Resistance Action Commitee (HRAC) (2019). https://www.hracglobal.com/

Jensen PK, Kudsk P (1988) Prediction of herbicide activity. Weed Res 28(6):473–478. https://doi.org/10.1111/j.1365-3180.1988.tb00830.x

Kudsk P (1989) Experiences with reduced herbicide doses in Denmark and the development of the concept of factor-adjusted doses. In: Brighton crop protection conference—weeds, Brighton, pp 545–554

Kudsk P (2008a) Chapter 16: Optimising herbicide performance. In: Weed management handbook. John Wiley & Sons, New York, pp 323–344. https://books.google.com/books?id=jXbGt8ttluIC&pgis=1. Accessed

Kudsk P (2008b) Optimising herbicide dose: a straightforward approach to reduce the risk of side effects of herbicides. Environmentalist 28(1):49–55. https://doi.org/10.1007/s10669-007-9041-8

Kudsk P (2014) Reduced herbicide rates: present and future. In: 26th German conference on weed biology and weed control, pp 37–44. https://doi.org/10.5073/jka.2014.443.003

MAPAMA (2019) Cultivos herbáceos e industriales en España. https://www.mapa.gob.es/va/agricultura/temas/producciones-agricolas/cultivos-herbaceos/cereales/default.aspx

Minkey D, Moore J (1996) Estimating dose response curves for predicting glyphosate use in Australia. s.l., s.n., p 4

Montull JM, Soenderskov M, Rydahl P, Boejer OM, Taberner A, (2014) Four years validation of decision support optimising herbicide dose in cereals under Spanish conditions. Crop Protection 64:110–114

Netland J, Torresen KS, Rydahl P (2005) Evaluation of the weed module in the Danish decision support system "Crop Protection Online" adapted to Norwegian conditions. In Proceedings of the 13th EWRS Symposium, 19–23 June 2005. Bari, Italy: European Weed Research Society, CD-ROM

Renton M, Diggle A, Manalil S, Powles SB (2011) Does cutting herbicide rates threaten the sustainability of weed management in cropping systems. J Theor Biol 283(1):14–27. https://doi.org/10.1016/j.jtbi.2011.05.010

Rydahl P (2003) A web-based decision support system for integrated management of weeds in cereals and sugarbeet. EPPO Bull 33:455–460

Rydahl P, Pedersen L (2003) User interfaces and system architecture of a web-based decision support system for integrated pest management in cereals. EPPO Bull 33:473–481

Rydahl P, Berti A, & Munier-Jolain N (2008) Decision support systems (DSS) for weed control in Europe – state-of-the-art and identification of best parts for unification on a European level. In: Endure international conference, pp. 12–15

Rydahl P, Munier-jolain N, Masin R, Sattin M, Germiniani E, Lescovcek R, Bojer OM (2015) A generic Decision Support System for integrated weed management. Aarhus University, Denmark

Seefeldt SS, Jensen JE, Fuerst P (1995) Log-logistic analysis of herbicide dose-response relationships. Weed Technol 9(2):218–227

Sønderskov M, Kudsk P, Mathiassen SK, Bøjer OM, Rydahl P (2014) Decision support system for optimized herbicide dose in spring barley. Weed Technol 28:19–27. https://doi.org/10.1614/WT-D-13-00085.1

Sønderskov M, Fritzsche R, de Mol F, Gerowitt B, Goltermann S, Kierzek R, Rydahl P (2015) DSSHerbicide: Weed control in winter wheat with a decision support system in three South Baltic regions – field experimental results. Crop Prot 76:15–23

Streibig JC (2003) Assessment of herbicide effects. In: EWRS (ed) Herbicide interaction, pp 1–44

Streibig JC, Kudsk P, Jensen JE (1998) A general joint action model for herbicide mixtures. Pestic Sci 53:21–28

Torra J, Cirujeda A, Recasens J, Taberner A, Powles SB (2010) PIM (Poppy Integrated Management): a bio-economic decision support model for the management of *Papaver rhoeas* in rain-fed cropping systems. Weed Res 50(2):127–139. https://doi.org/10.1111/j.1365-3180.2010.00761.x

Yu Q, Han H, Cawthray GR, Wang SF, Powles SB (2012) Enhanced rates of herbicide metabolism in low herbicide-dose selected resistant *Lolium rigidum*. Plant Cell Environ 36(4):818–827

# Chapter 14
# AvenaNET and VallicoNET: DSS for *Avena sterilis* and *Lolium rigidum* Control in Spanish Dryland Cereal Crops

José L. González-Andújar

**Abstract** AvenaNET and VallicoNET are web-based DSS developed for *Lolium rigidum* (ryegrass) and *Avena sterilis* spp. *ludoviciana* (winter wild oat) control in Spanish dryland cereals. This chapter describes the rationale, structure and evaluation of these DSS. Both systems present a common structure that contains an interface, a database and a bioeconomic model. The interface has been kept as simple as possible, and it requires simple agronomic, biological and economic data. The databases store information on the available herbicides for the control of *A. sterilis* and *L. rigidum*. The bioeconomic model contains a detailed life cycle structure including integrated management strategies, weed-crop competition and economic submodels. Both DSS followed an evaluation process consisting of the verification of the functions contained in the system, its ergonomics and the evaluation in field conditions. The validation results revealed that the performance of both systems was satisfactory.

**Keywords** Winter wild oat · Ryegrass · Decision Support System (DSS) · Web-based DSS · Weed management · Herbicide · Bioeconomic model · Decision-making · Validation

## 14.1 The Problem

Cereals are staple crops with a wide geographical distribution, around 16% of the world's useful agricultural area (Zimdahl, 2004). The production of cereals is affected by a series of biotic and abiotic agents that interact with the crop and decrease its yield (Fig. 14.1 Renton and Chauhan, 2017). More than 42% of cereal crop yields is decreasing due to these agents, one of the most important being the

J. L. González-Andújar (✉)
Department of Crop Protection, Institute for Sustainable Agriculture, Córdoba, Córdoba, Spain
e-mail: andujar@cica.es


G. R. Chantre, J. L. González-Andújar (eds.), *Decision Support Systems for Weed Management*, https://doi.org/10.1007/978-3-030-44402-0_14

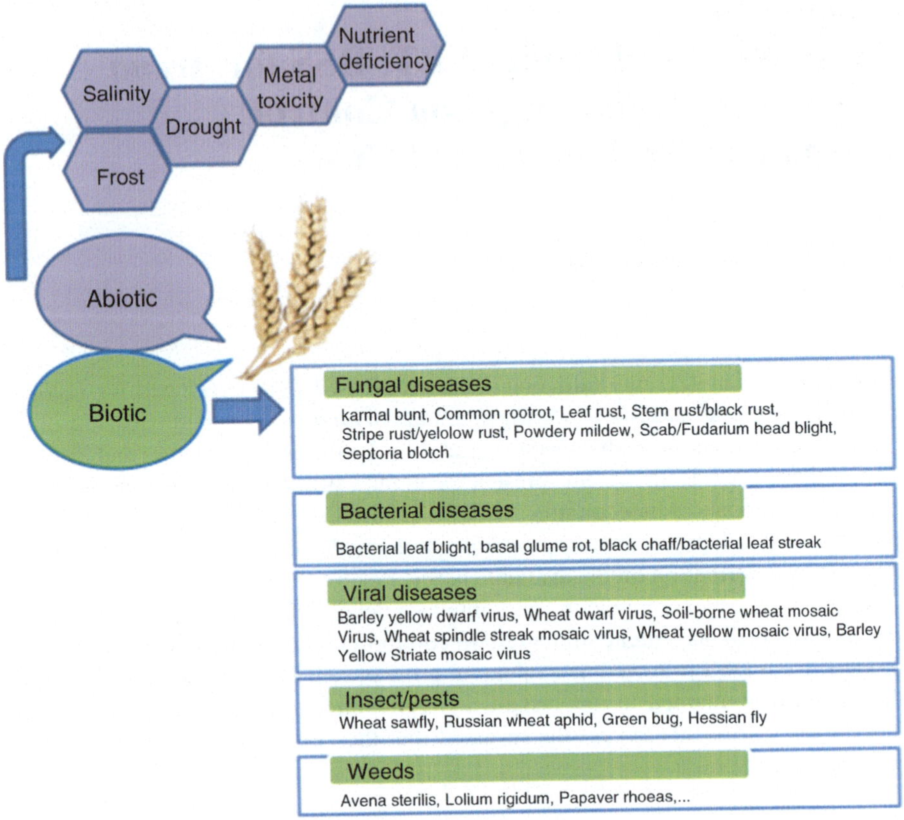


**Fig. 14.1** Biotic and abiotic factors responsible for the decline in wheat yield

presence of weeds. In Spain, the dominant weed species are winter wild oat (*Avena sterilis*) and annual ryegrass (*Lolium rigidum*) (Gonzalez-Andujar and Saavedra 2003).

Winter wild oats have a life cycle that is highly adapted to cereals (Fernandez-Quintanilla et al. 1997). The emergence period covers from the end of October to the middle of April. Winter wild oat is a strong competitor causing substantial losses. High *A. sterilis* densities, in the 300 panicles $m^{-2}$ range, have decreased cereal yields by up to 50% (Torner et al. 1991).

Annual ryegrass is a winter annual grass considered to be among the most troublesome weeds of cereal crops in the Mediterranean area (Reeves 1976; Gonzalez-Andujar and Saavedra 2003). The emergence period is mainly focused from late autumn to midwinter (Recasens et al. 1997). Yield losses in cereal crops can reach up to 80% depending on season and infestation (Izquierdo et al. 2003). The main method for control of both species is herbicide application. Even though herbicides may be effective and safe when properly applied, there is a growing concern for the environmental effects of using these chemicals, and the efficacy of this tactic is currently threatened by the evolution of herbicide-resistant populations (Heap 2020).

In order to assist Spanish farmers in decision-making for the control of *A. sterilis* and *L. rigidum* in wheat, two DSS were developed at the beginning of the twenty-first century—Avena-PC and Lolium-PC—designed for use in computers and which subsequently resulted in web-oriented DSS: AvenaNET and VallicoNET. Both DSS were developed for testing both the biological and economic performance of *L. rigidum* and *A. sterilis* integrated management strategies for dryland cereal systems in Spain.

## 14.2 VallicoNET and AvenaNET Description

The structure of VallicoNET and AvenaNET consists of an interface, a database and a model (see Chap. 2). In Fig. 14.2 the access page is shown, from which the user accesses the database and simulations of weed control strategies.

### *14.2.1 Interface*

The communication between the DSS and the user is of vital importance for the acceptance of the tools to help decision-making. One of the characteristics that it must have is the ease and simplicity of information entrance. Both systems have a similar data entry screen (Fig. 14.3). From the initial window, the user accesses through simulation of control strategies to the data entry screen that is divided into two large sections (Fig. 14.3). The first comprises the agronomic and biological parameters, while the second is designed for the economic parameters. There are three agronomic and biological parameters that the user needs to provide to the system. First, users need to carry out a sampling in order to evaluate the infestation density of *A. sterilis* and *L. rigidum* seedlings. Then they must include an estimate of the expected yield potential of the crop, based on expert knowledge, and finally

**Fig. 14.2** Main page for *Avena sterilis* (AvenaNET) and *Lolium rigidum* (VallicoNET)

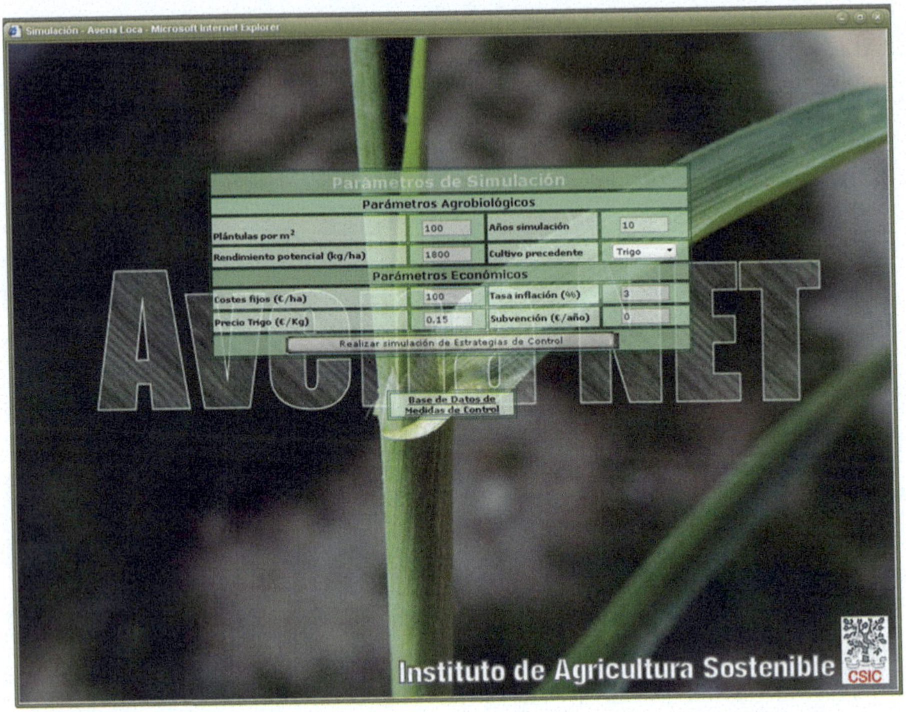


**Fig. 14.3** Data entry screen for AvenaNET

they have to indicate the previous crop. In addition, the user must enter economic information, such as fixed costs, expected crop price in the campaign, inflation rate and the expected subsidy. Finally, in the last section called 'simulation years', they must indicate the time horizon (in years) to simulate. The output windows (Fig. 14.4) provide the results which consist of all possible management strategies, ordered by priority, according to the economic output (€/ha). These results can be delivered in an Excel file.

## *14.2.2 Database*

The database is an important component of the DSS where all the information for giving recommendations is located. VallicoNET and AvenaNET integrate information on the available herbicides: active substances; commercial name; recommended application rate and their effectiveness, allowing their modification and deletion; and the introduction of new herbicides (Fig. 14.5).

**Fig. 14.4** Output of VallicoNET in short-term (left panel) and a long-term (10 years) (right panel) horizons

## 14.2.3 The Model

Both DSS are based on bioeconomic models composed of three submodels (see Chap. 2): population dynamics, weed-crop competition and economics.

### 14.2.3.1 Population Dynamic Submodels

The life cycle-based model structure is similar to other models proposed in the literature (Doyle 1991; Gonzalez-Andujar and Fernandez-Quintanilla 1993, 2004; Holst et al. 2007; Gonzalez-Andujar 2008) (Fig. 14.6).

The dynamics of the seed bank ($S$, Seeds m$^{-2}$) at time $t$ is described by:

$$S_{t+1} = (1-g)(1-m)S_t + sf(1-p)P_t \tag{14.1}$$

where each year a fraction $m$ of seeds experiences natural mortality, while a fraction $g$ emerges. Density of plants that emerge and survive until the adult stage is indicated as $P_t$. A fraction $s$ survives until reproduction. Each surviving plant will produce on average $f$ viable seeds representing the seed rain that returns to the seed

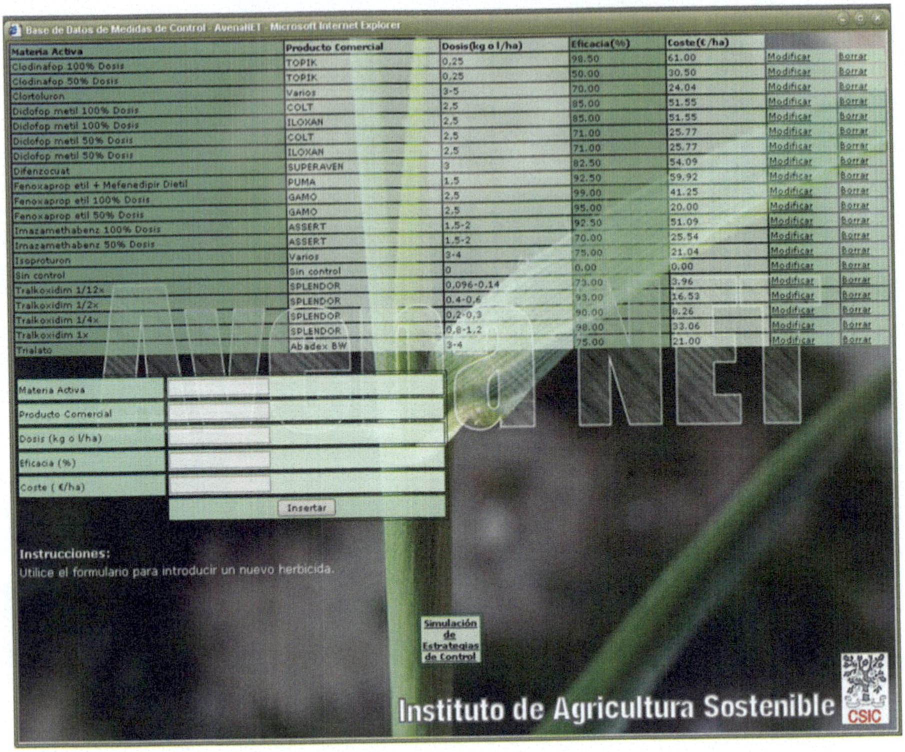


**Fig. 14.5** Herbicide database of AvenaNET

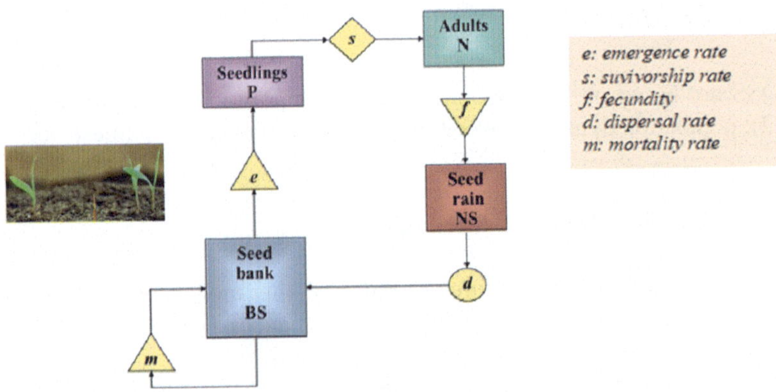


**Fig. 14.6** Basic annual weed demographic model

bank. A fraction, $p$, of the total seed rain is assumed to be losses due to biotic and abiotic factors.

The effect of weed plant density on fecundity (density-dependent factor) $f$ is introduced by:

$$f = f_0 / (1 + aP_t) \tag{14.2}$$

A fraction $c$ of emerged plants is killed by weed control. The density of seedlings that survive weed control is:

$$P_t = (1 - c) g S_t \tag{14.3}$$

The control measures introduced are of a chemical and cultural type (e.g. delay in sowing). AvenaNET and VallicoNET contemplate a total of 41 and 17 control strategies, respectively.

#### 14.2.3.2 Weed-Crop Competition Submodel

Many empirical models of weed-crop competition that relate the loss of yield as a response to the density or biomass of weeds ($X$) have been developed (Cousens 1985; Jamaica-Tenjo and Gonzalez-Andujar 2019). In this case, the popular hyperbolic model was chosen (Cousens 1985):

$$Y_{L=} \frac{Ix}{1 + \frac{I}{A}x} \tag{14.4}$$

In this model, $Y_L$ represents percent yield loss as a result of competition; $I$ is the percent yield loss per unit weed when $x \to 0$; and $A$ is the asymptotic yield loss when $x \to \infty$.

The above equation can be expressed in terms of cereal yield, $Y$ (kg ha$^{-1}$), contemplating the reduction with respect to the maximum weed-free yield ($Y_{wf}$, kg ha$^{-1}$) in the following way:

$$Y = Y_{wf} \left[ 1 - \frac{Ix}{100\left(1 + \frac{I}{A}x\right)} \right] \tag{14.5}$$

#### 14.2.3.3 Economic Submodel

To assess the DSS feasibility, a net economic return (NER, € ha$^{-1}$) was calculated as:

$$\text{NER} = YP - H - C \tag{14.6}$$

where $Y$ is yield (kg ha$^{-1}$), $P$ is cereal price (€ kg$^{-1}$), $H$ is herbicide cost (€ ha$^{-1}$) and $C$ (€ ha$^{-1}$) are variable costs associated with cereal production (tillage, fertilizing, seeding, etc.) except weed control practices.

The time path of expected returns over an extended period is given by the annualized net return, ANR (€ ha year$^{-1}$). For an n-year period, annualized net return was defined as:

$$\text{ANR} = (\Sigma\text{NER}_t\left(1+i\right)^{-t}\}\{i/[1-\left(1+i\right)^{-i}\} \quad (14.7)$$

where $i$ ($i > 0$) is the annual discount factor and $t$ is year.

## 14.3 Running the Model

Below is an example of the use of ValllicoNET (similar to AvenaNET). The user enters the data on the initial data entry screen (Fig. 14.3). In this case, the agronomic and biological parameters are (1) the weed density of infestation (established in 100 seedlings m$^{-2}$), (2) the expected potential yield (set in 1800 kg ha$^{-1}$) and (3) the previous crop (wheat monoculture), and the economic parameters are (4) the fixed costs (€ 100 ha$^{-1}$), (5) the estimated price of wheat (€ 0.15 kg$^{-1}$), (6) the expected inflation rate (3%) and (7) crop subsidy (none). The simulation periods considered are short term (1 year) and long term (10 years), with the objective of comparing both periods of time.

Figure 14.4 shows the output of the example used. The strategies considered are presented as being ordered according to their economic result. In our case, the best long-term strategy (see Fig. 14.4; right panel) is a mixed strategy of herbicide (diclofop-methyl) and delayed sowing. In the short term (Fig. 14.4; left panel), there are three strategies that offer similar economic results, two that include the use of herbicide (diclofop-methyl) along with cultural strategies (delayed sowing and crop planting density) and another that includes the herbicide at half of the recommended dose. Only the herbicide strategy (diclofop-methyl) together with delayed sowing remains solid over the time. It some cases, it is possible that the best economic strategies in the short and long term do not match.

## 14.4 Evaluation and Current Status

In order to have operational validity, DSS must undergo an evaluation process (Houseman 1994) consisting of the verification of the functions contained in the system, the ergonomics of the system that is related to its 'friendliness' for the end user and the evaluation in field conditions.

VallicoNET and AvenaNET were evaluated in their PC versions (LOLIUM-PC and AVENA-PC) in commercial fields of winter wheat from agronomic, economic and environmental points of view (Gonzalez-Andujar et al. 2010, 2011). Different

**Table 14.1** Results of evaluation by technicians of some aspects of the relevance of VallicoNET and AvenaNET for their practical use

| Questions | Rate |
|---|---|
| Do you consider it necessary to have computer tools to help decision-making? | 7.08 |
| Indicate in what degree would it be useful in your professional performance? | 6.66 |
| Do you consider that the suggestions or recommendations provided by the DSS are useful for the advisory work? | 6.15 |
| Do you consider that the DSS would allow you to make faster decisions? | 6.15 |
| Do you consider that the use of DSS would allow you to think about a greater number of variables when making decisions? | 6.25 |
| Would you trust the decisions of the DSS? | 6.53 |

The participants assessed each of the criteria in a continuum 1–10 scale, corresponding to the following responses: 1 as unsatisfactory and 10 as extremely satisfactory

trial setups were conducted from 2003 to 2005 in wheat fields in northern, central and eastern Spain. In general, the evaluations showed that both systems provide a flexible tool for potentially recommending less herbicide while providing similar weed control and crop yields to those obtained with the standard farmer practice. LOLIUM-PC recommendations provided, on average, substantial reductions in herbicide use (57% lower) in relation to the standard farmer practice (full dose) and slight reductions (14% lower) in relation to a half-dose treatment (Gonzalez-Andujar et al. 2011). AVENA-PC provided a substantial herbicide saving in relation to the standard farmer practices (65%) and 31% lower in relation to the half-dose treatment (Gonzalez-Andujar et al. 2010).

In addition, surveys were conducted with 14 experienced technicians to evaluate different ergonomic aspects of the systems. They were asked to mark in a table-like questionnaire the following criteria: (1) usefulness of the information provided by the DSS, (2) user-friendliness and (3) learning easiness. Average results for both DSS are shown in Table 14.1. The results were positive (but not enthusiastic) and highlighted the need for tools such as DSS to help decision-making (Table 14.1).

Both systems remain academic products, due to the fact that neither have never been released to potential users. Among the main drawbacks, lack of resources, updating and maintenance can be mentioned. For the current use, an upgrade should be undertaken considering that in the past decade, there has been an important change in the herbicide availability for cereals in Europe and the consideration of new non-chemical management strategies. Likewise, a new evaluation process should be undertaken, particularly with regard to new management strategies for *L. rigidum* and *A. sterilis* herbicide resistant.

## 14.5 Conclusions

The complexity of the information needed to manage weed populations within a context of sustainable agriculture requires the use of computer tools such as DSS that allow farmers and technicians to optimize decision-making. ValllicoNET and AvenaNET are two DSS that have been developed for the control of *A. sterilis* and *L. rigidum* in Spanish dryland cereals. They have both shown their potential usefulness as agronomic tools for helping decision-makers. Despite this, current lack of financial support and a continuous need for systems' update are among the main necessities that limit successful transfer of DSS-oriented models to farmers and other stakeholders.

## References

Cousens R (1985) A simple model relating yield loss to weed density. Ann Appl Biol 107:239–252
Doyle CJ (1991) Mathematical models in weed management. Crop Prot 10:432–444
Fernandez-Quintanilla C, Navarrete L, Torner C, Sanchez del Arco MJ (1997) *Avena sterilis* en cultivos de cereales. In: Sans X, Fernandez-Quintanilla C (eds) Biologı'a de las Malas Hierbas de España. Phytoma España, Valencia, Spain, pp 4–17
Gonzalez-Andujar JL (2008) Weed control models. In: Jørgensen SE, Fath BD (Editor-in-Chief) Population dynamics: Vol. 5. Encyclopedia of ecology, 5 vols. Elsevier, Oxford, pp 3776–3780
Gonzalez-Andujar JL, Fernandez-Quintanilla C (1993) Strategies for the control of *Avena sterilis* in winter wheat production systems in central Spain. Crop Prot 12:617–623
Gonzalez-Andujar JL, Fernandez-Quintanilla C (2004) Modeling the population dynamics of annual ryegrass (*Lolium rigidum*) under various weed management systems. Crop Prot 23:723–729
Gonzalez-Andujar JL, Fernandez-Quintanilla C, Bastida F, Calvo R, Gonzalez-Díaz L, Izquierdo J, Lezaún JA, Perea F, Sánchez Del Arco MJ, Urbano JM (2010) Field Evaluation of a Decision Support System for herbicidal control of *Avena sterilis* ssp. *ludoviciana* in winter wheat. Weed Res 50:83–88
Gonzalez-Andujar JL, Fernandez-Quintanilla C, Bastida F, Calvo R, Izquierdo J, Lezaun J (2011) Assessment of a decision support system for chemical control of annual ryegrass (*Lolium rigidum*) in winter cereals. Weed Res 51(3):304–309
Gonzalez-Andujar JL, Saavedra M (2003) Spatial distribution of annual grass weed populations in winter cereals. Crop Prot 22:629–633
Heap IM (2020) International survey of herbicide resistant weeds. Available at: http://www.weedscience.org [Accessed Apr 1, 2020]
Holst N, Rasmussen IA, Bastiaans L (2007) Field weed population dynamics: a review of model approaches and applications. Weed Res 47:1–14
Houseman CI (1994) Sotfware quality—what are the developers' responsibilities? Investigación Agraria: Producción y Protección Vegetal. Fuera de serie n° 1: 1–7
Izquierdo J, Recasens J, Fernandez-Quintanilla C, Gill G (2003) The effects of crop and weed densities on the interactions between barley and *Lolium rigidum* in several Mediterranean locations. Agronomie 23:529–536
Jamaica-Tenjo DA, Gonzalez-Andujar JL (2019) Modelos empíricos de competencia cultivo-maleza. Revisión Bibliográfica. ITEA-Información Técnica Económica Agraria 115(4):289–306. https://doi.org/10.12706/itea.2019.007

Recasens J, Taberner A, Izquierdo J (1997) *Lolium rigidum* Gaud. en cultivos de cereales. In: Sans X, Fernandez-Quintanilla C (eds) Biologia de las Malas Hierbas de España. Valencia, Phytoma España, pp 49–64

Reeves TG (1976) Effect of annual rye grass (*Lolium rigidum* Gand.) on yield of wheat. Weed Res., 16:67–76

Renton M, Chauhan BS (2017) Modelling crop-weed competition: why, what, how and what lies ahead? Crop Prot 95:101–108

Torner C, Gonzalez-Andujar JL, Fernandez-Quintanilla C (1991) Wild oat (*Avena sterilis*) competition with winter barley: plant density effects. Weed Res 31:301–307

Zimdahl RL (2004) Weed-crop competition: a review. Wiley-Blackwell, Chichester, 211pp

# Chapter 15
# A Simulation Model as the Core for Integrated Weed Management Decision Support Systems: The Case of *Avena fatua*-Winter Wheat in the Semiarid Pampean Region of Argentina

**Franco A. Molinari, Aníbal M. Blanco, Mario R. Vigna, and Guillermo R. Chantre**

**Abstract** This chapter describes a mathematical simulation model for the multiannual assessment of Integrated Weed Management (IWM) strategies. The model allows to simulate the competitive interaction between an annual weed species and a grain crop. From the weed's side, the following processes are represented: (1) demographic dynamics on a daily basis considering the numeric composition of the different phenological states, (2) intra- and interspecific competition, (3) seed production and (4) the effect of different control methods. Regarding the crop, the following variables are computed: (1) leaf area index (LAI), (2) competition on the weed and (3) expected yield as a function of weed competition. The model was developed on Microsoft Excel® with Visual Basic complements. Results are

F. A. Molinari (✉)
Departamento de Agronomía, Universidad Nacional del Sur,
San Andrés 800, Bahía Blanca 8000, Buenos Aires, Argentina

Comisión de Investigaciones Científicas de la Provincia de Buenos Aires (CIC),
Bahía Blanca, Buenos Aires, Argentina
e-mail: franco.molinari@uns.edu.ar

A. M. Blanco
Planta Piloto de Ingeniería Química—PLAPIQUI, Universidad Nacional del Sur—CONICET, Bahía Blanca 8000, Buenos Aires, Argentina
e-mail: ablanco@plapiqui.edu.ar

M. R. Vigna
INTA EEA Bordenave, Bordenave 8187, Buenos Aires, Argentina

G. R. Chantre
Departamento de Agronomía, Universidad Nacional del Sur - CERZOS/CONICET,
Bahía Blanca, Buenos Aires, Argentina
e-mail: gchantre@criba.edu.ar


G. R. Chantre, J. L. González-Andújar (eds.), *Decision Support Systems for Weed Management*, https://doi.org/10.1007/978-3-030-44402-0_15

provided for the wild oat (*Avena fatua*)-winter wheat (*Triticum aestivum*) system, a typical system of the south-west area of the semiarid Pampean region of Argentina. The model was calibrated and validated with experimental data collected along 4 years. Several multi-year scenarios were generated to evaluate the effect of different IWM strategies against common herbicide-based practices. Finally, possible improvements to the model and some guidelines towards the development of a long-term DSS for weed management are provided.

**Keywords** Integrated weed management · Agricultural system modelling · Weed competition · Wild oat · Barley · Wheat

## 15.1 Introduction

Weeds are the main cause of crop yield loss since the early stages of agriculture. It is widely acknowledged that in the short term, herbicide-based control methods play a major role in maximizing agricultural systems' productivity, but their continuous application produces negative environmental effects. In this regard, the combination of both preventive (e.g. legal, cultural) and proactive (e.g. chemical, mechanical, physical and biological) measures has been proposed as 'a way-through' for mitigating environmental and society-derived effects (e.g. chemical food residues, soil and water contamination, biodiversity loss, ecotoxicity, etc.). Such activity is known as Integrated Weed Management (IWM). However, it should be highlighted that the success of IWM resides on the application of knowledge-based principles rooted mainly on weed biology and ecology. Thus, a precise understanding of demographic variables (e.g. phenological stages) and processes (e.g. interspecific weed-crop competition) will help to develop new approaches for long-term management. As suggested by Ghersa and Holt (1995), the ability to predict weed and crop phenology is an essential aspect for designing sustainable management programs.

IWM can benefit from model-based Decision Support Systems. Several examples of DSSs have been presented in previous chapters of this volume (Chaps. 11–14). According to our knowledge, no DSS to aid in IWM in the medium term has been developed so far for Argentinian systems. Thus, the objective of this chapter is to contribute to fill this gap by proposing a simulation model for multiannual weed management planning, as the basis for a DSS. In this sense, this chapter integrates some algorithmic developments produced in our research group in the last years together with agronomic and biological insight gathered from extensive field experimentation along the last decades.

From a practical agronomic perspective, DSS-oriented models should include crop-weed demography, as well as the intervening ecophysiological elements that will finally define crop yields and weed population dynamics. The present model allows to simulate the competitive interaction between an annual weed and a grain crop. From the weed's side, the following processes are represented: (1) demographic dynamics on a daily basis considering the numeric composition of the different phenological stages, (2) intra- and interspecific competition, (3) seed production and (4) the effect of different control methods. Regarding the crop, the following variables are computed: (1) leaf area index (LAI), (2) crop competition over the weed and (3) expected yield as a function of weed-crop competition.

The wild oat-winter wheat system, typical of the South-West (SW) area of the semiarid Pampean region of Argentina (SPRA), was adopted as study case to illustrate the capabilities and overall performance of the model as the core of a DSS. Wild oat (*Avena fatua* L.) (AVEFA) is one of the most conspicuous weeds in winter cereal crops producing large yield losses and reducing harvest quality (Martín and Scursoni 2018). In the SW of Buenos Aires province, in the SPRA, *Avena fatua* displays irregular field emergence patterns due to large interannual variability of precipitations, seasonal temperature fluctuations and a species-specific adaptation to the local environment (Chantre et al. 2012, 2018). In the SPRA, *A. fatua* is present in 60% of cereal plots (Scursoni et al. 2014) producing yield losses of 20–25% in wheat crops (Scursoni and Benech-Arnold 1998; Scursoni et al. 2011). AVEFA seeds are also grain contaminants producing a significant reduction in selling price (Martín and Scursoni 2018).

The model was calibrated and validated with experimental data collected along 4 years at the Argentinian National Institute of Agricultural Technology (INTA) EEA-Bordenave, Buenos Aires, Argentina (37°46′08.0″S 63°05′30.5″W). Several multi-year scenarios were simulated to evaluate the effect of different IWM options against herbicide-based practices.

## 15.2 Simulation Model Description

The present model is of general purpose with potentiality for being adapted to different annual weed-crop systems. The model simulates weed management scenarios within a multiannual planning horizon (tactic-strategic) using a daily time step for variable calculation within the agronomic season. A daily time step permits a high level of detail being compatible with the actual frequency of weather data availability and forecasts. The multiannual simulation allows the visualization of medium- and long-term effects of weed management decisions in the field.

For most processes, our approach follows the typical equations widely adopted in previous weed modelling studies reported in the literature (Gonzalez-Andujar 2013; Lodovichi et al. 2013). In a few cases, however, some innovation was introduced to represent with more detail some specific processes to better represent the undergoing biology.

## 15.2.1 Model Insights

In Fig. 15.1, a general diagram of the simulation model considering an annual weed-winter cereal system is presented. On the left, a thermal timescale representing weed demographic stages is shown. At the base of the diagram, both chronological and thermal timescales for crop growth-development are included. At the top, a clear distinction between fallow and crop growth-development periods is made. In the latter, the crop sowing time, leaf area index (LAI) curve and critical competition period (CCP) are depicted.

The weed life cycle is represented in a simple fashion defining the most representative stage variables for an annual weed. The following stages are considered:

$s = 0$: seed bank
$s = 1$: seedling (one true leaf)
$s = 2$: early vegetative (two to four true leaves)
$s = 3$: advanced vegetative (tillering)
$s = 4$: initial reproductive (flowering)
$s = 5$: advanced reproductive (senescence)

The variables and parameters of the model are detailed in Tables 15.1, 15.2, 15.3 and 15.4.

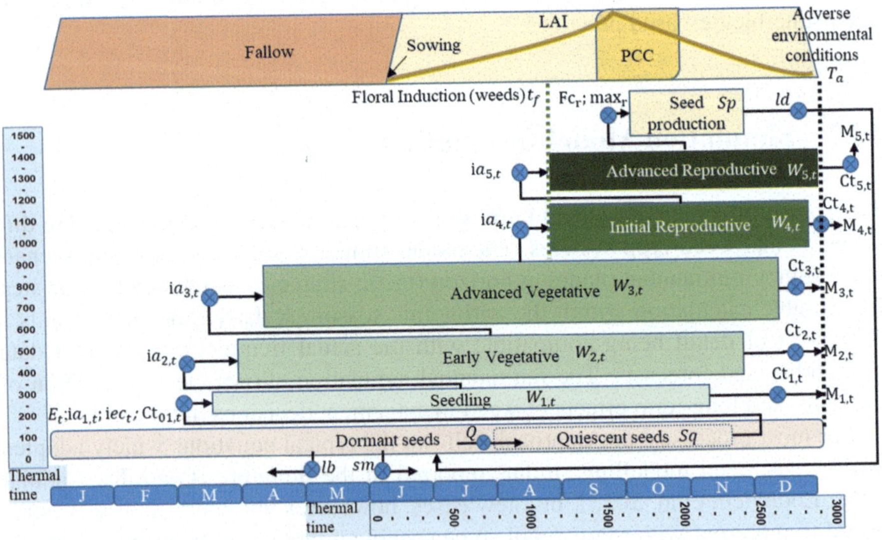


**Fig. 15.1** General diagram illustrating the simulation model, considering a weed (winter autumn)-winter cereal system throughout an agricultural campaign (calendar year). (See Tables 15.1–15.3 for a description of terms)

**Table 15.1** Model variables

| Variable | Description |
|---|---|
| $t$ | Julian day |
| $y$ | Year |
| $s$ | Weed phenological stage |
| $W_{s,t}$ | Accumulated weed density in $s$ and $t$ |
| $I_{s,t}$ | Incoming cohorts of individuals in $s$ and $t$ |
| $O_{s,t}$ | Outcoming cohorts of individuals of $s$ and $t$ |
| $M_{s,t}$ | Accumulation of individuals eliminated by control actions plus those affected by thermal/hydric stress in $s$, $t$ |
| $\theta_t$ | Accumulated TT |
| $T_t$ | Average daily temperature in $t$ |
| Sq | Quiescent (non-dormant) seeds |
| $S_y$ | Quiescent seeds from $y$ |
| $\mathrm{Sp}_y$ | Total weed seed production in $y$ |
| $E_t$ | Proportion of emerged seedlings |
| $\mathrm{ia}_{s,t}$ | Weed intraspecific competition in $s$ and $t$ |
| $\mathrm{Wk}_{s,t}$ | Weighted weed density in $s$ and $t$ |
| $\mathrm{ie}_t$ | Crop-weed interspecific competition |
| $\mathrm{LAI}_t$ | Leaf area index of the crop in $t$ |
| $r$ | Reproductive cohorts' group |
| $W_r$ | Weed density in $r$ |
| Yld | Expected crop yield (proportion of weed-free yield) |
| WC | Weed-crop interspecific competition |
| $\mathrm{MC}_{s,t}$ | Number of individuals eliminated by control actions in $s$ and $t$ |
| $\mathrm{Mstress}_{s,t}$ | Number of individuals affected by thermal/hydric stress in $s$ and $t$ |
| $\mathrm{Ct}_{s,t}$ | Weed mortality rate due to control in $s$ and $t$ |
| $\mathrm{Ct}_{01,t}$ | Weed mortality rate over a pre-seedling stage by residual herbicides in $t$ |

*TT* thermal time

#### 15.2.1.1 Seed Bank Dynamics

The weed seed bank dynamics considers the seed production from those individuals that reach the advanced reproductive stage ($s = 5$). Seed losses due to abiotic (field emergence) and biotic (predation, mortality and longevity) factors are accounted. The seed bank is considered to be composed by two seed fractions, quiescent (non-dormant) and dormant seeds. Thus, the daily fraction of the seed bank capable of producing new seedlings each day is estimated from the former fraction. The dynamics of the seed bank is represented by the following equations:

$$\mathrm{Sq} = \sum_{-1}^{y=-L} \left( S_y \right) \tag{15.1}$$

**Table 15.2** Weed demographic parameters

| Parameter | Description |
|---|---|
| ns | Number of phenological stages |
| Tb | Base temperature for TT |
| $Th_s$ | TT required for transition from $s$ to $s + 1$ of a given cohort |
| $t_f$ | Floral induction day |
| $L$ | Seed bank longevity (in years) |
| $Q_y$ | Seed fractions produced in year $y$ that are quiescent in year 0 |
| sm | Seed bank annual mortality rate |
| ld | Seed loss rate at natural dispersal |
| lb | Seed loss rate by biotic factors during the first fallow (predation, mortality) |
| $K$ | Agroecosystem's carrying capacity |
| Iam | Intraspecific competition mortality rate |
| $f_s$ | Competition factor for stage $s$ |
| nr | Number of simulated groups of reproductive cohorts |
| $Fc_r$ | Individuals' fecundity of $r$ |
| $max_r$ | Maximum seed production of $r$ |
| $t_a$ | Day of adverse environmental conditions (stress) |
| $m_{stresss}$ | Weed mortality rate by adverse environmental conditions in $s$ |

*TT* thermal time

**Table 15.3** Crop parameters

| Parameter | Description |
|---|---|
| $t_e$ | Crop emergence date (Julian day) |
| $t_m$ | Crop maturity date (Julian day) |
| CCP | Critical competition period |
| $Sf_t$ | Susceptibility of the crop in day $t$ |
| LAIhc | Value of LAI representing a highly competitive situation |
| Cs | Standard crop density |
| Ca | Actual crop density |
| Myl | Maximum yield loss proportion (high interspecific competition) |
| $A$ | Crop-derived constant |
| $K$ | Weed competitiveness constant |

$$S_y = \mathrm{Sp}_y\left[(1-\mathrm{ld})(1-\mathrm{lb})Q_y\left(1+(\mathrm{sm}\cdot i)\right)\right] \tag{15.2}$$

where Sq represents the number of quiescent seeds at the beginning of the crop season at year 0 (seeds per $m^{-2}$) and $L$ is the seed bank longevity (in years). The expression integrates quiescent seeds from $L$ years previous to the beginning of the current season. $Sp_y$ is the total seed production corresponding to a given year $y$; ld is the seed loss rate during seed dispersal; lb is the seed loss rate due to biotic factors

**Table 15.4** AVEFA demographic parameters for the study case

| Parameter | Description | Value | Reference |
|---|---|---|---|
| ns | Number of simulated phenological stages | 5 | EK |
| Tb | Base temperature for TT accumulation | 0 °C | Martín and Scursoni (2018) |
| $Th_1$ | TT required for transition from $s = 1$ to $s = 2$ | 70 °Cd | Martín and Scursoni (2018) |
| $Th_2$ | TT required for transition from $s = 2$ to $s = 3$ | 280 °Cd | Martín and Scursoni (2018) |
| $Th_3$ | TT required for transition from $s = 3$ to $s = 4$ | 400 °Cd | EK |
| $Th_4$ | TT required for transition from $s = 4$ to $s = 5$ | 300 °Cd | EK |
| $t_f$ | Floral induction day | 258 Julian days | EK |
| $L$ | Seed bank longevity | 3 years | Scursoni (2001) |
| $Q_{-1}$ | Proportion of seeds produced in year −1 that are quiescent in year 0 | 0.7 | Scursoni (2001). |
| $Q_{-2}$ | Proportion of seeds produced in year −2 that are quiescent in year 0 | 0.2 | Scursoni (2001). |
| $Q_{-3}$ | Proportion of seeds produced in year −3 that are quiescent in year 0 | 0.1 | Scursoni (2001). |
| sm | Seed bank annual mortality rate | 0.0732 | Scursoni (2001). |
| ld | Seed loss rate at natural dispersal | 0.67 | Scursoni (2001) |
| lb | Seed loss rate by biotic factors during the first fallow | 0.2075 | Scursoni et al. (1999) |
| $K$ | Carrying capacity | 250 ind $m^{-2}$ | EK |
| Iam | Intraspecific competition mortality rate | 1 | EK |
| $f_1$ | Competition factor for $s = 1$ | 0.15 | EK |
| $f_2$ | Competition factor for $s = 2$ | 0.3 | EK |
| $f_3$ | Competition factor for $s = 3$ | 0.6 | EK |
| $f_4$ | Competition factor for $s = 4$ | 1 | EK |
| $f_5$ | Competition factor for $s = 5$ | 0 | EK |
| nr | Groups of reproductive cohorts | 2 | Adapted from Scursoni (2001) |
| $Fc_1$ | Individual fecundity of Group 1 | 187.2 seeds $ind^{-1}$ | Adapted from Scursoni (2001) |
| $Fc_2$ | Individual fecundity of Group 2 | 19.3 seeds $ind^{-1}$ | Adapted from Scursoni (2001) |
| $max_1$ | Maximum seed production of Group 1 | 12,000 seeds $m^{-2}$ | Adapted from Scursoni (2001) |
| $max_2$ | Maximum seed production of Group 2 | 2500 seeds $m^{-2}$ | Adapted from Scursoni (2001) |
| $T_a$ | Day of adverse environmental conditions | 1 Julian day | EK |
| $m_{stress1}$ | Weed mortality rate by adverse environmental conditions in $s = 1$ | 1 | EK |

(continued)

**Table 15.4** (continued)

| Parameter | Description | Value | Reference |
|---|---|---|---|
| $m_{\text{stress2}}$ | Weed mortality rate by adverse environmental conditions in $s = 2$ | 1 | EK |
| $m_{\text{stress3}}$ | Weed mortality rate by adverse environmental conditions in $s = 3$ | 1 | EK |
| $m_{\text{stress4}}$ | Weed mortality rate by adverse environmental conditions in $s = 4$ | 1 | EK |
| $m_{\text{stress5}}$ | Weed mortality rate by adverse environmental conditions in $s = 5$ | 0 | EK |

*EK* expert knowledge; *TT* thermal time

during fallow; $Q_y$ is the proportion of seeds produced in year $y$ that are quiescent in $y = 0$; sm is the seed bank annual mortality rate.

#### 15.2.1.2 Field Emergence Modelling

The estimation of the weed seed bank proportion emerging each day is estimated through mathematical models based on quantitative environmental variables (i.e. temperature, precipitation or soil-derived environmental-based indexes). Typical emergence models include, for example, nonlinear regressions and artificial neural networks (Gonzalez-Andujar et al. 2016). For further details on available options for weed emergence modelling the reader is referred to Chap. 5.

#### 15.2.1.3 Phenological Stages ( $W_s$)

The weed demography is simulated through daily cohorts. Each cohort is represented by the individuals which emerge simultaneously each day. The individuals of each cohort go through different phenological stages during the weed life cycle (see green rectangles in Fig. 15.1).

The weed demographic balance is given by:

$$W_{s,t+1} = W_{s,t} + I_{s,t} - O_{s,t} - M_{s,t} \qquad \forall s, \forall t \tag{15.3}$$

where $W_{s,t}$ is the accumulated density (individuals per m$^{-2}$) at a given phenological stage ($s$) in a given day ($t$); $I_{s,t}$ are the incoming cohorts (i.e. the group of cohorts that enters $s$ in $t$); $O_{s,t}$ are the outcoming cohorts of $s$ in $t$; $M_{s,t}$ is the sum of individuals eliminated by control actions plus those affected by thermal/hydric stress in $s$ and $t$.

The logic adopted to simulate the daily dynamics of simulated cohorts is summarized in Figs. 15.2 and 15.3. As observed in Fig. 15.2, the dynamics starts with the cohort A as incoming cohort of stage $s$ in $t = 1$ ($I_{s,1}$). Cohort A requires a given thermal time (TT), say Th$_s$, to allow transition from $s$ to $s + 1$ (see the flow from $I_{s,1}$

| $t$ (julian day) | $I_S$ (incoming cohorts in s, i.m$^{-2}$) | $W_s$ (phenological stage s, i.m$^{-2}$) | $O_s$ (outcoming cohorts of s, i.m$^{-2}$) | $I_{(s+1)}$ (incoming cohorts to s+1, i.m$^{-2}$) | $W_{(s+1)}$ (phenological stage s+1, i.m$^{-2}$) |
|---|---|---|---|---|---|
| 1 | A | | | | |
| 2 | B | A | $Th_s$ | | |
| 3 | C | A+B | A $Th_s$ | A′=A × $ia_{(s+1),3}$ | |
| 4 | | B+C | $Th_s$ | | A′ |
| 5 | | B+C | B + C | B′+C′=(B+C)×$ia_{(s+1),5}$ | A′ |

**Fig. 15.2** Descriptive diagram describing cohorts' flows between phenological stages. (See Tables 15.1–15.3 for a description of terms)

| $t$ (julian day) | $I_S$ (incoming cohorts to s, i.m$^{-2}$) | $W_s$ (phenological stage s, i.m$^{-2}$) | $O_s$ (outcoming cohorts of s, i.m$^{-2}$) |
|---|---|---|---|
| 1 | D | | |
| 2 | E | D | |
| 3 | | D+E-$M_{s,7}$ | D′=D × (1-$Ct_{s,3}$) |
| 4 | | E-($M_{s,7}$-D × $Ct_{s,7}$) | E′=E × (1-$Ct_{s,3}$) |

**Fig. 15.3** Descriptive diagram describing control dynamics and cohorts' flows. (See Tables 15.1–15.3 for a description of terms)

to $O_{s,\,3}$). After TT requirements are fulfilled, cohort A is affected by the intraspecific competition ($\text{ia}_{(s+1),\,3)}$), and then a new cohort, A′, conforms the following stage ($s + 1$). Moreover, for the case of additional cohorts (B and C), emerging in different days ($I_{s,\,2}$ and $I_{s,\,3}$ respectively), the transition to the following state occurs simultaneously ($O_{s,\,5}$) in the considered example. To sum up, $\text{Th}_s$ required for transition from $s$ to $s + 1$ is the same for all initial cohorts, but the rate of TT accumulation may be different for each one, depending on the weather conditions (see arrows).

In Fig. 15.3, in $t = 3$, a control action is performed over the individuals in $W_{s,\,3}$ (D + E). Cohort D is affected by the rate of control ($\text{Ct}_{s,\,3}$) generating the D′ cohort. In $t = 4$, the same logic applies to E.

The estimation of the incoming cohorts at $s = 1$ in $t$ ($I_{1,\,t}$) is obtained as:

$$I_{1,t} = \text{Sq} \cdot E_t \cdot \text{ia}_{1,t} \cdot \text{ie}_t \cdot \left(1 - \text{Ct}_{01,t}\right) \quad \forall t \tag{15.4}$$

where $I_{1,\,t}$ are the incoming cohorts at $s = 1,t$ (i.e. group of cohorts which enter $W_{1,\,t}$); Sq is the amount of quiescent seeds; $E_t$ is the proportion of emerged seedlings in $t$; $\text{ia}_{1,\,t}$ is the weed intraspecific competition (as a survival rate) at $s = 1$, $t$; $\text{ie}_t$ represents the interspecific competition of the crop over the weed in $t$; and $\text{Ct}_{01,\,t}$ is the weed mortality over the transition between germinating seeds and seedling stage, in $t$.

For each phenological stage, a given thermal time accumulation ($Th_s$) is required to allow the transition to the next phenological stage. For TT accumulation $\theta_{s,t}$, the following equation (Begon et al. 2006) is used:

$$\theta_{s,t} = \sum_{n}^{i=1} \left( T_{(t+i)} - \text{Tb} \right) \qquad \forall s, \forall t \tag{15.5}$$

where $T_{(t+i)}$ is the mean daily temperature in $t$ + i and Tb is the base temperature of the weed species under study.

This equation starts with $I_{s,t}$ and operates on a daily basis until the environmental requirements for transition to the next phenological stage after n days are fulfilled. The following expression describes this process:

$$\left\{ \begin{matrix} O_{s,(t+n)} = I_{s,t}\left(1 - \text{Ct}_{s,t}\right) \\ I_{(s+1),(t+n)} = O_{s,(t+n)}\,\text{ia}_{s,t} \end{matrix} \right. \text{if}\,\theta_{s,t} \geq \text{Th}_s \forall s, \forall t \tag{15.6}$$

where $I_{(s+1),(t+n)}$ is the incoming cohorts at $s$ + 1 in $t$ + n; $O_{s,(t+n)}$ is the outcoming cohorts of $s$ in ($t$ + n); and $\text{ia}_{s,t}$ is the weed intraspecific competition (as a survival rate) at $s$ in $t$.

Equation (15.6) applies for all phenological stages except for stage 3 where an additional environmental condition (minimum floral induction time) is required. Thus, as depicted in Eq. (15.7), weed individuals at advanced vegetative stage must reach the floral induction time as an additional requirement for transition to $s = 4$.

$$\left\{ \begin{matrix} O_{3,(t+n)} = I_{3,t}\left(1 - \text{Ct}_{3,t}\right) \\ I_{4,(t+n)} = O_{3,(t+n)}\,\text{ia}_{4,t} \end{matrix} \right. \text{if}\,\theta_{3,t} \geq \text{Th}_3 \quad \& \quad t \geq t_{\text{f}} \forall t \tag{15.7}$$

where $t_{\text{f}}$ is the time of floral induction (Julian day).

#### 15.2.1.4 Weed Intraspecific Competition

In this work, intraspecific competition of the weed was considered to depend on the carrying capacity of the system ($K$) and on the density of the present individuals weighted according to their specific phenological stage ($\text{Wk}_{s,t}$). This is indicated in Figs. 15.1 and 15.2 and described by Eqs. (15.8) and (15.9):

$$\text{ia}_{s,t} = \left\{ \begin{matrix} 1, & \text{Wk}_{s,t} < K \\ 1 - \text{iam}, & \text{Wk}_{s,t} \geq K \end{matrix} \right. \forall s, \forall t \tag{15.8}$$

$$\text{Wk}_{s,t} = \sum_{\text{ns}}^{i=s} \left\{ W_{i,t} \cdot f_i \right\} \qquad \forall s, \forall t \tag{15.9}$$

where $ia_{s,t}$ is the weed intraspecific competition in $s$ and $t$; iam is the intraspecific competition mortality rate; $K$ is the agroecosystem's carrying capacity; $Wk_{s,t}$ is the weighted weed density from actual stage $s$ to ns in $t$, ns being the number of phenological stages simulated; $W_{i,t}$ is the accumulated density (individuals per $m^{-2}$) in $i$ and $t$; and $f_i$ is the competition factor of i.

#### 15.2.1.5 Crop-Weed Interspecific competition

The competitive effect of the crop over the weed ($ie_t$) is simulated as a survival rate on $s = 1$ (Fig. 15.1). The magnitude of $ie_t$ depends on the crop sowing density and on the leaf area index (LAI). $LAI_t$ can be easily obtained with specific crop simulation tools, such as the Decision Support System for Agrotechnology Transfer (DSSAT) software (Jones et al. 2003).

For the calculation of $ie_t$, the following equation is implemented:

$$ie_t = 1 - \left\{ \min\left[ \left( \frac{LAI_t}{LAIhc} \right) \left( \frac{Cs}{Ca} \right); 1 \right] \right\} \qquad \forall t \tag{15.10}$$

where $ie_t$ is the proportion of individuals that survive interspecific competition; $LAI_t$ is the LAI in $t$; LAIhc is a value of LAI representing a highly competitive situation; Cs is the standard sowing density; and Ca is the actual sowing density. The relationship between LAI and $C$ determines the mortality of the weed individuals. The minimum function establishes that $ie_t$ is constrained between 0 and 1.

#### 15.2.1.6 Weed Seed Production

Regarding seed production, several different groups of 'reproductive cohorts' were adopted. Such a modelling approach allows considering differences in the fecundity and maximum seed production of individuals influenced by the diverse environmental conditions that occur along the season (Fig. 15.1). The following equation represents weed seed production:

$$Sp = \sum_{nr}^{r=1} \left\{ \min = \left( W_r \cdot Fc_r; \max_r \right) \right\} \tag{15.11}$$

where Sp is the total seed production at the end of the season; $r$ corresponds to a given 'reproductive group'; nr is the number of simulated 'reproductive groups'; $W_r$ is the density of individuals in $r$; $Fc_r$ is the fecundity of $r$; and $\max_r$ is the maximum seed production of $r$.

#### 15.2.1.7 Weed-Crop Interspecific Competition

The interference of the weed on the crop is taken into account for the expected yield calculation. In this work, the expected crop yield equation proposed by Pannell et al. (2004) was adopted:

$$\mathrm{Yld} = \frac{\mathrm{Cs} + a}{\mathrm{Cs}} \cdot \left[ \frac{\mathrm{Ca}}{\left(a + \mathrm{Ca} + (k \cdot W)\right)} \cdot \mathrm{Myl} + (1 - \mathrm{Myl}) \right] \tag{15.12}$$

where Yld corresponds to the expected crop yield (as a proportion of the weed-free yield); Cs is the standard crop density; $a$ is a crop-dependent constant; Ca is the actual crop sowing density; $k$ is a constant which accounts for weed competitiveness; $W$ is the weed density of individuals which survive control methods; and Myl is the maximum yield loss proportion due to high interspecific competition.

Although Eq. (15.12) considers diverse weed control options, crop sowing densities, maximum yield loss, etc., the following important factors are not explicitly considered: (1) phenological stages of weed individuals, (2) critical competition period (CCP) and (3) interspecific competition before weed management interventions. In order to account for the complex competitive effect of the weed over the crop along the whole season, a new variable was introduced (WC) to replace $W$ in Eq. (15.12). The estimation of WC is as follows:

$$\mathrm{WC} = \frac{\sum_{t=t_e}^{t_m} \sum_{s=1}^{ns} \left\{ (W_{s,t} \cdot f_s) \mathrm{Sf}_t \right\}}{(t_m - t_e)} \tag{15.13}$$

where WC is the sum of weed competitive effects (over the crop) at the end of the season; $t_e$ is the crop emergence time (Julian day); $t_m$ is the time of crop maturity (Julian day); $f_s$ is the competition factor of stage $s$; and $\mathrm{Sf}_t$ is a factor that represents the crop susceptibility to competition in $t$.

### 15.2.2 *Weed Mortality (M)*

The following equation describes weed losses due to weed management and stress mortality:

$$M_{s,t} = \mathrm{MC}_{s,t} + M\mathrm{stress}_{s,t} \qquad \forall s, \forall t \tag{15.14}$$

where $M_{s,t}$ is the sum of individuals eliminated by control actions in $s$ and $t$ ($\mathrm{MC}_{s,t}$) and the number of individuals affected by thermal/hydric stress in $s$ and $t$ ($M\mathrm{stress}_{s,t}$).

#### 15.2.2.1 Management Simulation

Simulation of weed management is performed according to the following equation:

$$\mathrm{MC}_{s,t} = W_{s,t} \cdot \mathrm{Ct}_{s,t} \forall s, \forall t \tag{15.15}$$

where $\mathrm{Ct}_{s,t}$ is the weed mortality rate due to control interventions in $s$ and $t$ and $\mathrm{Ct}_{s,t}$ represents the proportion of individuals controlled by different management methods. Within the chemical options, two possible scenarios are considered: (1) weed mortality occurring after post-emergence control and (2) residual pre-emergence effect. For post-emergence control, different mortality rates apply on individuals at each phenological stage (from $s = 1$ to $s = 5$) at the time of the chemical intervention ($t = \phi$). For the residual effect, mortality is accounted over the transition between germinating seeds and seedling stage ($\mathrm{Ct}_{01,t}$) between $t = \phi$ and $t = \phi + \Omega$, where $\Omega$ is the adopted residual time span.

#### 15.2.2.2 Weed Mortality Due to Stress

Equation (15.16) considers the environmental adverse conditions that reduce the survival of the weed individuals:

$$\begin{cases} M_{\mathrm{stress}_{s,t}} = W_{s,t} \cdot m_{\mathrm{stress}_s}, & \text{if } t \in T_\mathrm{a} \\ M_{\mathrm{stress}_{s,t}} = 0, & \text{otherwise} \end{cases} \quad \forall s, \forall t \tag{15.16}$$

where $m_{\mathrm{stresss}}$ is the mortality rate due to adverse environmental conditions on stage $s$ in the Julian days belonging to set $T_\mathrm{a}$. Set $T_\mathrm{a}$ contains the specific days where adverse weather conditions take place.

## 15.3 Study Case: AVEFA-Winter Wheat System in the South-West Area of the Semiarid Pampean Region of Argentina

### *15.3.1 Model Parameters*

The proposed study case is the simulation of AVEFA in competition with winter wheat (*Triticum aestivum* L.), a typical system in the south-west area of the semiarid Pampean region of Argentina. For all the purposes, weather data records generated by the weather station of INTA at Bordenave were adopted (https://inta.gob.ar/documentos/informacion-agrometeorologica).

#### 15.3.1.1 AVEFA Parameters

The emergence of AVEFA was simulated with the neural network model proposed in Chantre et al. (2018), which was specifically tuned and validated for the region under consideration.

#### 15.3.1.2 Wheat Model Parameters

The LAI of wheat crop was simulated using DSSAT. The data generated by simulation were approximated in the present work through the following simplified equation (Eq. 15.17):

$$\mathrm{LAI}_t = \begin{cases} \alpha^{(\omega * D_t)}, & \mathrm{If}(D_t < \mathrm{G1}) \\ \beta - \gamma \cdot D_t + \delta \cdot D_t^2 - \varepsilon \cdot D_t^3, & \mathrm{If}(D_t \geq \mathrm{G1}) \\ 0, & \mathrm{If}(D_t \geq \mathrm{G2}) \end{cases} \tag{15.17}$$

where $\alpha$, $\omega$, $\beta$, $\gamma$, $\delta$, $\varepsilon$ are the parameters adapted from DSSAT, $D_t$ is the accumulation of the crop's thermal time (°Cd) at day $t$, G1 is the accumulated thermal time at the day of the largest LAI, and G2 is the cumulated thermal time required to reach physiological maturity in wheat. The parameters of Eq. (15.17) are detailed in Table 15.5.

### 15.3.2 *Weed Management*

In Table 15.6, the parameters corresponding to chemical and mechanical control actions to simulate the AVEFA-wheat system are provided.

### 15.3.3 *Calibration and Validation*

In order to correctly estimate expected yield (Yld), parameters $a$ and $k$ (Eq. 15.12) (which depend on the crop variety and on the competitive ability of the weed over the crop) should be tuned for the system under consideration. In this work, these parameters were calculated by correlation of experimental data resulting from field experiments conducted at EEA-INTA Bordenave (Bordenave, Buenos Aires, Argentina) along 4 years (Lopez and Vigna, n.d.) totalizing a pool of 41 data points ($N = 41$).

From the available data set, 70% of the data points were randomly chosen for parameter estimation. The parameter estimation problem was implemented in Excel using the Solver tool to seek for the values of $a$ and $k$ that minimize the RMSE

**Table 15.5** Wheat parameters for the study case

| Parameter | Description | Value | Source |
|---|---|---|---|
| $t_e$ | Crop emergence day | 174 Julian days | Cronotrigo@ |
| $t_m$ | Physiological crop maturity day | $t_m = f(G2)$ | Adapted from DSSAT |
| G1 | TT at maximum $LAI_t$ | 1116 °Cd | Adapted from DSSAT |
| G2 | TT for physiological crop maturity | 2260 °Cd | Adapted from DSSAT |
| CCP | Critical competition period | 300–320 Julian days | Cronotrigo@ |
| $Sf_t$ | Susceptibility of the crop between day 0 and CCP | $Sf_t = $ Min ($LAI_t$, 1) | EK |
| $Sf_t$ | Susceptibility of the crop during CCP | 5 | EK |
| $Sf_t$ | Susceptibility of the crop after CCP | 1 | EK |
| $\alpha$ | $LAI_t$ parameter | 0.1138 | Adapted from DSSAT |
| $\omega$ | $LAI_t$ parameter | 3.71 $E^{-3}$ | Adapted from DSSAT |
| $\beta$ | $LAI_t$ parameter | 47.98 | Adapted from DSSAT |
| $\gamma$ | $LAI_t$ parameter | 0.08012 | Adapted from DSSAT |
| $\delta$ | $LAI_t$ parameter | 5.02 $E^{-5}$ | Adapted from DSSAT |
| $\varepsilon$ | $LAI_t$ parameter | 1.07 $E^{-8}$ | Adapted from DSSAT |

*EK* expert knowledge

**Table 15.6** Control strategies used, detailing type of control, application time, control method (usual commercial formulation), residual time span ($\Omega$) and mortality rate of control over a pre-seedling stage ($Cr_{01,t}$) and at seedling ($Ct_1$), early vegetative ($Ct_2$), advanced vegetative ($Ct_3$), initial reproductive ($Ct_4$) and advanced reproductive ($Ct_5$)

| Type of control | Application time | Control method | Residual span ($\Omega$) (days) | Residual effect ($Ct_{01,t}$) | $Ct_1$ | $Ct_2$ | $Ct_3$ | $Ct_4$ | $Ct_5$ |
|---|---|---|---|---|---|---|---|---|---|
| Non-selective | Fallow | Glyphosate SL 48% (2 L $ha^{-1}$) | – | – | 1 | 1 | 1 | 0.5 | 0 |
| Non-selective residual | Fallow | Glyphosate SL 48% + flumioxazin CS 48% (2 L $ha^{-1}$ + 0.120 L $ha^{-1}$) | 6 | 1 | 1 | 1 | 1 | 0.5 | 0 |
| Tillage | Fallow | Ploughing | – | – | 1 | 1 | 1 | 1 | 0 |
| Selective | Post-emergence | Pinoxaden EC 6%, 0.8 L $ha^{-1}$ | – | – | 1 | 0.9 | 0.7 | 0.2 | 0 |
| Selective residual | Post-emergence | Flucarbazone-sodium WG 70%, 80 g $ha^{-1}$ | 20 | 0.5 | 1 | 0.9 | 0.3 | 0 | 0 |

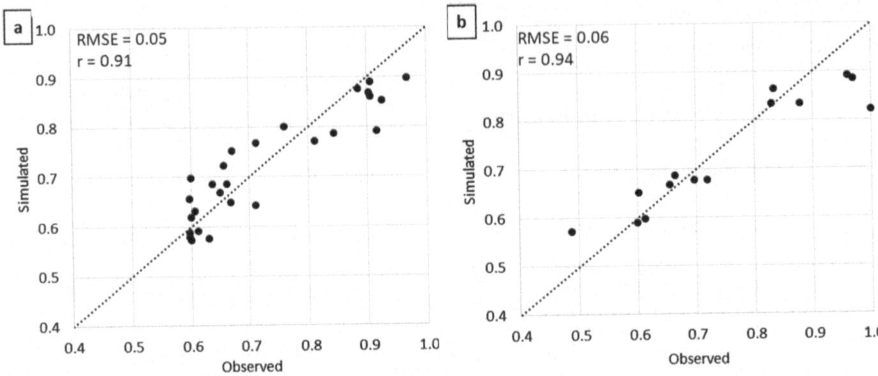


**Fig. 15.4** Expected wheat yield (Yld), calculated vs. observed obtained with the AVEFA-wheat model. (**a**) Train set (70% data). (**b**) Validation set (30% data). *RMSE* root mean square error

(Fig. 15.4a). The remaining 30% of the data points were simulated with the calculated parameters ($a$ = 100, $k$ = 5) obtaining the results shown in Fig. 15.4b, which are considered satisfactory for the purposes of this study.

## 15.3.4 Simulation Results

Following, the results of multiannual simulations of the AVEFA-winter wheat system are shown to illustrate the main characteristics and outcomes of the proposed model. Three cases are presented, each one combining different management strategies (including chemical, mechanical and cultural interventions). Chemical controls adopted in this work include selective and non-selective herbicides with or without residual effects. In the following simulation experiments, usual commercial formulations are provided as examples of typical treatments.

### 15.3.4.1 Case I: Cultural + Mechanical Management

Wheat is sown at June first each year, at a sowing density of Ca = 200 pl $m^{-2}$. The specific adopted variety possesses a LAIhc = 1 and a Myl = 20% whose values correspond to a high competitive wheat variety (Buck Napostá) (Lopez and Vigna, n.d.). A mechanical control (ploughing) was performed each year, previously to sowing.

In Fig. 15.5, the effect of the intraspecific competition can be observed as a reduced number of weed individuals incorporating to the system between April and May of year 1. Before the wheat is sown, the ploughing produces a steep elimination of AVEFA individuals. Then, the interspecific competition generated by the highly competitive crop produces a reduced incorporation of weed seedlings in the early stages of the crop development.

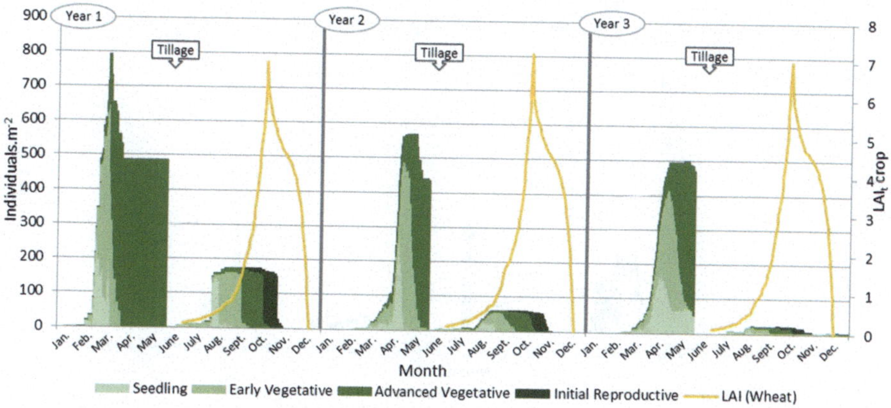


**Fig. 15.5** Multiannual AVEFA dynamics showing in different shades of green the relative composition of each phenological stage from a high non-dormant seed content (4500 seed m$^{-2}$). The combined effect of mechanical control (ploughing) and cultural strategy (high competitive wheat variety) is simulated. The development of the wheat is observed through the $LAI_t$ (yellow line). Arrows indicate control dates

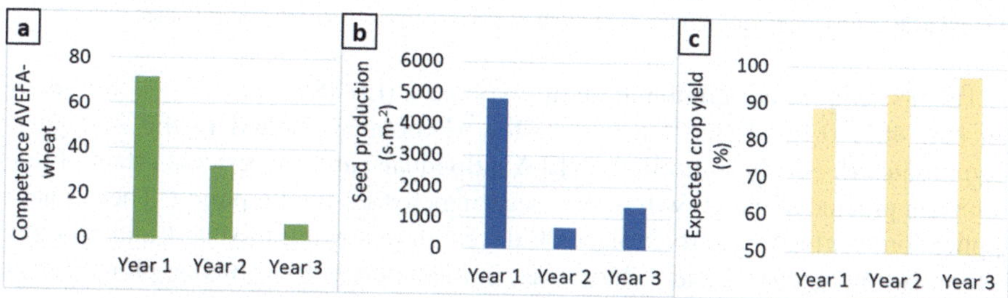


**Fig. 15.6** Competition of AVEFA over winter wheat (WC) (**a**), AVEFA seed production (Sp) (**b**), expected wheat yield (Yld) (**c**), for the 3-year series and combination of mechanical control (ploughing) and cultural measure (high competitive crop variety)

Obtained results show a progressive reduction in the AVEFA-wheat competition in the third year (Fig. 15.6a). Seed production decreased in the second year but slightly increased in the third one (Fig. 15.6b). Regarding the crop, good yields are observed considering the large initial weed infestation producing an increase of 8.7% in the third year, regarding year 1 (Fig. 15.6c).

#### 15.3.4.2 Case II: High Use of Chemicals

Wheat is sown at June first each year, at a sowing density of Ca = 200 pl m$^{-2}$. The specific adopted variety possesses a LAIhc = 6 and a Myl = 60%. These values correspond to a low competitive wheat variety (Cooperación Nanihue) (Lopez and Vigna, n.d.). The results of the applied chemical and mechanical interventions are presented in Fig. 15.7.

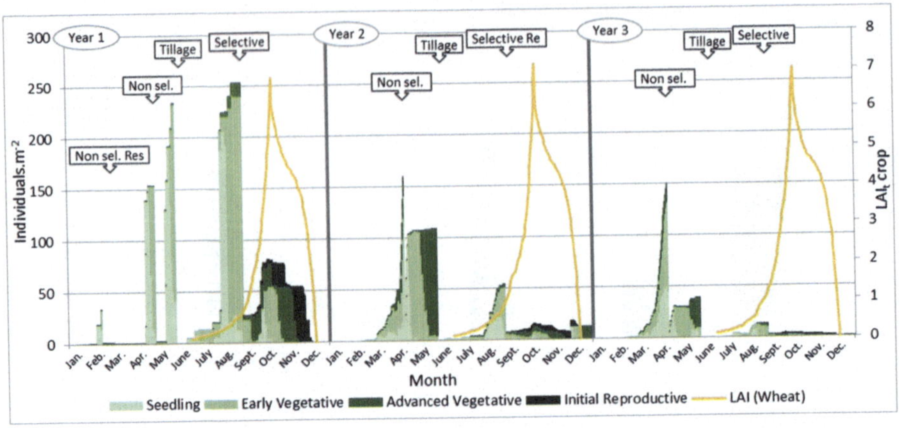


**Fig. 15.7** Multiannual AVEFA dynamics showing in different shades of green the relative composition of each phenological stage from a high non-dormant seed content (4500 seed $m^{-2}$). Each year, the combined effect of one mechanical control (ploughing) and several chemical controls is simulated: non-selective residual (glyphosate LS 48%, 2 L $ha^{-1}$ + flumioxazin SC 48% 0.120 L $ha^{-1}$), non-selective (glyphosate LS 48%, 2 L $ha^{-1}$), selective (pinoxaden CE 6%, 0.8 L $ha^{-1}$) and selective residual (Flucarbazone-sodium WDG 70%, 80 g $ha^{-1}$). The development of the wheat is observed through the $LAI_t$ (yellow line). Arrows indicate control dates

The first control is a combination of glyphosate (LS 48%, 2 L $ha^{-1}$) + flumioxazin (SC 48% 0.120 L $ha^{-1}$) during fallow (February) applied in the first year. Glyphosate effectively controlled AVEFA individuals, and the residual effect of flumioxazin precluded the growth of new seedlings for a 2-month period. Then, a new application of glyphosate took place 30 days before sowing in year 1 and 45 days before sowing in years 2 and 3. Moreover, immediately before sowing, ploughing is also performed in each year to eliminate the remaining individuals. In post-emergence of wheat, applications of pinoxaden (CE 6%, 0.8 L $ha^{-1}$) are performed in the first and third years combined with a residual herbicide (Flucarbazone-sodium WDG 70%, 80 g $ha^{-1}$) in year 2. The application of the selective herbicides performed an efficient control of AVEFA individuals avoiding competition during the CCP (October 25th to November 14th, CronoTrigo©).

Results show a progressive reduction of the AVEFA-wheat competition in the third year (Fig. 15.8a). Seed production remained low in years 2 and 3 (Fig. 15.8b). Crop yield was significantly reduced in year 1 due to large weed infestation, showing an increment of 16% and 21% in years 2 and 3, respectively (Fig. 15.8c).

#### 15.3.4.3 Case III: Cultural Management + Chemical Control

Wheat is sown at June first of each year, at a sowing density of Ca =400 pl $m^{-2}$. The specific adopted variety possesses a LAIhc = 6 and a Myl = 60%, corresponding to a low competitive wheat variety (Cooperación Nanihue) (Lopez and Vigna, n.d.). As

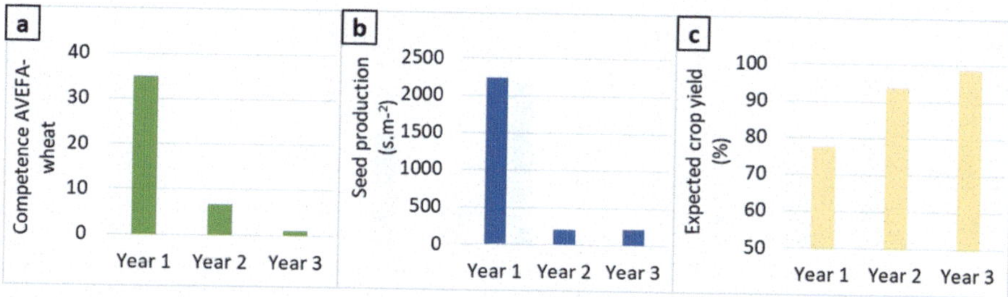


**Fig. 15.8** Competition of AVEFA on wheat (WC) (**a**), AVEFA seed production (Sp) (**b**) and expected wheat yield (Yld) (**c**). Each year, the combined effect of one mechanical control (ploughing) and several chemical controls is simulated: non-selective residual (glyphosate LS 48%, 2 L ha$^{-1}$ + flumioxazin SC 48% 0.120 L ha$^{-1}$), non-selective (glyphosate LS 48%, 2 L ha$^{-1}$), selective (pinoxaden CE 6%, 0.8 L ha$^{-1}$) and selective residual (Flucarbazone-sodium WDG 70%, 80 g ha$^{-1}$)

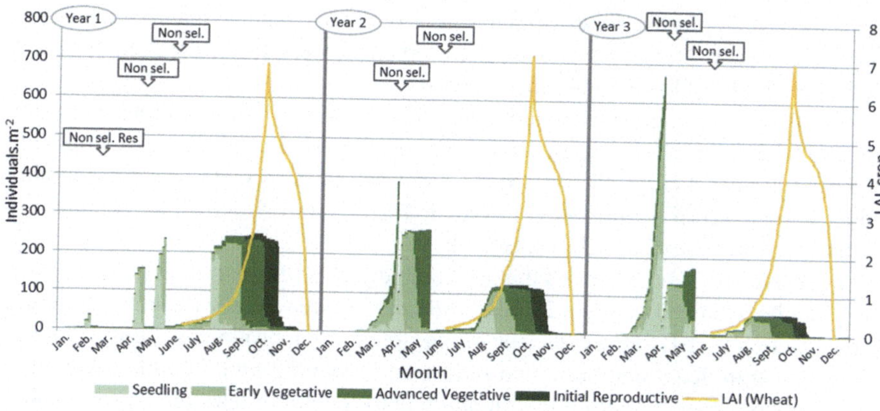


**Fig. 15.9** Multiannual AVEFA dynamics showing in different shades of green the relative composition of each phenological stage from a high non-dormant seed content (4500 seed m$^{-2}$). Each year, the combined effect of one cultural management measure (100% increase in crop density regarding standard crop density) and several chemical controls is simulated: non-selective residual (glyphosate LS 48%, 2 L ha$^{-1}$ + flumioxazin SC 48% 0.120 L ha$^{-1}$) and non-selective (glyphosate LS 48%, 2 L ha$^{-1}$). The development of the wheat is observed through the LAI$_t$ (yellow line). Arrows indicate control dates

a cultural management option, crop density is doubled compared to previous cases. Several chemical controls were also applied. In particular, two applications of a non-selective herbicide were simulated during fallow. A non-selective herbicide with residual effect was added in year 1 to cope with the large initial infestation.

In the fallow of the first year, the effect of a mixture of glyphosate LS 48%, 2 L ha$^{-1}$ + flumioxazin SC 48% 0.120 L ha$^{-1}$ can be observed, followed by a second application of glyphosate LS 48% (Fig. 15.9). Both applications coincide with those of Case II (Fig. 15.7). Different from Cases I and II (Figs. 15.5 and 15.7), an

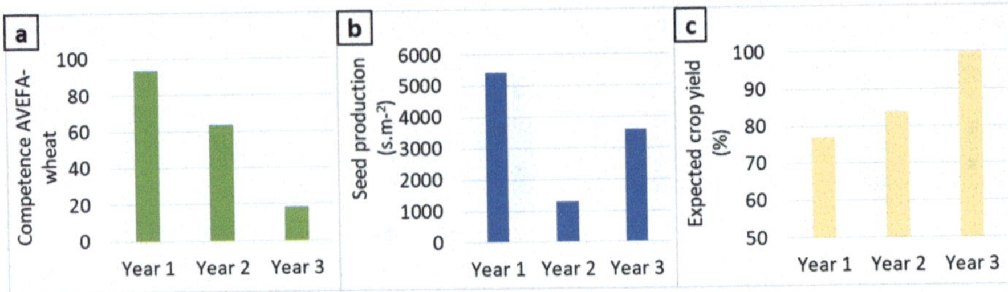


**Fig. 15.10** Competition of AVEFA on wheat (WC) (**a**), AVEFA seed production (Sp) (**b**) and expected wheat yield (Yld) (**c**). Each year, the combined effect of one cultural management measure (100% increase in crop density regarding previous cases) and several chemical controls is simulated: non-selective residual (glyphosate LS 48%, 2 L $ha^{-1}$ + flumioxazin SC 48% 0.120 L $ha^{-1}$) and non-selective (glyphosate LS 48%, 2 L $ha^{-1}$)

application of glyphosate is performed instead of the mechanical intervention before sowing each year. The effect of doubling crop density produces a reduction of new AVEFA individuals from September onwards.

Results show a progressive reduction of AVEFA-winter wheat competition in the third year (Fig. 15.10a). Seed production decreased in year 2 but moderately increased in year 3 (Fig. 15.10b). Similar to Case II, crop yield was considerably reduced due to AVEFA infestation in year 1 (Fig. 15.10c) with a yield increment of 7% and 33% in years 2 and 3, respectively.

In general, it can be observed that acceptable yields (close to or even larger than 90%) can be achieved in years 2 and 3 with all the three adopted control strategies (Cases I, II and III). However, regarding weed seed production, only Case II achieved a significant and sustained reduction in years 2 and 3 (<200 seeds $m^{-2}$) which suggests a more controllable situation in forthcoming seasons. This is a rather expected result since Case II is the most intense control strategy including a mechanical intervention before sowing, together with a selective treatment during the crop growing season. This is probably the most expensive and environmental impacting strategy. In Cases I and III, although the seed production was reduced with respect to year 1, the lack of application of the selective herbicide in September translates into a larger seed production which impacts harvest quality and contributes to seed bank replenishment.

Finally, it should be mentioned that in CASE II, where ploughing is performed before sowing, the application of chemical controls during fallow has no effect on the model outcome since the weed is completely killed by the mechanical treatment. However, a chemical application during fallow is a common practice in order to keep soil water content and in this case also to help to illustrate the effect of the residual action. In Case III, on the other hand, where a non-selective chemical treatment is performed before sowing, a previous chemical intervention precludes that some plants became large enough to escape these pre-sowing applications.

## 15.4 Conclusions and Future Work

To be used for practical IWM DSS, the most relevant agronomic variables should be estimated along rather extensive periods, typically several years. Additionally, significant uncertainty is present in agronomic systems, in particular associated to weather parameters' forecasts (daily temperatures and rainfall). In this context, stochastic studies, which demand intensive simulation, make more sound than just deterministic or scenario analysis. Finally, if the management option space is to be automatically explored, lots of simulations are usually required. Therefore, a simulation model should possess a level of detail compatible with reasonable computation times. Fortunately, current computer systems and software produce very rapid computations allowing therefore a very deep level of detail in the underlying DSS model. In fact, the described model approaches weed dynamics and weed-crop interaction with a larger level of detail than the usually found in similar studies. The most relevant features of the proposed modelling framework are the following:

1. *Interspecific competition (crop-weed)*: The interest of the adequate representation of this process is basically related with accurately quantifying the effect of some cultural control measures which have a paramount importance in IWM. The competition of the crop on the weed, although modelled in a simplified way, considers both the degree of competition of the crop variety and the crop density through the LAI. These factors affect the most susceptible state of the weed along the whole crop development period.
2. *Interspecific competition (weed-crop)*: It is simulated as crop yield loss, calculated as a function of a representative weed density. A somewhat innovative formula was developed for its calculation which considers both the amount of the weed in the corresponding growth stage and the period that remains in competition with the crop. Moreover, the maximum yield loss in competition allows simulating the effect of more or less competitive crops (varieties).
3. *Intraspecific competition*: This allows representing scenarios of high infestation and scarce control measures. Through the somewhat novel proposed approach, it is possible to simulate the competition of the weed plants within their own population. The effect of the competition of older individuals on younger plants can be easily tuned through several weighting factors.
4. *Seed bank:* Besides the typical elements of the seed bank dynamics (emergence, mortality, seed rain, etc.), the following phenomena are also modelled in this system: dormancy and longevity. Multiannual simulations allow quantifying the consequences of the different management strategies on this variable of paramount importance for the agroecosystem.
5. *Phenological stages*: The classification of the weed life cycle in its most relevant growth stages allows producing more detailed simulations. The daily demographic balance of the weed in each phenological stage allows a detailed modelling of processes such as inter- and intraspecific competition, at the expense of a large number of parameters to be provided/tuned.

6. *Seed production:* Simulations provide seed production per plant. Plants are classified into reproductive groups depending on the conditions that experienced along growth and development. Each group possesses different fecundity. In this way, it is possible to investigate how different management strategies impact on the weed dynamics and consequently on seed production.
7. *Weed management*: The model incorporates the effect of the control actions through mortality rates that affect in different magnitudes weed plants in different growth stages. This allows the quantification of the impact of a certain action based on the relative susceptibility of the plants present the day the control is performed. The approach also considers the residual effect of herbicides on weed seedlings. This segregated way of modelling mortality hopefully provides a more precise computation of the effect of herbicides' action on the agroecosystem, at the expense of a more parameterized model.

However, in order to build a practical DSS from the proposed model, additional features are required. Specifically, the following complements should be considered in future developments:

(a) *Economic evaluation*: The inclusion of a module for economic evaluation would allow the calculation of the cost of the whole management strategy as the integration of the costs of the individual control actions along the planning period, together with the income due to the selling of the produced cereal, which in turn will depend on the corresponding yields and qualities.
(b) *Environmental impact*: This module is intended to quantify the impact of the different control actions, in particular of the use of chemicals, through appropriate indexes such as EIQ, externalities, and other methodologies available in the literature.
(c) *Resistance*: Weed resistance to herbicides is a complex topic and a very challenging process to model. A possible approach to incorporate resistance quite straightforwardly in our modelling framework is to consider the presence of a resistant population coexisting with the susceptible one. Each population may be represented by different mortality rates to herbicides, producing therefore different demography responses to each chemical treatment, whose extension and impact could be hopefully captured with long-term simulations.
(d) *Optimization module*: In order to automate the exploration of control options, an optimization engine will be adapted aimed at optimizing the IWM problem regarding the two main objective functions: economic benefit and environmental impact.

## References

Begon M, Townsend CR, Harper JL (2006) Ecology: from individuals to ecosystems, 4th edn. Blackwell Publishing Ltd, Malden

Chantre GR, Blanco AM, Lodovichi MV, Bandoni AJ, Sabbatini MR, López RL, Vigna MR, Gigón R (2012) Modeling *Avena fatua* seedling emergence dynamics: an artificial neural network approach. Comput Electron Agric 88:95–102. https://doi.org/10.1016/j.compag.2012.07.005

Chantre GR, Vigna MR, Renzi JP, Blanco AM (2018) A flexible and practical approach for real-time weed emergence prediction based on Artificial Neural Networks. Biosyst Eng 170:51–60. https://doi.org/10.1016/j.biosystemseng.2018.03.014

Ghersa CM, Holt JS (1995) Using phenology prediction in weed management: a review. Weed Res 35:461–470

Gonzalez-Andujar JL (2013) Population dynamics: weed control models. In: Reference module in earth systems and environmental sciences. Elsevier Inc., Amsterdam. https://doi.org/10.1016/b978-0-12-409548-9.01018-6

Gonzalez-Andujar JL, Chantre GR, Morvillo C, Blanco AM, Forcella F (2016) Predicting field weed emergence with empirical models and soft computing techniques. Weed Res 56:415–423. https://doi.org/10.1111/wre.12223

Jones JW, Hoogenboom G, Porter CH, Boote KJ, Batchelor WD, Hunt LA, Wilkens PW, Singh U, Gijsman AJ, Ritchie JT (2003) The DSSAT cropping system model. Eur J Agron 18:235–265. https://doi.org/10.1016/S1161-0301(02)00107-7

Lodovichi MV, Blanco AM, Chantre GR, Bandoni JA, Sabbatini MR, Vigna M, López R, Gigón R (2013) Operational planning of herbicide-based weed management. Agric Syst 121:117–129. https://doi.org/10.1016/j.agsy.2013.07.006

Lopez R, Vigna MR (n.d.) INTA EEA Bordenave. Studies on bioecology and grass weed control in cereal crops in southwest Buenos Aires. Annual reports 1989–1994.

Martín AN, Scursoni JA (2018) *Avena fatua* L. escapes and delayed emergence in wheat (*Triticum aestivum* L.) crops of Argentina. Crop Prot 103:30–38. https://doi.org/10.1016/j.cropro.2017.08.020

Pannell DJ, Stewart V, Bennett A, Monjardino M, Schmidt C, Powles SB (2004) RIM: a bioeconomic model for integrated weed management of *Lolium rigidum* in Western Australia. Agric Syst 79:305–325. https://doi.org/10.1016/S0308-521X(03)00089-1

Scursoni JA, Benech-Arnold R (1998) Rendimientos de cultivos de trigo y cebada y crecimiento de *Avena fatua* en distintas condiciones de fertilización nitrogenada. IV Congr. Nac. Trigo. II Simp. Nac. Cereal. Siembra Otoño-Invernal 3.48

Scursoni J, Benech-Arnold R, Hirchoren H (1999) Demography of Wild Oat in Barley Crops: Effect of Crop, Sowing Rate, and Herbicide Treatment. Agronomy Journal 91(3):478–485

Scursoni JA (2001) Demography of *Avena fatua* L. in production systems of beer barley (*Hordeum vulgare* L.) under different agronomic management practices. University of Buenos Aires.

Scursoni JA, Martín A, Catanzaro MP, Quiroga J, Goldar F (2011) Evaluation of post-emergence herbicides for the control of wild oat (*Avena fatua* L.) in wheat and barley in Argentina. Crop Prot 30:18–23. https://doi.org/10.1016/j.cropro.2010.09.003

Scursoni JA, Gigón R, Martín AN, Vigna M, Leguizamón ES, Istilart C, López R (2014) Changes in weed communities of spring wheat crops of Buenos Aires province of Argentina. Weed Sci 62:51–62. https://doi.org/10.1614/WS-D-12-00141.1

# Index




G. R. Chantre, J. L. González-Andújar (eds.), *Decision Support Systems for Weed Management*, https://doi.org/10.1007/978-3-030-44402-0

The manufacturer's authorised representative in the EU is Springer Nature Customer Service Centre GmbH, Europaplatz 3, 69115 Heidelberg, Germany. If you have any concerns regarding our products, please contact ProductSafety@springernature.com

ted and bound by CPI Group (UK) Ltd, Croydon, CR0 4YY
22/09/2026
02221060-0016